Hermann Poppelbaum

NEW LIGHT ON HEREDITY AND EVOLUTION

NEW LIGHT ON HEREDITY AND EVOLUTION

Other Works in English by the same Author

Man's Eternal Biography, Adonis Press, New York (1945)

Man and Animal, Anthroposophical Publishing Company, London (1960)

A New Zoology, Philosophic-Anthroposophic Press, Dornach (1961)

NEW LIGHT ON HEREDITY AND EVOLUTION

by
Hermann Poppelbaum

Translated by
Norman Macbeth

ST. GEORGE PUBLICATIONS
Spring Valley, New York

Authorized translation by agreement with the Philosophisch-Anthroposophischer Verlag, Dornach, Switzerland

Published in Switzerland under the title *Entwicklung Vererbung und Abstammung*

FIRST EDITION

ISBN #0-916786-15-3

Library of Congress Catalogue Card Number 77-79731

Manufactured in the United States of America

Cover Design
By
Florin Lowndes

Contents

I

Maintenance of Form and Transmission of Form

The three phenomena of evolution, heredity, and descent are closely bound up with each other. They pose innumerable problems for the scientist. He must recognize that the phenomena are *biological*, having to do with life, and that these problems must be solved all together or not at all.

In the nineteenth century these three themes took on a new aspect. Analytical investigation was applied, and the answers worked out by it achieved enormous popularity. Evolution, natural selection, experimental embryology, Mendelism, all grew into separate branches of science. Nowadays a single student can hardly peruse the results of all the research. But the broad outlines are known to the layman and influence his view of the world. He may even have the impression that the principal problems have been solved.

The layman will not welcome the suggestion that he must think everything through afresh. Even the specialist is not eager to do this. But that is what we aim to do in this book. It is written in the conviction that a new attack will be profitable, if only because the basic positions have been too much consolidated. We also believe that Rudolf Steiner's anthroposophy will be of immense aid in throwing new light on the questions that have been raised by various students in this field.

* * * * *

Let us postpone the consideration of descent, which is our hardest problem, and see how evolution and heredity can be looked at as a single problem.

Growth is a central phenomenon of life that never loses its fascination no matter how often we see it unroll in the growing, blossoming, and fruiting of a plant, in the changing phases of the butterfly, or in the stages of a human life. Always it raises the question of how a single whole can appear in successive phases that either end in death or lead back to their beginning. Nothing like this is found in the inorganic world.

In contrast to this miracle of continual becoming, the fleshing-out of the living body with matter appears as a riddle of the second rank. True, the substances in a living being are different from those in lifeless nature. They are highly complicated, intricately formed and mixed, and they are in a watery half-fluid condition that holds them in continual flux. But we cannot escape the impression that it is the living form that rouses the substances to life and maintains them in that condition. If the form releases the substance, if the substance escapes from the boundaries of the form, then it becomes lifeless again and resumes the sluggishness from which it was protected during "the time of its life". Substance is only flotsam in the stream of the forming-process.

Substance does not build the form; the form holds the substance in an ever-varying balance, for which von Bertalanffy coined the phrase "flow-equilibrium". This equilibrium is in continual motion and never allows the substance to become stationary. The great discoveries of biochemistry have only deepened this riddle of perpetual fluctuation. It is one of the prime phenomena of life.

But we must hasten to add that the form itself is nothing static or permanent. It wrestles continually with its own lability as well as with substance. It never remains the same; it is continually renewed and continually changed. As the substance is worked over, the form fluctuates constantly; now it appears in greater opulence, now in greater simplicity, sometimes even reaching to meagerness.

If we ascribed this fluctuation in the form to a lack of precision in the life-process, we would be slighting its significance as a manifestation of life. On the other hand, however, we find that the characteristic forms of a developing embryo or larva cannot be connected with or traced back to the shape of the protein or protoplasm found in it. There is no transition from the screw-like shape of the living molecule (as three-dimensionally portrayed) to the simplest multicellular early stages of the new form. These stages are the peculiar attribute of precisely the species that "has" them. One can recognize this species early in organisms that have front and back, top and bottom, their own earthly or cosmic orientings. It is one of the wonders of life that a being can maintain its own form in the face of all the unpredictable changes and dangers of the outer world. Every living thing exists on a sort of "in spite of" basis. We must probe into the depths where its struggle is waged.

This struggle against external vicissitudes has been noticed by many students, but as a rule they do not go on to ask what it is that maintains the resistance. If we answer that it is life itself, it sounds too pat and it may be a tautology. But here spiritual science supplies an answer for which our own considerations have already prepared us. It suggests that what furnishes the resistance is *the species itself*, which vigorously asserts its own form in the face of all the substance streaming through it and gives its own stamp to the bodily expression of the forming forces. Species and formative-forces-body here coincide. They cannot come to rest, nor do they wish to do so.

Spiritual science has much more to say about this body of formative forces. For the normal consciousness (which can think it, but not see it) it is traceable in the fluctuations and variations of the form. Within certain limits this consciousness can fix the formative-forces-body as the picture commonly called the "habitus"; and it can let this habitus-picture "play", thus portraying the deviations that are natural for the picture. Not by chance does the biologist speak of the "play" of variations.

Here it is important that we should steep ourselves in the mystery of form and growth. Then we find something that does not come to rest in a definite way of filling space or in a definite lapsing of time. This that does not come to rest creates one form after another, out of itself, voluntarily. Goethe described it well when he said that this voluntary giving was "easy and joyful".

We should note how, in the course of time since Goethe's day, we have lost the feeling for the living species-form and for the voluntary quality of the variations. We began (and Darwin brought this to the highest point) to seek, in the conditions of environing nature, external reasons for the variations. It became impossible for us to think that a species was not forced to create variations by something outside itself. We did not see that special circumstances may elicit a variation, but do not produce it. The transplanting of a buttercup into a mountainous region calls forth a new shape from the bundle of formative forces that we have brought along. The forces respond to the outer circumstances as to the crook of a finger; they are not submitting to coercion. The deviation is not thrust upon the species against its will.

Another misunderstanding as to the voluntary quality of variations is the doctrine of the "blurred copy". Investigators who embrace this line of thought seem to accuse nature of carelessness in her handiwork.* It is sad to see how far a hypothesis can stray from life when one loses the path to nature and how unjustly a scientist can judge nature when he does not take the trouble to follow her.

This brings us to the phenomenon of *heredity*. This consists

---

*Julian Huxley, in *Evolution in Action* (1953), speaks of "incompletely self-copying genes" as a cause of variations. Even Graham Cannon (*Lamarckism and Modern Genetics*, page 91), who is very skeptical in these matters, sees variations as imperfections in the complicated chemistry of the cell.

in the transmission of the form as a whole, although many investigators have tried to shift the focus to the transmission of specific traits. Despite these efforts, the important thing to grasp is that the *transmission* of the form proceeds from the *maintaining* of the form, which is the more comprehensive phenomenon. The series of steps leading from the general to the particular can be seen in nature as follows:

(a) The individual maintains its own form amid the flux of circumstances, and can restore it when it is damaged (regeneration). In certain cases it can even melt up the blemished work and start over again.

(b) The individual entrusts the maintenance of its form to a single part of its system, such as a bud (hydra) or a sprout (the budding stolon of marine Hydrozoa).

(c) The maintenance of the form for the future is entrusted to an organ or system especially set aside for this purpose (the reproductive system and gonads).

(d) As a last stage, the transmission of the form as the function of the germ-glands is made more complicated by the appearance of bisexuality, with the necessary consequence that the gametes* produced by each sex must be combined.

Only in the fourth case is reproduction combined with the separation and reuniting of the sexes. Thus our series clearly shows that this phenomenon is subsidiary rather than central. Therefore it is completely misleading to use sexual reproduction as the basis for interpreting the facts of heredity and to allow it to determine our whole approach to the problem.**

---

*The gamete or germ cell is a reproductive cell whose nucleus, and often cytoplasm, fuses with that of another gamete (constituting fertilization), the resulting cell (zygote) developing into a new individual.

**Julian Huxley, in Chapter 5 of *Evolution in Action*, carries sex all the way back to the earliest one-celled animals; truly a miraculous feat of the imagination.

We should not ask how heredity comes about through the collaboration of two sexes. The true question is how we can explore the region in which the transmission of form in general is regulated. The answer can only be—through growing out of the spatial and temporal boundaries of the individual. For this purpose there is no need for a splitting of gametes, or a fertilization, or a complicated apparatus for dividing the germ. We must only extend the process of restitution, which otherwise appears only after an injury, to a coming generation. If we have grasped the operation of the formative-force-body in the regenerating individual, then we can see the phenomenon of heredity as a metamorphosis in time of the process of restitution.

This is not merely a new semantic formulation. It is a clue that enables us to grasp a very important connection. For example, one may go on to study the phenomenon of grafted hybrids (chimeras), the so-called vegetative bastards. Chimeras are the apparently integrated individuals that can arise through grafting two different species, such as medlar and hawthorn or tomato and nightshade. The closely related formative-force-bodies (two Rosaceae in the first case and two Solanaceae in the second) can tolerate each other for a while, but in the end they split up. One partner then treats the other as a stranger and expels it. Etienne Wolff* induced the embryonic lung-bronchioli of mice and chickens to grow together and even to build a common duct. After a few days, however, they separated from each other.

* * * * *

We do not grasp the importance of the formative-force-body unless we note that it is not a vague way of speaking of a new factor that must be added to the body in order to make it

---

**La Science des Monstres*, Paris 1948.

live. We are speaking of a reality that has a very definite and knowable relation to the material bodily organization. We are not marking a boundary beyond which something incomprehensible begins. We remain in the realm of facts that can be presented to our thinking by description and reasoning.

Here we may derive unexpected assistance from Rudolf Steiner's remark that the formative-forces-body is not to be understood through the usual three-dimensional concept of space. We must search for the polar opposite of conventional space; a space that is not empty and undifferentiated, a space that has an inner structure.* This clue puts the investigator on a fruitful path. Many puzzling phenomena can now be reexamined in the light of a non-Cartesian anti-space structure. Riddles approach solution; for example, the problem of how the substances ingested as food by man or beast can be taken hold of by the formative-forces-body in the digestive tract. Substance is here freed from its usual fetters, divorced from its normal relation to gravity through the colloidal consistency of the plasma; but it is also brought under the influence of an entirely different region, one that is not centered in matter but works out of the periphery of the earth. Thus substance acquires its new characteristics not from its own structure but in accordance with the position that it has in the organism. As it is displaced and moved about in the organism, it must change its nature.

Here we come to an axiom of the new conception of evolution: The properties of a substance are not determined by its composition, and it does not strive to retain these properties no matter where it may be; rather, it changes itself in accordance with the displacements that it undergoes in the life-processes of the organism. Not only does it lose some of its

---

*This space has been vividly described by G. Adams, "Space and Counterspace", an essay appearing in *The Faithful Thinker* (Hodder & Stoughton, 1961) and by L. Locher *Raum und Gegenraum* (Dornach, 1958).

weight through being converted into a fluid, but it acquires new and living qualities that are a counterpart of its physical qualities. These are the materials and forces that build what is living.

The literature on the formative-forces-body is extensive. We can only refer to it generally. Here we will merely try to achieve new formulations that may serve to further the progress of biology.

* * * * *

If we want to understand the relations of the organs to each other, we see what a profitable path we have been led to by our new approach to space. In the old view these relations were of a neighborly character; juices were exchanged, and communication was carried on by means of bridges in space (blood vessels and nerves). But now we see that organs commune with each other even when they are far apart as measured with a ruler. They are adjacent in qualitative space, etheric neighbors. For example, there is a goodly number of inches between the kidneys and the eyes, but the facts force us to see this distance as small because the relations of these organs are "near", as shown by the fact that both are affected by correlated diseases. The polarity between colon and forebrain* is another example of superspatial connections. In a more general way, the organs above the heart have secret correlations to those below the heart.** These relationships do not depend on the messenger substances that are so diligently sought for in modern physiology. Indeed, if we assume (or even discover) that specific substances maintain the intercourse between distant organs, we are only led to the delicate question of precisely how these supposed messengers find

---

*Mentioned by Rudolf Steiner in his first course for physicians (1920).

**First discovered in 1909 by Wilhelm Fliess.

their way to the right spot and demonstrate their kinship. It may be said that they are attuned to each other, but it is just this vague suggestion that we would replace by the new idea of etheric adjacency and structural space.

If the organism lay in normal Cartesian space, it would have to be directed in some way and we would be faced with the problem of locating the control center. If we asserted that this

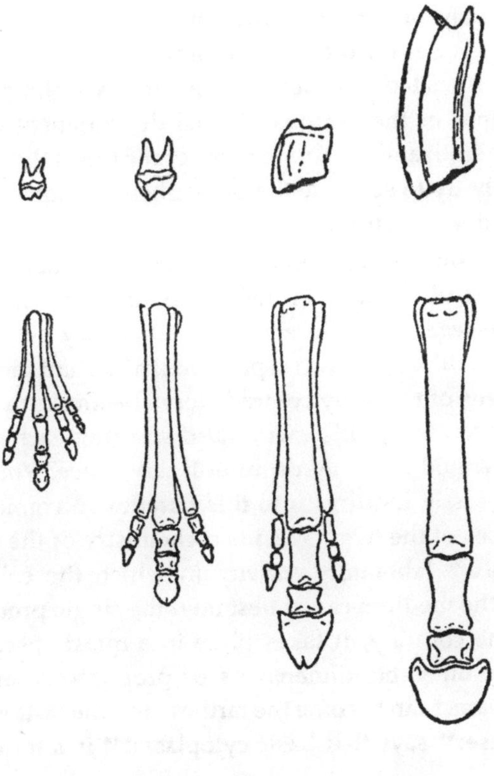

Figure 1

Correlation between teeth and limbs as shown by the gradual increase in size of the upper molar teeth and the skeleton of the hand in four stages of the development of the horse since the Eocene.

was in the organism as a whole, the usual concept of a steersman would not be satisfied. But if we assume that organs having special positions also have special contacts with each other, many a riddle can be solved. We will even be led to discover relationships that are now unknown. For example, light is thrown on the otherwise mystifying embryonic correlations between teeth and forward extremities if we look at the growing child and the physiological condition of the teeth. Rudolf Steiner remarked that awkwardness in the hands hastened the decay of the teeth (dental caries), and he even suggested exercises for the hands as a way of improving the teeth. In the so-called genealogy of the horse, we glimpse the same principle in the curious parallel development of the length of the teeth and the proportions of the limbs (Figure 1). We must only try to separate the basic phenomenon from its Darwinian superstructure.

If the organism lies in differentiated space, i.e. space that is full of bends and twists, then the space "in" the single cell must also be regarded differently. Then we may be able to solve the great riddle of how it is possible, in one and the same cell, for dozens of mutually contradictory chemical processes to take place "side by side" without disturbing each other, which they would certainly do in ordinary space. Von Bertalanffy* expressly inquires into this paradox in connection with processes of the liver. The microchemistry of the cell is not a miniature laboratory activity in which the colloidal structure of the medium can at best hold the single processes apart. On the contrary, it takes place in a quasi-space that possesses innumerable dimensions or proportions, and in which the nearest can become the farthest and the farthest the nearest. Wieser** says that basic cytoplasm*** is a milieu in

---

**Gefüge des Lebens*, 1939, page 80.

***Gewebe des Lebens*, Bremen 1959, page 61.

***The cytoplasm is all the protoplasm of a cell excluding the nucleus.

which enzymes move freely and in which proteins can dissolve and recombine. This is necessary, because the protoplasm of the cells is a colloid that fluctuates between the dissolved and gelatinous conditions, i.e. a more fluid and a more solid consistency. This is where "the protean quality of protoplasm is revealed; solid enough to create forms, yet fluid enough to permit the transport of matter". Certainly! But precisely these fluctuations in the condition of the protoplasm presuppose an organization that directs the plasma and that cannot reside in conventional space.

Looking at all these relationships, we begin to see how the organism lies in the "form-field", which is the expression we will use for the peculiar nature of etheric space. The "blueprint" of an organism does not result from the chemistry of the various components of the protein, etc.; it images an extra-spatial order that gives form and position to the organs and also determines the earthly-cosmic layout of the organism as a whole. The enzymes, hormones, etc. that move about in the organism are not shaping-causes; rather, they are mere indicators of the relationships in the form-field at a particular spot. The total structure of the living being proceeds from the superspatial form that is developed in etheric space. This is an attempt to make the formative-forces-body comprehensible.*

The hormones are themselves effects. Their positions show where the form-field, in producing them, has already been active. Our most important finding is that the creation of form, and also the transmission of form (the central fact in all the phenomena of heredity), must be ascribed to the workings of the formative-forces-body in our new qualitatively structured space. Let us recognize that here we see the concrete image of what is abstractly called "species".

---

*It will not be difficult for the student to find the improvements in this formulation as compared to the author's first attempt in *The Formative-Forces-Body of Living Beings* (1922).

* * * * *

As we showed earlier, the transmission of the form is independent of the mechanics of reproduction, and particularly of sexual reproduction. This latter must have an origin entirely different from that of the transmission of the form, since this is quite possible without any such arrangements. Modern genetics has interpreted the facts of heredity in such a way as to give enormous importance to the events of reproduction, but again we must assert that here a derived phenomenon has been confused with a more universal one. In using the facts of fertilization and cross-breeding as the pivot of its interpretation, genetics has committed a methodical error of the first order. Newton did the same thing for optics, with grave results, when he took a subsidiary phenomenon (the spectrum with green in the middle) as the basis for his explanations, rather than the more basic phenomenon (the blue and yellow margins of color on the edges of an object viewed through a prism) discovered by Goethe.

In the still prevalent (though much modified) idea of *heredity-carriers*, what we object to is the unbiological element. It is not the actual facts of heredity but our own habits of thinking that drive us to assert the existence of tangible *genes* as the agents for the transmission of traits. The latest research has made it clear that all of the disk-shaped segments of the chromosomes that are called genes actually consist of one and the same chemical substance (desoxyribonucleic acid, commonly referred to as DNA). We will discuss the significance of this material later. But it is hard to see why anyone should assert that this rather monotonous material can "represent" all hereditary traits, unless it is assumed that this representation resides in something other than the chemical structure that this substance "has".

We can even say that the concept of "traits", which plays such an important role among the geneticists, springs from an entirely unbiological attitude. Among the breeders of domestic animals, traits (good points, bad points) are much discussed. But for a person who thinks biologically they are unjustified abstractions. Curiously enough, the abstract "trait" has entered into an uneasy alliance with the Mendelian abstraction "gene", so that the two are linked in all explanations despite some of their modern modifications.* The theoreticians still speak constantly of the transmission of qualities through carriers that can be localized in matter.

But the riddle of heredity does not depend on any substance, even the most complicated chemical compound; it is a peculiar stamp that appears of itself within the flow of the generations. Precisely how it works is a problem for the new research that we hope to stimulate. We know that the role of matter is only that of a door-opener, not that of a moving cause. It is still often asserted that the riddle of heredity has been solved by the discovery of the genes in the chromosomes, but this has already been made obsolete by studies showing the wide discrepancy between the genes as actually seen in space and the influence that they are supposed to exert on the developing organism.

This discrepancy is not produced by our fragmentary insight; it follows from an objective incommensurability. We have demonstrated this above. Now we must turn our attention to the problem of the entry of the formative forces. Obviously these forces, no matter how we may picture their antispatial nature, need certain footholds in order to launch their work on the substances. They need definite points of entry from the superspatial into the visible and spatial entity that is to develop, whether egg, embryo, or seed. Roux, in his experiments with tadpoles, killed one cell during the early two-cell stage by singeing it with a pin. The dead cell, as long as it clung to the intact cell, prevented this from growing into a whole embryo, so that only one half developed. But if the dead

---

*See Chapter II below.

half was removed, a complete larva was able to form. Thus it is clear that every defect in substance will hinder the normal entry of the pattern of formative forces. By inhibiting the orderly operation of these forces, the experimentor can induce defects and distortions, such as cyclopia (the tendency of the eyes to fuse together), brainlessness, fusing of the limbs, etc.*

In searching for the points of attack of the formative forces, we find some significance in the physical state of the plasma. The colloidal nature of the cytoplasm signifies not a spatial order only, but a matrix that is adjustable in all directions, as though equipped with elastic dimensions. This makes Bertalanffy's flow-equilibrium possible, although it does not explain it. Both the colloidal state and the flow-equilibrium are among the conditions that are necessary if life is to communicate itself to substance; but they are not an explanation of life.

Bearing in mind the significance of the footholds or points of entry for the formative-forces body, we can understand the great success that Spemann and his school have had in implanting tissue-fragments from one embryo into other embryos. If the fragment is planted in a position other than the one from which it was taken, it works as an "inducer" in that the organizing of the surrounding part of the embryo seems to proceed from it. This, of course, only strengthened the view that there are substances that act as inducers or evokers of certain formations. But it must not be forgotten that even in the normal development of the germ mere *contact* between neighboring organs in a rudimentary state acts as an inducer, i.e. the new formation radiates from that point. Thus, for example, the formation of the neural tube (the principal part of the central nervous system) is induced at the point where it touches the mesodermal chord; while the lens of the eye

---

*Etienne Wolff: *La Science des Monstres*, Paris, 1948.

appears at the point where the optic vesicle, growing outwards from the brain, comes into contact with the embryo's exterior skin (ectoderm). No material secretion is needed for the evocation; a mere spatial contact is enough to give the signal for the entry of the formative forces. Naturally this is not a blind impulse; we are dealing with orderly relations of position and size, which have their own growth-rates, etc. Thus if the optic cup has been artifically reduced in size, contact with the skin will produce a correspondingly small lens.* An implanted organizer cannot create forms at will; it can only produce a substratum for the entry and activity of the "legitimate" form-forces. Neighboring tissues have their inducing influences everywhere, but the surroundings must be ready to react; in fact the surroundings must actually help to open the door for the formative forces.

* * * * *

Thus the arising of a living shape depends on a substratum that is accessible to the activity of the superspatial form. We use the word "substratum" deliberately, because it does not lead to the idea of a producer (a cause), yet it carries the hint of special aptitude that we have in mind when we speak of a substratum suitable *for* this or that phenomenon. When there is a real causing, as in physics, what is caused can be deduced (and usually calculated) from the given conditions. But with the entry of the formative forces, the connection between the door-opening function and the quality of what enters is so loose that, although we are sure that there is a connection, we cannot correlate the results with the conditions in any exact way. It is only the building of a basis, the helping something to manifest; not a causing or a producing.

---

*Wieser: *Gewebe des Lebens*, page 86.

Let us pause to reflect how many things appear in a new light when viewed in this way. One example is the human capacities of thinking, feeling, and willing, which need what Rudolf Steiner called "bodily counterparts" before they can come to birth in man. The sense-nerve system, the rhythmic system, and the metabolic-limb system must be developed so that these activities can function physically. They must reach into the bodily counterparts in order to work on earth. What these three capacities would be without this "extension into the physical" we can learn only through supersensible research. But the door-opening function of the three systems is analogous to that by which the formative-force organism enters into the living substance of the body. What we must see here is a new way of working together, with the aim of fetching down into the physical something that would otherwise remain hidden. What arises through this fetching down into visibility is an intensification of life, an enhanced presence, occurring through manifestation in a deeper-lying realm.

According to Rudolf Steiner's spiritual research, the basis for human idea-formation (often called object-consciousness) is a barely perceptible secretion of uric acid in the cerebral cortex. This secretion, however, has nothing to do with the content of what is thought. Thus a well defined and very simple substance is responsible for forming the basis. Returning to our own case, as the living form pours in, the corresponding activity is entrusted to the living protein, a substance that is in continuous flux and is impermanent by nature. But the protein is aided by a special form-transmitter, which is ribonucleic acid with its dual modifications in nucleus and plasma.*

******Desoxyribonucleic Acid in the Nucleus, Ribonucleic Acid in the Protoplasm;*** **paper by Dr. K. H. Simon (Ithaca, New York) in "Naturwiss. Rundschau", August 1960.**

What is the situation with the oftmentioned difference between the roles of nucleus and plasma for life itself? Although they look very different from each other when stained for microscopic examination, the two regions of cytoplasm and nucleus are closely bound up in life. Biologically it would be absurd to think of the nucleus as physiologically separated from the plasma. Recent studies have shown that the membrane of the nucleus takes from the plasma the substance needed by the nucleus, and that the plasma itself is constantly receiving substances expelled by the nucleus. The organels of the cell, such as the nucleolus etc., contain substances that penetrate the membrane when they are in lively action; while others, lying immediately adjacent to the membrane, such as the newly discovered "microsomes", are obviously sites for protein-synthesis in the cell. Wieser* asks why it should be assumed that the enzyme systems appearing in the plasma should leave the heredity-carriers in the nucleus undisturbed, when it is known that enzymes are constantly exchanged between nucleus and plasma. We too ask how it could be possible for either the substance or the apparatus of heredity in the nucleus to remain entirely undisturbed and unchanged by this stream of lively activity?

All the events in the plasma and the nucleus are so dependent on each other that the plasma could apparently, if necessary, provide all the substances needed for the nucleus. This makes it hard to cling to the old belief that everything happening in the cell is directed by the nucleus.

We must also reckon with the fact that the stocks of nucleic acids stored in the chromatine are continually broken down and built up as long as the cell is active in its metabolism. Only when the cell is preparing to split, and is therefore not active in its metabolism, do we find the genes in their orderly rows on the loops of the chromosomes. What does this do to the

**Gewebe des Lebens*, page 68.

idea of genes held ready in the chromosomes for all the needs of reproduction and heredity? Are we to cleave to their continuity, or must we admit that the genes are subject to the law governing all living substances, which decrees that they must be in constant flux, always being broken down and rebuilt? How does this affect the doctrine that the genes are already marked out as the present representatives of the traits that will develop later?

Thus we see that the physiology of nucleus and plasma yields no material guaranty for the retention of a permanent structure. If there is any imprinting of form in the organic field, its production and transmission are not the monopoly of any specific material.

Wolfgang Wieser, the great student and critic of the latest biological experiments, comes to the same result.* He asks: "Beyond the activity of single genes and enzymes, where is the steering of the whole process of development?" . . . . . "As we penetrate into the molecular region of biological organisms, we find difficulties of new dimensions, which hint at the existence of new and perhaps impenetrable ordering principles at the root of all life."

Difficulties of new dimensions, because one of the most important ordering principles has been neglected. This is the entry of the superspatial formative-forces-body, which is different for each species. The system of forces guiding the creation of forms, as E. S. Russell called it,** lies in antispace as an individual formative-forces-body. The reader who likes paradoxes can contrast the outstretched body of the substances to the "instretched" body of the formative forces.

Living development and the transmission of forms are tied to this necessary component or "limb" of all living beings.

**Gewebe des Lebens*, page 103.

***The Directiveness of Organic Activities*, Cambridge, 1945.

II

The Overcoming of Mendelism

We have tried to grasp how fertilization occupies a special place among the phenomena, and we were impressed by the fact that, in a causative way, the inheritance of form has nothing to do with fertilization. In order to understand the direction taken by modern genetics, we must again speak of what is so easily forgotten, i.e. that fertilization is a late arrival on the scene and is only the last of a number of methods by which form is transmitted in the organic world.

Fertilization must, in turn, be seen as a special form of a more general event, which is the reunion of previously separated elements or beings or merely parts of beings. With his large view of nature, Goethe recognized this separation and reunion in the vessels that transport blood or the juices of plants. He saw these so-called "anastomoses" as a prime phenomenon. For him the anastomoses were a first sign of what later, on an entirely different level, appears as the separation and reunion of the sexes.

The pollinizing of a blossom stands roughly midway between the vascular anastomoses and the separation and reunion of the sexes. In blossoms, especially those of shrubs and trees, the products of the stamens and pistils come together after a very short separation. The outer shape of the plant is not altered by this, even though a tree may have blossoms of both types (monoecious) or of only one. Thus the carriers of the opposed products are not modified. In a maple, for example, one cannot at once tell whether it will have male or female blossoms.* The way in which the form of a tree rejects distinct

*For this reason it has always been doubtful for deeper students, such as Goethe's contemporary Schelver, whether one should really ascribe sex to plants.

male or female earmarks shows us that in the plant kingdom sexually transmitted heredity is still something alien and enforced.

In the animal kingdom, once we get past the lower stages, sexual reproduction gradually predominates. The activity of sprouting no longer leads directly to reproduction; it is pressed down into the organs, where it brings forth multiplication, segmentation, and the forming of sex-cells. Sexual differentiation of form generally begins in an early stage of life, let us say toward the end of the larval period.

One common feature that can be observed, from the coupling of algae-threads up to the highest inverebrates, is that the products of sex-cell formation (zygotes*) are remarkably well equipped to survive unfavorable seasons. Examples are the resting spores in algae, winter-eggs in small freshwater crustaceans, frost-resistant eggs in many insects. It is really only our human intellect that leads us to call winter and drought unfavorable; often they are necessary in the scheme of things as a special impulse that prepares for further growth in the spring.

The reunion of substances or beings that have long been separated from each other can strengthen the life-forces for the future. One who seeks the peculiar meaning of the reunion of the sex-products should not underestimate this since, as we have already shown, the mere production of new individuals does not need the division and reunion of the sexes. In the reunion of the separated parts, indeed even in the mere separation, there is a latent element of desire that is still unknown to the plant but expresses itself in many ways among animals; e.g. the wedding rites found even among the molluscs, the forming of swarms among worms, dwindling of the organs of motion in female scale-insects, degeneration of

---

*Strictly, the zygote is the fertilized egg before it undergoes further differentiation, but the term is sometimes extended to refer also to the individual developing from that egg.

the male to a tiny parasite that is firmly attached to the female (among fishes) or lives in her inner organs (Bonellia worm).

In Rudolf Steiner's spiritual science the experience of desire is bound up with the so-called astral body of the animal.* This is the real distinction between the transmission of form in general and the particular variety of it now before us. The transmission of form, originally a task of the formative-forces-body and accomplished solely by it, is now connected with the element of desire and proceeds further under its spell. From now on heredity, entangled with the sexual element, is an entirely different process.

Thus far spiritual science. Can we confirm this from the phenomena? In Rudolf Steiner's descriptions (which rest on spiritual vision) of very early times, the separation of the sexes is an impulse originating in the human kingdom and migrating thence into the animals and plants.** This makes it doubly clear that the division into sexes is of secondary importance.

The modern doctrine of heredity rests on Mendel's classical experiments and on the interpretation in terms of gametes, etc. The crucial point in the Mendelian theory is that the formation and reunion of the sex-cells are taken as the cornerstone of the interpretation. Here we must reiterate that the transmission of form, as a general phenomenon, does not need this interpretation. Our first chapter showed that, through the concept and the reality of the formative-forces-body, it might be possible to give a non-Mendelian explanation of the facts of heredity.

But how did the Mendelian interpretation come to birth? Through taking the idea of alleles***, or pairs of factors, as the

---

*See the author's *A New Zoology*, Chapter 2.

**Rudolf Steiner: *Occult Science, an Outline*.

***Alleles are genes that occupy the same relative position (locus) on homologous chromosomes and undergo pairing during meiosis.

pivot. Instead of studying the transmission of the form as a whole, attention was focused on the way in which certain traits are reproduced in crossbreeding. Every student learns this today. He is told: If two individuals differing in only one trait, such as white or red color in the blossoms, are mated with each other, the first hybrid generation will contain only red individuals. The further course of events, however, shows that the "white" trait is only slumbering and that in later crosses it will reappear in a predictable percentage of cases. If two individuals of this first hybrid generation are bred together, they will produce 1/4 purebreeding reds, 2/4 mixed-breeding reds, and 1/4 purebreeding whites. This is Mendel's famous rule of segregation.

The most important point in this discovery is that the progeny divide into carriers of the red and white traits in *predictable* proportions. This predictability holds good for further crossings and also for back-crossings with the original form.

In the Mendelian notation this result can be symbolically presented as follows:

P (Parent Generation) R + w
F^1 (First Hybrid Generation) Rw
F^2 (Second Hybrid Generation) RR, Rw, Rw, ww (Proportion 1:2:1).

Let us note however, that we cannot predict whether a specific individual will show the red trait (pure or mixed) or the white one. We can calculate only average ratios among the progeny. Thus the Mendelian law is only a statistical statement. The symbolic notation in the form of factors should never trick us into forgetting that this is a mere shorthand postulation; to wit, that every trait can be represented as a factor. The phenomenon itself does not lead us to this manner of expression. The equation is only a neat way of avoiding lengthy descriptions and expressing "the course of heredity" in the tersest possible form.

In the years since Mendel's law was rediscovered (1900), the art of working with these factors and predicting the results of future crosses has been developed to a very high point. It would have remained a narrow specialty, however, had not the results of cell-research (cytology) given it substantial assistance. The two sciences are now interlaced in the closest and most fascinating way. Just because sexual reproduction was viewed as the central fact in heredity, it was easy to become enchanted by the distribution of traits among hybrids and to forget that crossbreeding is only a special case to be viewed alongside the breeding of unmixed types. Students had long been familiar with the remarkable manoeuvers of the threadlike germ-substances (chromatin) during the division of the cell. It had often been said that nature would go to so much trouble only if she regarded this chromatin substance as very important, and that she was obviously determined that it

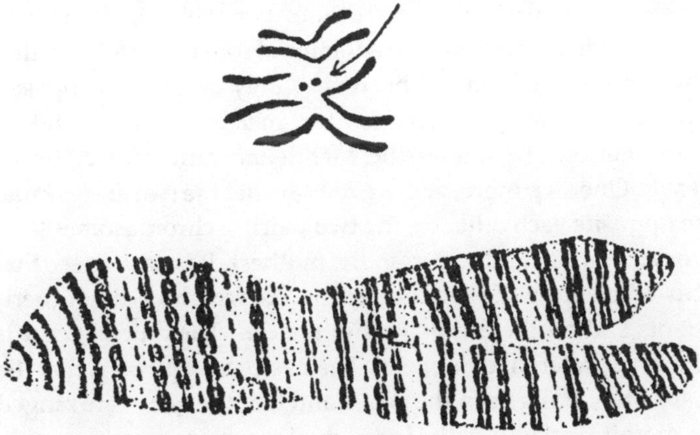

Figure 2

Above: Double set of chromosomes in ovaries of Drosophila.

Below: Two giant chromosomes in the salivary glands of the larva. The banding shows the completely analogous construction of the two partners out of similar chromomeres (genes?). After Bridges.

should be equally distributed among the daughter cells. The number of these nuclear loops (chromosomes), and even their individual forms, had been found to be constant for each animal or plant species.* Among plants there were often several hundred, among animals and men a smaller number. The number is always even, since one set is obtained from each parent.

As can be verified in any textbook, all these facts add up to an impressive mass of evidence for the proposition that the chromatin-substance must be regarded as the bearer of heredity. The total number of such bearers (genes) is called the genome.** The genes are conceived as distributed along the chromosomes or packed into them (Figure 2).

For this approach, the 50% reduction in the number of chromosomes that always occurs before the formation of sex-cells (in the so-called reduction-division, or meiosis) is an important confirmation. Again the textbooks show how it has come to pass that the separation of the alleles is thought to occur in the moment of the reduction. From this it follows that the gametes can contain bearers of only one of the opposed factors of a trait-pair. This in turn means that in a random combination of gametes the Mendelian ratio of 1:2:1 must result. One step more, and we can say that the two alleles must lie opposite each other on the two pairing chromosomes (one from the father and one from the mother). It is easy to see that this whole train of thought rests on the crossbreeding experiments and on the events within the cell. The crucial point is the interpretation of meiosis. The basic conception is that the hereditary traits lie in the genes and that the genes are ranged in definite loci within the individual chromosomes.

It may be difficult to grasp some of the details of this line of

*This rule, however, has recently been shown to suffer many exceptions.

**The genome or genom is the set of all *different* chromosomes found in each nucleus of a given species. A haploid nucleus has one genome.

thought, which is presented here in a simplified and incomplete way, but it is easy to see what remains in the mind of the reader who has to some extent followed the argument. He will believe that the materials standing proxy for the living qualities are manipulated by the chromosomal mechanism of the germ-apparatus. The scientist who is trying to grasp the attitudes that underlie a method must realize that the Mendelian segregation, which is actually a phenomenon of nature, has first been taken to pieces by the interpreters and then ascribed to an apparatus whose mechanistic operation hides in the invisible.

It is typical of the inclination of our time that laymen and scientists alike feel satisfied with the idea of bits of matter representing the single traits that appear later. There is the same tendency as in physics (where it may be justified) to explain a process from the relations of the particles that enter into it. The entire process thus becomes the sum of the behavior of all particles. This is, in the true sense, an additive method of explanation.*

Those who are interested in the fate of modern thinking were alerted by Rudolf Steiner to the significance of such trends of interpretation. This is just the contrary of our procedure in Chapter I, where we aimed at the whole. The choice between the two paths comes at the point where the researcher decides that he will explain a phenomenon out of its supposed elements or that he will try to explain the behavior of the parts out of the whole phenomenon. The modern doctrine of heredity has chosen the first alternative. The biology founded by Rudolf Steiner must try to pursue the second. It must view the entirety of the organism as a primary fact.

To grasp the direction in which genetics has been moving, one must be aware that it has never been able to break away

---

*Von Bertalanffy long ago showed the unbiological character of the additive method; see his *Modern Theories of Development* (Harper, 1962).

from the fundamental axiom that the traits have loci in the chromosomes,* although we shall soon see that many a fact speaks against this. It has clung, too, to the abstract concept of the trait, or the trait-pair, although there is so much contrary experience that desperate auxiliary assumptions have been necessary to keep the whole doctrine from collapsing.**

The reader should always bear in mind that the phenomena themselves do not require us to derive our views of heredity from the formation of gametes and the details of fertilization. We have repeatedly shown that these events in the cell can have no direct relation to the fact of heredity, because heredity can take place in the absence of all sex. Even the abstract concept of the "traits", as we showed above, has been drawn from the jargon of breeders, who are principally concerned with crossing particular plants or animals and who customarily speak of traits or points. But in nature, heredity is a procedure that occurs within the boundaries of a species. Most crosses of two species are infertile. Also, the traits that interest breeders are largely superficial ones. Points such as the plumage of pigeons or the coat and color of domestic mammals do not reach into the depths of the organization or affect the structural plan. The breeder's points are the external trappings of the structure. They are not basic characteristics, but merely "marks".***

It is striking that many of the hereditary qualities that have

---

*This was recently reiterated by Julian Huxley in *Evolution in Action* (1953), Chapter 1, where he says: "The relative position of the genes can be mapped. . . . Genetic geography is linear, and you can point to the exact position of a gene along the length of a particular chromosome."

**Chapter 7 of Garrett Hardin, *Nature and Man's Fate* (1959), contains a fairly typical list of such desperate assumptions, although there they are bravely referred to as "subtleties".

***Graham Cannon of Manchester has noted this distinction, but he has not been taken seriously because of his radical opposition to Darwinism. See *The Evolution of Living Things*, Manchester University Press 1958, espcially Chapter 5.

been studied by the Mendelians belong to this superficial order (the various stuntings that follow the Mendelian rules are an exception). Where the traits impinge upon the structure, they disturb it; e.g. winglessness, stunting of limbs, lack of color in the eyes, and all the innumerable distorted mutations of *Drosophila melanogaster*. If we learn to distinguish superficial points from structural features, we see that almost all the "Mendelizing" traits belong to the former category. Julian Huxley adds that the progeny of the eyeless mutants of *Drosophila*, if bred pure for eight or ten generations, are again endowed with almost normal eyes.* He ascribes this, however, to a counter-mutation that repairs the earlier damage.

This distinction between trait and structural feature has been little heeded because of the Mendelians' obstinate generalization of the concept of traits. This has gone so far that in their calculations the lack of a trait is actually treated as a trait (Bateson). This trick is used to supply the alleles that are needed for the formula.

* * * * *

To summarize again, the gene and the trait are both unbiological abstractions by which we argue ourselves away from life. But now the reader may ask the opposite question—whether we can escape from such abstractions if we are to understand heredity. Is it possible to formulate the Mendelian law of segregation without referring to the manoeuvers of genes and chromosomes? Stripped of all "genetic" afterthoughts, how would a simple description of the phenomena sound? Here is the answer. "Two contradictory traits, brought together by crossbreeding, will remain merged in the first hybrid generation but will break apart in the second, so that half of the grandchildren escape from the

---

*"The Listener", October 1959, No. 23.

mixture and become purebreds while the others repeat the game."

This is the way a breeder might express it, but it shows that the essence of Mendel's law can be stated without any mention of the mechanism of genes and chromosomes. In his book *The Theory of Knowledge Implicit in Goethe's World-Conception,** Rudolf Steiner showed how important it could be for science if the familiar laws of physics were formulated in a simple way. The so-called exact mathematical formulation is only a variant that is not necessarily more precise than another form. Mendel's law presented as a formula is only pseudo-mathematical, a mere coquetry.

Our formulation yields as important insight into the actual facts of heredity, viz. that the formative-forces-bodies of two crossbred forms try in the second generation to break away from the clash of contrasting traits and to get rid of each other. This succeeds in a predictable number of cases. This brings us to the noteworthy concept of the elasticity of the formative-forces-body, which appears in the repulsion of homologous but mutually contradictory qualities. Now we can see that the famous "autonomy" of traits (already assumed by Mendel) is an exceptional condition that is brought to an end by nature herself after a certain amount of time.

Thus it is possible at one and the same time to remain true to the phenomena and to achieve a deeper grasp of heredity. What is expressed in the idea of the elasticity of the body of formative forces can be tested in the crossbreeding experiments. The percentage of individuals that escape into purebreeding is a numerical expression of the power of the purifying recoil.** What the Mendelians call the purity of the ga-

---

*Anthroposophic Press, 1940.

**This percentage, i.e. the power of the recoil is much higher than is commonly realized. Mendel himself pointed out that, assuming four offspring in each generation, the tenth generation would produce 1023 purebreeding reds (RR), 1023 purebreeding whites (ww), and only 2 mixed breeders (Rw).

metes and the random combination in fertilization are merely devices that permit calculations according to the laws of probability. Such assumptions do not belong in the description of the phenomenon itself, but among the premises for a mathematical way of expression. They are correct only under the supposition that the location of the genes in the gametes is known. The observed facts, however, speak against one or the other of these premises. A basic objection to Mendelism is that the decisive agent is really not the genes, but chance, and that the law of segregation rests on the regular operation of the laws of chance. The famous ratio can be reproduced by tossing two pennies in the air: in a quarter of the cases, both will be heads; in another quarter, both will be tails; while the remaining two quarters will have one heads and one tails. This is how Mendel's figure of 1:2:1 arises.

One more remark on the Mendelian doctrine of the purity of the gametes. This looks like an indirect confession that in every crossing there is a sort of elastic recoil, even though it is supposed to happen in the gametes. The truth about the "elastic" nature of the etheric body can barely peep through this atomistic approach.

Although the Mendelians assert that they have accurately located many genes, there is still no way in which they can explain just how a gene, through its mere presence among the other genes, can exert influence on a predetermined and far removed point in the organism.* There is a great gulf between the supposedly constant and localized gene and the later-appearing trait. There is no visible connection between the ghostlike idea of the representation of all traits in the genome

---

*Again we quote Julian Huxley, *Evolution in Action*, Chapter 1: "The gene complex exerts its effects through the mechanism of individual development. There is always a longer or shorter chain of processes intervening between the genes in the chromosomes and the characters of the functioning animal or plant. Unfortunately, the precise way genes act during development is still very imperfectly understood."

and the actual appearance of the living animal. Yet this is absolutely essential if we are to understand the process of heredity, which consists precisely in the recurrence of each trait in its predetermined place. The developments in genetics during the past three decades clearly show that, if the basic idea of representation is to be maintained, the facts will require profound modification of all ideas about the nature of the genes and the traits. The original assumption of "one gene for one trait" has long since been abandoned. Let us see how the Mendelists, despite these difficulties, have striven to rescue their position and to continue to reckon with the genes as the "producers" of the hereditary qualities.

First it was necessary for them to assume that, in many cases, a single trait was caused by *several* genes, which had to collaborate to produce it. Since these genes often had very different locations in the chromosomes, their collaboration was not easy to understand, especially when the operation of a single gene was shrouded in complete darkness. As early as 1910 Nilssen-Ehle was forced to reckon with cumulative multiple hereditary factors in order to apply the Mendelian formulas to the results of his crossbreedings of wheat. Few students perceived at that time that this practice was beginning to undermine the classical concept of the gene. But today it is clear that it destroys one of the central assumptions on which Mendelism was originally based, the principle of the independence of each single gene and of the random combination of genes during fertilization. Both these postulates arose with Mendel himself.

Even commoner is the opposite case, where a single gene is said to produce several different traits (pleiotropy). We remain in perfect ignorance as to how one gene can work in a number of different directions or, stating it another way, how nature could have been led to entrust contradictory tasks to one and the same gene.

If the Mendelians seem to be reaching for straws, it is

because they want to cling at all costs to a concept that runs contrary to nature.

The same question arises in connection with the desperate assumption that there are "lethal" genes that cause death in all gametes into which they happen to enter. There is, of course, no way to observe the murderous effect; it is merely calculated by hindsight from the resulting figures.* In order to reconcile the actual figures on segregation with the gene-concept, it was necessary to assume the existence of lethal genes through which certain combinations were eliminated when the gametes were formed. But when an assumption has been disproved by the results, it is hardly scientific to try to save it by inventing lethal genes. What are we to think of a device that has no purpose other than to adjust contradictory results by a wave of the wand? Even if there were some truth in this subterfuge, we would have to wonder about nature herself, which would be producing and fostering genes whose sole function was to kill a large number of her other laborious productions before they could even come to life.

This recitation shows how the gene-concept departs further and further from reality and how the scientists who want to save it are forced into a more and more difficult attitude toward nature. But let us glance at a different set of observed facts, which shows from another side what thin ice the gene-hypothesis rests on. We refer to the phenomenon of polyploidy, or the presence of several complete sets of chromosomes in each nucleus. The best known cases are the giant varieties of delphinium and other garden flowers that have no abnormalities except their size. This is already strange and raises questions about the "determining" role of the genes. But the most remarkable thing is that these giant varieties undergo a spontaneous reduction, reverting to the normal

---

*Garrett Hardin, in the list referred to in the note on page 30, hopefully remarks: "*How* the gene kills is a matter for further biochemical research".

number of chromosomes after a few generations. This demonstrates that the living being, under its own power, can arrange one or several reductions in order to get rid of superfluous sets of chromosomes, and all without in any way touching upon heredity or segregation. This leads us to ask why it should be assumed that the reduction of the chromosomes during normal gamete-formation plays such an important and complex role.

Almost all the cells of the individuals of any species have the same number of chromosomes. Thus a full set of supposed hereditary dispositions is supplied to all tissues, even the most specialized, and not to the sex-cells alone. This speaks against rather than for the role of the chromatin as the producer of all hereditary traits. But even if every cell were supplied with all characters, we would only have to conclude that the actual qualities of the organs and tissues would have to be called forth each time by a different agent. This would abolish the genes' monopoly.

But it cannot be proved that the original gene-concept was correct in asserting that the genes were divided among the daughter-cells exactly in accordance with their future roles. If the older authors were right in this opinion, then "every disposition would have to follow a predetermined course through the cells in order ultimately to emerge as a trait purified from all other influences".* In other words, at all times all the contents of the chromosomes would have to be divided little by little among the separate regions and tissues of the body. The disposition for eye-color, for example, would have to be set free to work in the cells of the iris; the disposition for the shape of the head, in the skull-region, etc. Nothing supports such a hypothesis. The divisions of the nucleus are everywhere equal, thus giving the daughter-cells

*Wieser: *Gewebe des Lebens*, page 92.

all possible dispositions—if they really contain any dispositions at all.

Let us note again these two facts: first, that all the chromosomes are distributed to all parts of the embryo; second, that when there are extra sets of chromosomes, the organism can revert to the normal number by inserting its own processes of reduction. These indicate that the chromosomes, instead of collaborating actively in the functions of the cells and tissues, have much more of a reserve or auxiliary nature. In this connection we should never forget that, when an active cell splits, the chromosomes are always completely dissolved and reformed as the next division is prepared for.

The genes' monopolistic control of events in the cytoplasm can be refuted in still another way. For a hundred years it has been known that male honey-bees (drones) have only half a set of chromosomes, whereas the females (workers and queens) have the full so-called "diploid" set.* The only differerence in the nuclei of the two sexes is whether they have a half or a full set of chromosomes.** But note the enormous differences in organization and external appearance between the males and the females. Surely there is something wrong when we first ascribe these fundamental differences to the units of heredity, then assert that the chromosomes are identical except that one sex has twice as many. What does this do to the great hope that differences among mature animals can be correlated with differences among the genes? R. A. Beatty,*** after examining many cases of parthenogenetic haploidy in mammals, found no individual de-

---

*Diploid = having two homologous sets of chromosomes in each nucleus.
Haploid = having a single set of unpaired chromosomes in each nucleus; characteristic of gametes.

**M. J. D. Wight, *Animal Cytology and Evolution* (Cambr. Univ. Press, 1945) suggests that in the order of Hymenoptera all males are haploid, hence the bees are not exceptional.

****Parthenogenesis and Polyploidy in Mammalian Development*, Cambridge, 1957.

viations that could be traced to the number of sets of chromosomes, even in cases where there were more than two sets (polyploidy). If such polyploid embryos reach maturity, which they seldom do, they show none of the abnormalities of form that might be expected of them.

These results demonstrate how weak the argument is in favor of the autonomy of the genes.

We ought really to begin to doubt the functions of the genes when, through the microscope, we see the sequence of their positions on the loops of the chromosomes. The famous pictures of the chromosomes in the salivary glands of the fruit-fly (Figure 2) are striking, but even if the disks that we see there are really the genes (which many believe to be the case), they act like mere building-materials, prudently stored for use but completely redistributed when use actually occurs. One need only study the names that have been given to the various platelets. They read like a catalog that has been organized on a principle entirely different from that which later governs the qualities that are said to arise from the genes. It is obvious that the growth of the final form has nothing whatever to do with the arrangement of these platelets. This compels us again to postulate a system of forces that take hold of these building-materials and distribute them properly through the growing entity. The genes themselves can, as the need arises, be doubled in number, distributed in their carriers, reduced again, or completely eliminated, if this is necessary to prevent chaos from arising. How can one speak of autonomy in connection with such passive material, quite apart from the fact that the physiology of the cells shows that the various elements therein have a high degree of interdependence? Before every cell-fission the chromosomes arise anew from the divided chromatin of the active cells. The assertion that, because they always take on the same structure, the newly-formed chromosomes are the same old formations, is pure dogma and postulation.

Here we come again to the problem of the relations of nucleus and plasma. In recent decades there has been a vast increase in our knowledge of the finer physiological processes in the cell. It is now certain that the living plasma maintains the life of the nucleus, although this in turn feeds into the plasma certain materials that are influential there. The structure of the nucleus, whether in the loops or in the chromatin scaffolding of the so-called resting nucleus, is also beyond any doubt maintained by the living plasma. Thus the supposed guarantors of the future qualities are themselves physiologically dependent on the life-process. In addition, let us make the following general observation: the achievements of the individual cells do not add up to the achievements of the whole organism; the achievements of the whole organism determine what will take place in the separate cells. Theory may demand that the content of the nucleus, as the impulse-giver, should be independent of processes in the body, but for the physiologist this is nonsense.

The monopoly of the genes is being undermined by contradictory facts. The idea that they were trait-producers standing on the alert in the chromosomes was always governed by an atomistic localizing method of thinking, and must now yield to the concept of superspatial entities such as are described in our first chapter.

* * * * *

It is significant that for some years already reputable biochemists, such as Professor Carl Lindegren of Southern Illinois University, have opposed the idea that the genes are the "first movers". Lindegren writes:*

> Recent advances in understanding the chemical structure of desoxyribonucleic acid, together with evidence

**Nature*, London 30 September 1955, pages 1244-45.

> of its capacity to mediate the transfer of genes from cell to cell, have led to the belief that the secret of life lies in the nucleic acid. Such optimism has its basis in the idea that the gene is the fundamental, 'self-perpetuating', primary, living particle. . . . The possibility that the gene is a terminal or recent addition to the cell has not been sufficiently considered.

After this, he points out that, although the city of Chicago could no longer exist without its electrical installations, these were added to the city in fairly recent times.

Lindegren goes on to mention, with copious citations from other authors (Watson, Muller, Michaelis, Renner, Harder, etc.), those semi-autonomous formations such as the chloroplasts, microsomes, and mitochondria that reside in the cell but lie outside the chromosomes. He does this to show that a portion of the life-events are not steered by the nuclear substances, but enjoy a certain amount of autonomy. As to the gene itself, he asserts that, according to modern views, it consists of the pure chemical compound DNA and therefore cannot multiply itself by its own power, but must, like all living things, rely on its environment. This environment, through the enzymes, furnishes all the necessary syntheses. The secret of life is not to be found in the structure of any given chemical, "but in the organization and association which make the whole cell capable of producing all the components necessary to the synthesis of all the cellular organelles when placed in an adequate milieu."

Truly, an adequate milieu is the crux. But here we cannot be content with the plasma alone; the plasma itself points to larger connections. Examination of the interdependence of events in the cells and in the embryo leads us to begin with to the enzymes, which are outside the nucleus but seem to cause all the important plasmatic processes. The enzymes bring these processes into an equilibrium that is continually upset by the formation of new enzymes. But through these disturbances, according to the modern view, arises the differentiation of the cells that produces an embryo. Differences in development are caused by differences in enzymes. Every cell of the early embryo is furnished with different genes. Chemical differences precede organic differences. "Thus the formation of the heart", says Wieser,* "is announced by the appearance and concentration of myosin in certain cell-complexes. This substance is one of the proteins, which are responsible for the contraction of muscles. It works as a sort of inducer, touching off the next stages of organ-formation. One could say that the heart-muscle beats before there is a heart,** and that the heart arises in response to the beat."

But now we must ask whether the enzyme-endowment is anything other than a symptom of the differentiation-process, which would take place in any event. The apportioner cannot itself be an enzyme. Nor can it be identical with DNA, for this substance can become the great organizer only if it is guided as to sequences and locations; it must be made active and transported to or stored in various places if it is to undertake any function in shaping and forming. It seems more to wait for directions from outside than to itch to supervise the sculptural work.

Furthermore, the better we understand the constitution of nucleic acid, the less we understand how this labile substance could issue the directions. If it is (as many believe) a form-building substance, then we must join Wieser*** in demanding to know "the code in which it issues the commands that regulate the timing of all the events in the process of growth". And what could the answer be except that the entity govern-

---

**Gewebe des Lebens*, page 101.

**As early as 1920 Rudolf Steiner asserted that the heart should not be regarded as a pump, because in the embryo the beating appears earlier than the organ.

****Gewebe des Lebens*, page 102.

ing all these events in space and time is the formative-forces-body?

If one obstinately insists that the cell-products govern development, then the following problem arises. Each cell, as it is formed, is allotted the same living substance; but this substance must behave in highly diverse ways, depending on what region of the embryo the cell is located in. Different materials would have to migrate from the plasma into the nucleus in order to activate whatever gene-complexes were necessary at any given location.* The nucleus's monopoly would be destroyed, and the search for the governing principle would have to be transferred to the protoplasm. But the enzymes of the plasma are under the same disability as the nucleus; they point to something above space and time, and cannot themselves be regarded an autonomous. Like the nucleus, the plasma has no monopoly. If it is then asserted that the activities of the nucleus and plasma are attuned to each other, we must ask where the tuning takes place. The tuning of two things requires a third thing, but here the third thing cannot be present in ordinary space because it reaches beyond both time and space. Researchers who, like Wieser, want to form exact concepts in this field, regard the genome and the cytoplasmic units (mitochondria, microsomes, and plasmagenes) as a genuine "regulating system" with feedback connections. Thus the principle of "feedback", which man has applied to modern automatic machines, is drawn in to explain organic events. But such a feedback device is really only a simplified version of the comprehensive image of interwoven processes that we call life. It presupposes a third set of forces, which keep the two parts of the system that are connected by the "regulator" running, so that the regulator can work. But these directing forces cannot be on the same level as what is directed. They must be a system that awakens

---

**Gewebe des Lebens*, page 103.

order, *protecting* the organism from the chaotic influence of the environment.

We have shown that there is no self-regulating plasma, but that the various inclusions such as the chloroplasts, the antigens,* and other self-multiplying particles, enjoy a certain measure of autonomy. This leads us to the supersensible formative-forces-body, which works into the spatial structure and uses the plasma as its instrument and raw material.

Perhaps the reader would like to pause here and ask whether, after all we have said, any task remains for the chromosomes. Having lost their monopoly, have they lost all?

The chromosomes are storehouses for readily available building materials. Their character as storehouses is revealed by the curious orderliness that prevails in them, which we can verify in the so-called chromosome-maps. This orderliness has, as we said before, nothing to do with the later arrangement of the traits in the developed animal. It is more like the arbitrary orderliness that we establish in the attic when we put things to rights; a warehouseman's work, concerned with stowing rather than using; an alphabetical index, having no purpose other than to help us find what we need. In such an index the most disparate things may stand next to each other, but anyone who was troubled by this would have missed the whole point of the arrangement. When we grasp this, however, we also perceive that the real functions of these stowed-away substances are an entirely different matter. To use another comparison, the chromosomes are like a number of well-packed suitcases. The way in which the various objects are packed in the suitcases has no connection with the sequence in which they will be used when unpacked. They cannot be used at all until they are unpacked. The real activity of the nuclear substances occurs in the phases during which no chromosomes can be seen, whereas during the course of the

*Antigen—Substance capable of stimulating the formation of an antibody.

fission they are as it were condemned to idleness like the transported objects (see Figure 3).

But we must lay great weight on the fact that the chromosomes are formed *anew* whenever the cell prepares to split. This is when a double set of chromosomes is formed. We should never forget that in the last phase of one division or in the first phase of the next division, half of each chromatin thread is a completely new creation out of the plasma. This half is as though newly minted by a die. The mysterious force that forms this contrasting duplicate must lie in the medium in which it arises.

It cannot lie in the chromosome itself while this is being doubled. It has been shown experimentally, by means of radioactive isotopes, that the nucleic acid of the chromosomes goes through a constant process of renewal, hence is only apparently continuous and stable. The consequence is that this substance, instead of being responsible for maintaining the structure of the living being, is itself involved in a constant exchange of its ingredients. We can ascribe no continuity to it as a structured substance. We have already demonstrated this above.

The chromosomes are like master-moulds, direct contact with which produces the enzymes (Wieser, page 58), which then steer the metabolism in the plasma. We must admire the work of the biochemists in showing that the "side-chains" of the arising protein molecules are added as though cast in these moulds, so that the resulting form fits as a key fits in a lock. Thus in the secret places where living substance is formed, there is a minting process that creates mirror-pictures. This is a glimpse of the mysterious self-increasing power of the living proteins. But again the decisive thing in the arising of life is not the form of the mould but the activity that sets the casting in motion and, at the right place and the right moment, changes the mould.

No substance, even the most complicated, can be the bearer of life, because life is the bearer of all substance.

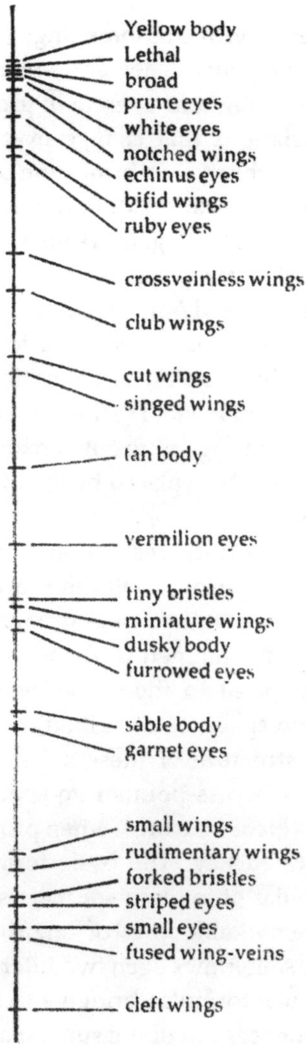

Figure 3

The "packing order" of the genes. (Map of Chromosome No. 1 in Drosophila. Twelve lethal factors are not shown.) After H. Nachtsheim.

Long experience with crossbreeding has forced a change in the concept of the genes. They are no longer viewed as lone wolves, but are thought to operate in groups, even in whole chains or associations that change over from one chromosome-half to the other.* This brings the concept closer to life, but at the same time shatters it. In the same way the forming anew of substance in the organic realm is no longer thought to be the handiwork of any particular material, but is ascribed to an activity that is spread out in space and that uses the physical structure of material merely as a form on which it can operate. The localized action of the particles was a materialistic dream, which had to yield to the concept of ordered collaboration. The originating of life out of matter was a completely false idea, and must be replaced by the idea that matter originates out of life.

Subject to these drastic restrictions, one can still speak of the role of the chromosomes. But they are now seen as products of the activity that they were formerly thought to cause. The enzyme-keys contained in them, when unpacked from their suitcases, point to the places where the life-bearing forces break into space, as described in Chapter I. The complexity of the structure of these substances hints that the formative forces' access-points into space must have a very special form; as great treasures, when properly guarded, need a very complicated key. The basic form, as shown by the structural formulas of modern specialists in the chemistry of albumen, is a remarkable spiral of longitudinal filaments with cross-binding, sometimes even two interlocking spirals. This suggests that such formulas bring us to a border area where the spatial condenses out of the superspatial and imprints on the emerging form traits that are not merely earthly but are

*In the Mendelists' so-called "crossing-over" we have an example of how the genes, in theory, work as closed societies, which is in flat contradiction to the classical concept of the gene as a lone wolf.

drawn from the movements of the stars; i.e. the loops that wind around an axis that moves forward on a line.*

In a last effort to understand the chromosomes, let us recall that early in the history of this branch of science one author, Valentin Haecker, asserted** that the much-discussed sex-chromosome (recognizable in man as a supernumerary, odd-numbered, or X-chromosome appearing in one or the other sex in the gametes) did not *determine* the sex, but merely *showed* what sex the gamete already had. This was called the index-hypothesis of the sex-chromosome. We will briefly review the course of Haecker's reasoning and two of his more closely studied examples.

In Aphis, one of the plant-lice, it is possible to tell in advance, during the formation of the spermatozoa, which of the daughter cells will or will not contain the heterochromosome, because those that will contain it are larger than the others and have mitochondria. The heterochromosome only confirms what one can already see, i.e. that half of the gametes will form females and the other half males. Thus the heterochromosome does not determine the sex, but only shows which it is. Similarly, in Phylloxera one can tell in advance that certain eggs will produce males, because they are smaller. They eliminate the heterochromosome in order to have a male set, but the elimination only confirms the sex that these eggs already have. Thus the sex decides whether the heterochromosome will be included or expelled from a certain egg, and not vice versa.

In these cases the heterochromosome is not the arbiter of sex, but an indicator (index) of the already determined course of development toward one or the other sex. Valentin Haecker cites several authors (Winge, Whitman and Riddle) who have

---

*Rudolf Steiner describes such lines as the true course of the sun with the planets spiraling around it.

**Valentin Haecker, *Allgemeine Vererbungslehre* (third edition, 1921).

connected sex with stronger or weaker metabolism and have therefore concluded that the heterochromosome is only an adaptation to a particular rate of metabolism, or is perhaps a means of keeping this rate constant. There have been cases where the sex was altered by artificially raising or lowering the rate of metabolism in the growing oocytes.* Such cases speak against the assumption that sex is determined by unequal (heterogametic) division. At the same time they bring the common explanation of the role of the chromosomes into serious doubt.

Thus the interpretations of chromatin and the chromosomes have gone through a remarkable number of changes, and these changes have by no means come to an end. For what is fair for the chromosomes as material phenomena is also fair for all other perceptible manifestations of nature. Every physically describable event in a living being must be regarded not as the thing itself, but as an index and symptom of its whole way of life. Physical substances and events will not thereby lose their importance for research, but they will be seen as pointers toward the real events. They will be manifestations of the impingement of imperceptible (supersensible) regions of reality. If we believe that the substances of the cells, tissues, and organs cause life, we close our eyes to the regions that are manifesting in them. But if we follow our clues, we can look into the actual workshops of life. Otherwise, they will not let themselves be "localized". Rudolf Steiner said to physicians that the efficacy of medicaments lay in processes rather than substances. It is not antimony that is therapeutic, but an "antimony-process" occurring where the substance is freed from its merely physical activity. Is it not possible that in the realm of life all substances go through such an emancipation and thereby become able to serve the phenomena of life?

*An oocyte is a cell that undergoes meiosis and thereby forms an ovum.

III
The Role of Heredity in Man

The unit that heredity has to manipulate in plants and animals is the species. Within the frame of the species occurs the transmission of form that we call heredity. In its concentration on the traits, which continually appear and disappear, the atomistic Mendelian view is apt to lose sight of this basic phenomenon of the "transmissibility of the species-form". In the preceding chapters we showed what can be done to shake up this view and bring the eyes back to the basic phenomenon. We also expounded the idea of the elastic formative-forces-body and demonstrated how, in the Mendelian segregation among hybrid offspring, we find a strong argument for the reality of this supersensible formative-forces-body. Indeed, in the facts of segregation we can, as Goethe would have said, see this elasticity with our own eyes.

The single individuals, whether plants or animals, receive from the hereditary stream as much of the common stock of forms as they can hold. Heredity is a pouring out of the species-form upon the individuals, who take up what flows into them, like vessels. Each separate being is a random receiver, comparable to a vessel overfilled from a spout. The smaller deviations of the individuals from each other (modifications) are not inheritable, hence are soon balanced out; at least, so long as no "mutation" occurs. But even a mutation, in the course of several generations, melts into the general species-form again, unless it is cultivated by breeders or, in exceptional cases, isolated by nature. When this happens, a new and special form can appear. But as a rule a mutation is drowned out in the same way as if it were a mere modification, although it may last a little longer before it disappears. In both animals and plants the species-form again restores itself as the center of gravity.

With Man* it is quite different. Here the individual deviation from the species-form has a fundamental significance. The species "Man", as a biological entity, has little reality; indeed, one can say that here the species is no longer something manifold and collective, but enters and disappears into the individual. Each man is a species of his own. This bold phrase of Rudolf Steiner's** puts our view of Man on an entirely new basis; his relationship to the species is exactly the opposite of that of the plant or animal. The single plant or animal is a sample of its species, and nothing more. The individuals are only random specimens of the basic form and have no permanent significance. (Only the theoretical Darwinists interpret the deviations as first steps toward a gradual change in the species.)

A person, however, is no such fleeting variation of the species-form "Man". As an individual he has an existence in his own right. Anyone seeking to know what "Man" is, must look at single men. Only they are real, in a spiritual sense, and only they demand recognition as separate beings. A multiplicity of men is always a sum of heterogeneous individuals, who may be added to one another but not melted into a collective entity.

Rudolf Steiner was the first to suggest that this view of Man can best be confirmed through biography. He uses the following words: "One who describes a person's real nature in a biography will understand that in this biography he has

---

*In a book on evolution, the German word *Mensch* must normally be translated as *man*. The reader may be startled to see this written with an initial capital, but the word occurs so often, in so many contrasts and combinations, that this practice may be advantageous when the species is referred to. We will try to avoid using "the human being". (Translator's note).

***Theosophy* (1904) Chapter 2 on "Reincarnation and Destiny".

something that corresponds to the description of a whole species of animals." In Man, the species-concept is transferred to the sphere of the individual, and we must rediscover the uniqueness of each person and its full scope. One could also say that each person is his own *type*, if this word had retained its real meaning instead of becoming trivial. The individual person is not a reprint of the general species-form; he is a separate spirit that appears in a body and establishes his own value. The special flavor in every life-history is the best means of showing the individual human stamp. Biological species always consist of many individuals, but each human species has only one member.

Putting the matter in another way, in the human individual the spiritual kernel, the ego, appears quite openly; whereas in the animal it remains hidden behind a multitude of individuals. In the two realms the "position" of this kernel towards the individual is quite different. Therefore in spiritual science the "group-ego" is spoken of among the animals and is seen as having a rather loose connection with the single animals; in contrast to which the "incarnating ego" of man steps right into the body and shapes it in its own image.* An animal is always a random variation of its species. A man is always unique, take him for all in all.

Now, the uniqueness of each human individual has, as Rudolf Steiner showed, a necessary consequence that we must not shrink away from. We must recognize that he has had repeated earth-lives. We cannot ascribe the differences between individuals to miracles of creation for each person, hence must view them as the result of development. The special coinage of the spiritual kernel (the ego) is best understood if we trace it back to earlier experiences. Rudolf Steiner

---

*In his book *Man and Animal*, the author has shown how this point reveals the basic difference between animals and man, and how it can be studied in five different aspects.

showed convincingly* that these earlier experiences must be conceived as taking place on earth rather than in some incorporeal existence; the main reason being that they have produced results which, like our childhood exertions in learning to write, have passed over into capacities in the permanent entity.

Just as experiences that we go through in youth coagulate into later capacities (or incapacities), so experiences in earlier incarnations can stay with us as abilities or short-comings returning with us and in us as part of our essence. Thus Man appears in later incarnations blessed or burdened with the results of earlier lives. (A more detailed explanation must be sought in the works of Rudolf Steiner or his students;** here our concern is only to lay a foundation for the traits of the individual person, which are entirely different from those created out of the storehouse of an animal or plant species.)

If one asks how the individual takes hold of the traits that he needs for his further development, the answer is that the human ego obtains them from its ancestors and parents. The body must be built up anew in each incarnation, after the ego has assembled, during the long interval between lives, the traits that it will have in the next life. The events preceding incarnation can, of course, be described only by spiritual investigation; but this reveals that the ego, long before it incarnates, begins to work with the line of future ancestors (as whose descendant it will later appear) and sees to it that the succeeding generations provide the needed hereditary properties. Thus the ego is anything but a casually selected vessel into which heredity pours its treasures. It is a point of crystallization that actively gathers from the generations the traits needed for its bodily development. This means that in animals and plants heredity is a passive event, whereas in Man it

Theosophy, Chapter 2.

**There is a good treatment in G. Wachsmuth: *Reincarnation as a Phenomenon of Metamorphosis*, Philosophic-Anthroposophic Press, 1960.

is a diligent selecting and combining. Man is the master of heredity, rather than its slave. Every incarnation is a masterpiece of selection out of the hereditary stock offered by a multitude of ancestors. Rudolf Steiner's works point out, however, that in this choosing man needs the aid and support of other beings.

It must be confessed that a bold rethinking is necessary if the received doctrine of heredity as a passive event is to be turned into the opposite, as required in the case of Man. But this is not impossible, and we are not vaguely pointing to a sort of "mystical" event. We are, however, greatly hindered by the fact that the customary view on the emergence of the individual out of the womb of heredity has made it difficult to conceive of the individual as *preceding* his heredity. But if it is hard to think of the traits being assembled by the ego out of the generations, it must be conceded that their coming out of dispositions in a bodily heritage is even harder to grasp. Yet the Mendelians take it for granted that the dispositions are somehow transported from their locations on the chromosomes to the places where they will develop into visible traits. This transportation is no easier to imagine than the ego's assembling-work. The fact that the ego can perform this work, selecting the traits with foresight, shows that it is connected with the will, which spiritual science asserts to be the ego's essential property. It does not merely look on; long before conception it actively asserts itself in combining the raw materials scattered through the generations. It follows from this that the ego is watching the generations before its own descent. This is a logical inference from the combination that later appears before the senses. But it can be supplemented by Rudolf Steiner's spiritual research. This asserts that the ego is actually influential in bringing the various ancestors together; in other words, that it plays a part in creating and guiding the hereditary streams among its ancestors. The combining of the threads is mainly (though not exclusively) the work of egos collaborating to bring about certain possibilities of incarnation. This leads to the further insight that obviously there are egos that work together in a neighborly way, being interested in similar combinations of traits. One could call them fishers in the same stream. They do not have precisely the same catches, but each profits from what the others have brought to merge into the stream. They work toward similar goals, hence the bodies that they produce are similar. This accounts for the resemblances in families, which are especially clear among brothers and sisters. These siblings have long worked together to produce the series of ancestors that culminated in their parents.

This means that brothers and sisters, for a long period before they are born, have seen eye to eye on how their ancestors should be combined; they have collaborated in a joint enterprise. The paradoxical result is that their bodily resemblance is a consequence of their long-standing connection through destiny. Those who belong together are not spiritually related because they have similar bodies; they have similar bodies because they have long been related in spirit and destiny.

We must add that this wisdom-filled knitting of the ancestral lines is not to be ascribed to the ego alone. This farsighted work depends largely on the profound wisdom of certain supersensible beings. The skeptic will, of course, resist the idea of such wisdom-beings,* but the skeptic himself is willing to ascribe a similar farsighted wisdom to the protein in the cells. This wisdom also transcends human understanding, because it is able to pursue many directions and purposes at one and the same time.**

---

*In spiritual science these are referred to as "Spirits of Wisdom". They are active in the formative-forces-bodies of all living beings.

**This is how it appears to the human understanding. In truth, however, nature pursues no limited goals. Her procedures can be called "multipurpose".

The idea of the ancestral lines being woven by the ego shows us why the special traits of a single man must be ranked with the differences between species. The developing body receives its individual features according to long-range plans, and its minting activity goes on through the first twenty years or so. From his heredity the child takes only a rough sketch of his form;* he must chisel out his full personality by his own efforts. The descending ego is always correcting as yet faceless deviations so as to bring the image of its personality to full expression. Therefore we see brothers and sisters, no matter how strongly the common hereditary model may shine through them, gradually diverging from each other through this individual chiseling of the features. The hereditary model is reshaped according to the special nature of each child.

Rudolf Steiner, who was uncompromising in this field, spoke of the replacement of the *hereditary model* by the ego's own handiwork, i.e. of the destruction of this model and the creation by the ego of something new. The ego-structure, he said, at first follows the hereditary model, but varies it at will, and ends by reconstructing it—a most remarkable procedure, falling somewhere between imitation and creating out of whole cloth. This is the most important key to the phenomenon of maturation occurring during the first two decades. One can see how the natural process of heredity is here raised into the human realm and intensified until it becomes a real transformation. This dissolves everything conservative and stable. Instead of merely stamping out new copies of the familiar forms, it leads to an entirely new shaping. Through being raised into the ego-realm, the spell of the past is broken; it has only one last service to perform for the present, viz. to prepare its way into the future.

Thus development in the human realm becomes a matter of *self-heredity*. Each man can say that he has not received his

---

*See Rudolf Steiner's pedagogical and medical courses.

traits from any ancestor or parent, because these had their own individuality as well as he. Therefore, no matter how paradoxical it may seem, he has received them from himself. In each succeeding life he is a further developed repetition of what he was in earlier incarnations. The phases of his development are drawn from the type that lies at the basis of his (and only his) individuality. This type has become identical with his spiritual kernel.

The sculptural work of the ego-type is not limited to the foresighted combining of ancestors through the generations. It continues during the embryonic period. This is why modern scientists have noted that the development of the human ego, from its beginning in the ovum through all the later phases, is governed by the special existence of "Man". Thus Portmann* remarks:

> As the sketch precedes the portrait, as its meager lines already contain the colored whole, and as each later stage adds essential features of the future work, so in the development of Man his special nature is present *from the beginning* in the decisive features; the human quality permeates the growth from the first moment. How many steps in development are arranged solely to produce two aspects that are peculiar to Man alone, his upright posture and his speech!

We must add, however, that every stage of growth, whether as embryo or after birth, does not reveal only the universal human image. This is, of course, permanently imprinted, but imprinted at the same time is the character of this particular person, of this special individual. The unique spiritual kernel works alongside the human image in general. In his appear-

---

*Adolf Portmann: *Zoology and the New Picture of Man*, Rowohlt's German Encyclopedia, Volume 20, Page 80.

ance, conduct, and destiny each person is a manifestation of *Man* and also of his own type. What belongs to the species is not allowed exclusive dominion, because then an egoless universal form would arise. This form is permeated with individual features that force the universal to recede. One could say that this universal form is like a basic presupposition (a premise) for the elaboration of the person's own shape and color. It is a preliminary design that is never fully executed.

The difference between Man and animal can perhaps best be expressed by saying that the single animal is always a mere variation pointing clearly to the model that lives in it; whereas a man emerges from the past and becomes "himself".

The growth of the individual as well as of the species is a continual rejection of "unsuitable" traits. For the species, a series of rejections is part of all advancement. To reach the stage of an amniote (reptile, bird, or mammal) means no-longer-to-be-an-anamniote (fish or amphibian); to become a mammal means no-longer-to-be-a-reptile; to become a man means no-longer-to-be-a-mere-hominid. The Darwinian doctrine of natural selection has no room for this phenomenon of rejection, because it regards development as a passive event (extinction of the unfit). But as soon as one sees an active growthlike process in the unfolding of the upward-leading forms, one can reasonably speak of tendencies (trends) of development. We will discuss this in Chapter IV.

Meanwhile it should be remarked that, in the heredity-processes in Man, this "rejection of the unfit" is something very real. It is not the same as in plants and animals. And if one accepts the active presence of the human individuality during the generations of the ancestors, then the process of rejecting unsuitable ancestral traits takes on more significance. For it is clear that this rejection is the counterpart of the "ancestor-selection" and has a wider import than the acceptance and adopting of desired traits. The converging hereditary streams of the predecessors offer much more than can be used. In preparing its body, the ego must get rid of far more than it needs for the construction of its unique form.

The activity of the incarnating ego is largely a sorting-out and rejection of the hereditary materials presented by nature. A person arises as his ego bridles the hereditary streams, either rejecting them entirely or choosing from them certain features to be built into his own form. We know that this makes a great demand on the reader, but we challenge him to entertain the thought that, wherever "mere heredity" prevails in a person, the shaping process is incomplete and really an illness. The passive transmission of form, which is necessary for the maintenance of an animal or plant species, is an illness in Man. *Where inherited traits appear without being altered, the ego is shirking its task.*

Such drastic statements make sense only when combined with other descriptions taken from spiritual science. They must be supplemented by ideas of the past of the ego and of its relationships with the so-called sheaths, the astral, etheric, and physical bodies. These ideas are suggested by a deeper study of the actual results of scientific work. Thus the word "hominization" which, in the modern study of fossils, has been commonly used for the gradual assimilation of the ancestral forms to the human form, is an empty phrase unless it means a process guided by definite forces. In spiritual science this is covered by the concept of the spiritual kernel, which intervenes very actively in the evolution of Man and gives it the decisive turn toward the human form. For the development of the individual man, the crucial concept is that of the unique ego and its prenatal activity, without which a complete description of the course of development is impossible. Lacking this concept, growth would have no "subject"; there would be nobody going through the development. We must reckon with this kernel right from the start, since it cannot be introduced later and the description will be a fragment without it.*

If the ego selects the parents before birth, the student will think of a number of further questions where Rudolf Steiner's indications are important. We cite these not because of their sensational content, but because they throw light on the way the hereditary garment is put on. In the period before conception, while the still unincarnated ego is striving toward the parents, there is a reciprocal *attraction* between the ego and the parents. The astral body of the descending person, says Rudolf Steiner, strives toward the mother, the ego toward the father; and only where both strivings are fulfilled, i.e. where a suitable pair of parents is found, can incarnation occur. Here we must repeat that such descriptions are discussible only in connection with others; they must not be taken out of context if they are to make sense. We have presupposed an understanding for the four-membered human constitution. On this basis it is easily understood that the astral body is more connected with the emotional imaginative fantasy-nature, and the ego more with the will-nature that governs the limbs. This makes it reasonable that the element of fantasy and thought should lead us to the mother, while the will connects us with the father. This is how Rudolf Steiner presented it.** Let us try to explore the consequences.

First of all, the two parents do not share equally in the forming of the child's nature. This is not because they trans-

*The family trees of royal houses give some clues as to how the incarnating ego is related to humanly arranged hereditary relationships. Georg Armborst's little book, *Genealogical Excursions through History*, contains a mass of interesting material on in-breeding (the so-called "loss of ancestors"). The family trees of Don Carlos, Frederick the Great, and Queen Isabella of Spain should be compared to see how degeneration arises through heredity. The individual element is here overwhelmed by the hereditary.

**Lectures of 14 November 1910 in Munich and 6 February 1911 in Düsseldorf.

mit different hereditary traits, but because one region in the child is father-influenced and another is mother-influenced. In the heredity of plants and animals, no such distinction is necessary; hence orthodox Mendelism has always assumed that both parents were of equal importance as trait-transmitters. Both are only conduits for the hereditary stock that is spread out over the progeny. In the strictest sense, however, this is not quite accurate. Many years ago we published a study of butterfly crossings in which the results varied as the mother-father combination was varied between the two strains. In the first weeks, the product of the crossing resembled the mother, and only later did the design and coloring of the father assert themselves; so that we had an intermediary hybrid.*

In Man, it seems to be much more important whether the father or the mother is the principal donor of heredity. The significant morphological difference between the two sexes is continued in the methods of transmission. There is a large and wholly unexplored field here for observation and reflection. Even now, however, there can be no doubt of the importance of these curious processes, hence Rudolf Steiner's indications could not be omitted here.

At this point we should again assert the new version of the "taking over of the form from the parents". In the case of Man, where a shaping ego must be reckoned with, this taking over is far from a passive process, even though a trait of the parental form or character may "reappear" in the children. We must always recall the sculptural activity of the ego, which utilizes a model or pattern but alters it freely and adopts only part of it without change. Combining, taking over, and adopting can be understood only as activities of the ego. We must conceive every human ego as equipped with its own specific capacity

---

*R. Goldschmidt and H. Poppelbaum, Zeitschift f. indukt. Abst. Vererb.-Lehre 1914.

IV

Precursors and Ancestors

One who seeks to portray the descent of living beings turns almost as a matter of course to the familiar figure of the tree. The trunk is rooted in the earliest ages of the earth and grows through the geological epochs. The branches are the various plant and animal families. There are older and younger branches. The degree of proximity of the single forms on the branches shows how closely they are related to each other.

One might expect to find a separate tree for the plant kingdom. But when one tries to work this out in practice, one discovers that the plant-tree and the animal-tree supplement each other to a remarkable extent. The major divisions of both kingdoms can be grouped together, all on one tree.* This work is very illuminating, particularly in showing how difficult it is to find a place for Man. One gradually realises that Man's course of development runs counter to that which leads to the specialized animals and plants. He really descends the trunk, in a direction opposite to the rising sap.

Until the last century there was little dissatisfaction with this picture of the tree as an explanation of the development from form to form. The progression came from the tree's forces of growth. Nature, it was commonly believed, gave birth to one form after another and did not have to be spurred on by any special force. She did it voluntarily, out of her own exuberance.

Only in the nineteenth century did anyone begin to suggest that some sort of external impulse was necessary to start the

---

*Gerbert Grohmann in his book *Pflanze* (1929) was the first to show the inner relationship between the major plant-groups and the phyla of the animals.

for this sculptural work. This is why the changes wrought in the inherited model are so diverse in different persons.

Thus at the close we come back to the mystery revealed by the spiritual-scientific study of human heredity—that merely allowing the hereditary process to occur, without any reshaping of what is inherited, points to a weakness in the ego, which has been charged with the reworking of what is transmitted. Such a feeble ego cannot properly *incorporate* what it inherits, hence this clings to the person and all his members as a burden, even as a real illness. In such a case one can see how, as though through a garment that does not fit the wearer, the inherited element peeks out everywhere, because it has not been worked into the new man. In the march through its incarnations, the human ego must learn not only to possess what it inherits, but to reshape it and work it over.

To be human means to make truly our own all that has been transmitted to us; to produce anew, in the light and through the power of the spirit-ego, all that is past; and to bring to rebirth, in the realm of our own responsibility, all that has been. It seems obvious to the unspoiled mind that beings standing higher than man must help in this work. This is confirmed in many ways through Rudolf Steiner's spiritual research.

tific information that portrays rBGH negatively (positively) in regard to health and environmental risks would increase (decrease) the expected individual difference (V^*). The predicted probability increase (decrease) should occur in particular for households at the margin for which this information would be sufficient to change their search behaviour.

Data

National-level supermarket scanner data for fluid milk demand were combined with information about the use of rBGH in milk production and product-specific labelling to evaluate labelling effects.[9] Over 13,000 supermarkets that either belong to national chains or operate independently in one of 64 metropolitan areas around the country were tracked by Information Resources, Inc. (IRI) over 13-week periods from January 1995 to December 1997 and over 4-week periods from January 1998 to December 1999.[10] IRI records sales quantity at the product code (UPC) level aggregated nationwide. Prices are temporally aggregated (within either a 13- or a 4-week period) and spatially aggregated and are based on list prices that do not take advertised sales into account.

The analysis focuses on beverage milk, excluding buttermilk and flavoured milk and only considers half-gallon and gallon containers. Prices and unit sales for fluid milk products offered by 11 different milk processors obtained from the IRI database were combined with information about the use of rBGH in milk production and the milk processor's labelling practice during the time period considered. This additional information was obtained through direct contact with processors. Eleven milk processors met the criteria for inclusion in the IRI database and also provided reliable information about rBGH and labelling characteristics.[11] These 11 firms produced 176 different milk products. They also accounted for 3.69% of nationwide supermarket skimmed and low fat milk sales and 2.23% of whole fat milk (as reported by IRI). Two of these 11 processors sold rBGH-free milk that was not labelled as such, seven sold fluid milk products that were labelled as rBGH-free and five sold conventional milk products.[12] None of the processors changed their policy with regard to rBGH use or labelling over the period 1995–1999. Milk products from all three categories of products – conventional, rBGH-free non-labelled and rBGH-free labelled milk – were available over the entire time period. The categorical data on product rBGH-characteristics were coded using two mutually exclusive dummy variables (*rBGHfreenonlabelled* and *rBGHfreelabelled*). Both of these dummy variables equal zero for conventional milk products.[13]

A market size variable (*marketsize*) that accounts for differences in the size of the market served by the 11 milk processors in this aggregated national-level data set was calculated. Annual state population estimates from 1995 to 1999 by the US Census Bureau (1999) for states in which the product is available were used to capture the number of potential consumers for a given milk product.[14]

9 A cooperative agreement with the USDA-ERS provided access to a commercial database.

10 IRI uses the food industry's definition of a supermarket: A grocery store with dairy, produce, fresh meat, packaged food and non-food departments, and annual sales of $2 million or more. Sales from health food stores, food cooperatives or natural food stores are not included.

11 Not all available milk processors are included in the IRI database, possibly because of limited size, local scope or unavailability in recorded supermarkets, and not all of the contacted milk processors were willing to provide reliable information about their rBGH policy.

12 Two milk processors produce items in more than one category, for example, rBGH-free labelled and conventional milk products.

13 Data on organic milk were initially also included in the analysis. However, organic milk sales were almost certainly affected by supply shocks associated with market penetration over the estimation period, resulting in implausible parameter estimates in the organic milk models. Thus, organic milk sales are excluded in the estimation models presented in the results section of this study.

14 Population estimates are expressed in terms of millions of people.

The data were organized into eight different fat content and container size categories to permit comparisons of homogeneous products. For instance, demands for whole milk products in gallon containers are estimated separately from demands for 2% milk in gallon containers and from demands for categories of milk in half-gallon containers. Two additional variables, the logarithm of quantity ratios between each milk product and its reference brand, and the price difference between each milk product and its reference brand, are also computed within each fat content and container size category. The reference brands used in the computation of these variables were derived generic private label product entries from the IRI database that were not identified with specific milk processors.[15] Within each fat content and container size category, the available generic private label entries were aggregated by computing simple price and unit sales averages. The generic brands used as reference brands are generally thought to include a substantial proportion of milk produced using rBGH, but are not labelled as such.

The panel data set used in the econometric analysis consists of 5293 observations. Each observation corresponds to a specific fluid milk product identified by its UPC that was sold in a specified 13-week or 4-week period from 1995 to 1999. Table 2.1a provides descriptive statistics for the variables included in the data set. Total sales of each brand ($unitsales_{mi}$) and total sales of reference brands ($unitsales_{mr}$) form the log quantity ratio, and prices for brand i (P_{mi}) and for the reference brand (P_{mr}) form the price difference variable.

Table 2.1b provides market share data for these fluid milk products across fat content and container size categories. Milk sales are distributed across these categories.

Table 2.1a. Descriptive statistics.

Variable	Observations	Mean	SD	Minimum	Maximum
$unitsales_{mi}$	5293	142,311	228,063	1	2,030,569
$unitsales_{mr}$	312	8,351,017	1.18×10^7	988,116	6.56×10^7
$\ln(unitsales_{mi}/unitsales_{mr})$	5293	−4.90	2.34	−14.91	−1.29
P_{mi}	5293	2.14	0.69	0.95	4.29
P_{mr}	5293	2.03	0.47	1.36	2.87
$P_{mi} - P_{mr}$	5293	0.21	0.43	−0.89	2.13
marketshare	5293	130.33	103.61	12.58	272.70
rBGHfreenonlabelled	5293	0.10	0.30	0	1
rBGHfreelabelled	5293	0.32	0.47	0	1

Table 2.1b. Market share of fluid milk products across fat content and container size.

	Observations	Mean unit sales	% of total unit sales
Half-gallon	3216	120,335.5	51.38
Fat-free	893	135,936.89	16.12
1%	799	94,011.279	9.97
2%	810	143,545.82	15.44
Whole	714	103,949.88	9.85
Gallon	2077	176,337.54	48.62
Fat-free	629	160,145.3	13.37
1%	372	148,949.87	7.36
2%	536	238,416.82	16.97
Whole	540	152,446.12	10.93

[15] Supermarket-specific private brands (e.g. Albertsons or Safeway store brands) could not be used as reference brands because they were not separately identified in the IRI database.

Econometric Specification

Following McFadden (1973) and Mathios (2000), conditional logit models were used to examine consumer purchases of fluid milk. The representative consumer's direct utility for purchase and consumption of fluid milk *i* given generally in equation (5) is specified linearly as:[16]

$$V_i^* = A_i\beta - \varepsilon_i. \tag{9}$$

In equation (9), the vector A_i indicates the attributes of milk brand *i*, and the vector β represents the weights the household places on them. The error term in equation (9) is assumed to arise mainly from randomness in attribute perception. Purchase and consumption of fluid milk *i* over alternative milk products indicates that:

$$A_i\beta - \varepsilon_i > A_j\beta - \varepsilon_j \text{ for all } j \neq i. \tag{10}$$

Under an *iid* logistic distributional assumption (Mathios, 2000), the probability that the *i*th fluid milk product (m_i) is purchased can then be written as:

$$P(m_i > 0) = \frac{e^{A_i\beta}}{\sum_{i=1}^{I} e^{A_i\beta}}, \tag{11}$$

where *e* the exponential function. The relative odds of the representative consumer choosing product *i* over some reference brand, m_r, is:

$$\frac{P(m_i > 0)}{P(m_r > 0)} = \frac{e^{A_i\beta}}{e^{A_r\beta}} \tag{12}$$

In the data set utilized in this study, the aggregate number of units of each fluid milk product sold in selected supermarkets is observed over either a 13-week or a 4-week period nationwide. Therefore, the left-hand side of equation (12) is redefined as unit sales of product *i* divided by unit sales of a reference brand. Redefining the left-hand side variable this way, and taking its logarithm, equation (12) can be rewritten as:

$$\ln\left(\frac{unitsales_{m_i}}{unitsales_{m_r}}\right) = (A_i - A_r)\beta' \tag{13}$$

Equation (13) forms the basis for the estimation equations used in the empirical analysis. The right-hand side of equation (13) is transformed into a linear function of the parameters for estimation.[17] In this formulation, the attribute difference vector $(A_i - A_r)$ denotes differences in attributes between the *i*th fluid milk product and the reference brand. This attribute difference vector includes price differences as well as information about whether a brand is conventional (and assumed to be produced with rBGH), produced without rBGH but not labelled as such or labelled as rBGH-free.

Milk products labelled as rBGH-free are more likely to be chosen by those consumers who, *ceteris paribus*, have a difference in utility between rBGH-free and conventional milk. The sign of the coefficient for milk that is rBGH-free but is not labelled as such is predicted to be smaller in magnitude than the coefficient for rBGH-free labelled products due to search costs. If search costs are higher than the difference in utility between rBGH-free and conventional brands, then unlabelled rBGH-free brands are only chosen by chance and the coefficient associated with the dummy variable indicating that they are rBGH-free but unlabelled should not be significantly positive.

The econometric estimation of fluid milk demand was carried out separately for each fat content (fat-free, 1%, 2% and whole) and container size (half-gallons and gallons) to allow for varying levels of substitutability between these products. Sample sizes for each of these fat content and container specific estimations were over 200.

Diagnostics

There may have been some important structural changes in the fluid milk data we evaluate. Organic fluid milk in gallons started to appear in supermarkets in April 1998, with steadily increasing aggregate sales. Additionally, data collection by IRI changed

16 The focus of this analysis is the choice of fluid milk brands based on their attributes. This focus, plus the assumption that the choice of $\underline{x}$ will be unchanged for different indirect utility functions, allows suppression of a constant term that relates to other goods consumed in this specification.

17 See also Mathios (2000) for a linear transformation of the conditional logit model.

from a 13-week to a 4-week period in 1998; this collection method change may affect demand estimates even though the dependent variable in the estimation equations is relative quantity. Chow tests (Chow, 1960) were conducted to investigate structural change in the market in 1998. The absence of structural change was rejected at conventional levels for approximately one-half of the fat content/container size categories. The regression results reported below split the sample time period for both half-gallon and gallon fluid milk demand estimation for consistency. Coefficient estimates for the split sample do not differ greatly for estimates over the entire sample.

Several tests for heteroskedasticity were performed on the data set. A general test for heteroskedasticity that does not specify particular variables (Breusch and Pagan, 1979) detected heteroskedasticity for all fat content levels and for both container sizes. More restricted tests for heteroskedasticity, using specific variables to define heteroskedasticity (White, 1980) failed to reject the null hypothesis.[18] Heteroskedasticity in the data appears to be introduced by a number of factors that cannot be easily separated. Consequently, the regressions were estimated in a generalized least squares (GLS) form and White-corrected standard errors are reported. Finally, no significant autocorrelation problems were detected in the sample, suggesting that the tests for structural change indicating a change in structure in 1998 result from the change in data collection procedures that occurred in 1998 rather than shifts in consumer behaviour over time.

The firms considered in the analysis comprise less than 5% of nationwide supermarket fluid milk sales in the IRI database. The share of one of these particular brands in a given region may be considerably greater than this nationwide proportion. Although the fluid milk supply curve at the (assumed to be perfectly competitive) farm level is expected to be quite flat, the derived retail fluid milk supply curves may not be. Of particular interest are price/quantity relationships for retail supply that relate to the costs of monitoring and enforcing rBGH-free labels. These costs are expected to arise at the milk-processing firm level, and are likely to differ across firms. Although we do not have useful measures for these firm-specific costs for fluid milk supply at the retail level, we instrument for them in the estimated demand models through firm-specific dummy variables for the milk processors included in the sample.

There are several alternatives to the instrumental variables approach that differ in the treatment of specific firm effects on the quantity of fluid milk purchases, including fixed effects models. Fixed effects models for demand without instrumenting for firm cost effects yielded very similar qualitative results to the instrumental variables approach. There were also generally no significant improvements in model fit when a random effects approach was utilized instead of fixed effects. Parameter estimates based on the instrumental variables approach are reported because of their potential to account for firm-specific cost differences in monitoring and enforcement for rBGH-free labels.

Regression Results

Tables 2.2–2.5 summarize instrumental variables regression results for the two separate time periods (1995–1997 and 1998–1999) and each fat content and container size combination. Estimated coefficients for price difference variables are negative and statistically significant in all fat content and container size combinations and both time periods. These coefficient estimates range from −5.56 to −2.32 for half-gallons and −3.93 to −1.0 for gallons, indicating that larger price differences between the product of interest and the reference brand leads to a decrease in the logarithm of the ratio of the quantity of its sales to the quantity of sales of the reference brand. Estimates for the *marketsize* variable coefficients are also positive and significant in most regressions, suggesting that, *ceteris paribus*, as the potential market for a product increases, sales increase.

[18] In applying White's approach, the population variable, month dummies and year dummies were used.

Table 2.2. Regression results for half-gallon, fat-free and 1% milk relative quantities.

	Fat-free		1%	
Independent variable	1995–1997	1998–1999	1995–1997	1998–1999
constant	−4.22***	−4.32***	−3.73***	−4.28***
	(0.23)	(0.09)	(0.14)	(0.08)
marketsize	0.002*	0.0007	0.004***	0.002***
	(0.001)	(0.0006)	(0.0007)	(0.0006)
rBGHfreenonlabelled	−0.72*	−1.67***	−2.41***	−1.60***
	(0.40)	(0.34)	(0.32)	(0.24)
rBGHfreelabelled	0.083	0.66***	−0.05	0.91***
	(0.35)	(0.20)	(0.23)	(0.17)
$P_{mi} - P_{mr}$	−2.32**	−3.34***	−3.08***	−2.96***
	(1.10)	(0.33)	(0.73)	(0.21)
Sample size	279	614	236	563
Degrees of freedom	274	609	231	558
F-statistic	2.49	35.78	51.65	92.72

Standard errors are corrected for heteroskedasticity and reported in parentheses.
*, ** and *** denote coefficients that are statistically different from 0 at the 10%, 5% and 1% level, respectively.

Table 2.3. Regression results for half-gallon, 2% and whole milk relative quantities.

	2%		Whole	
Independent variable	1995–1997	1998–1999	1995–1997	1998–1999
constant	−3.76***	−4.70***	−4.56***	−4.41***
	(0.18)	(0.16)	(0.23)	(0.11)
marketsize	0.002	0.004***	0.004***	0.002**
	(0.001)	(0.0007)	(0.001)	(0.0008)
rBGHfreenonlabelled	−2.88***	−1.98***	−2.0***	−2.71***
	(0.45)	(0.38)	(0.52)	(0.44)
rBGHfreelabelled	0.07	0.82***	0.37	0.26
	(0.20)	(0.18)	(0.30)	(0.17)
$P_{mi} - P_{mr}$	−4.65***	−3.75***	−5.56***	−4.17***
	(0.55)	(0.27)	(0.67)	(0.11)
Sample size	274	536	227	487
Degrees of freedom	269	531	222	482
F-statistic	24.56	59.21	24.24	62.23

Standard errors are corrected for heteroskedasticity and reported in parentheses.
** and *** denote coefficients that are statistically different from 0 at the 5% and 1% level, respectively.

The parameter estimates for *rBGH-freelabelled* are of central interest in this study. While these coefficients are only statistically significant for fat-free and whole milk in gallons for the first time period, they are significant and positive for almost all container sizes and fat content categories for the second time period, when the data were collected over more frequent intervals. Significant positive coefficient estimates range from 0.66 to 0.91 for half-gallons and from 0.57 to 1.57 for gallons. These results are consistent with the predictions of the

Table 2.4. Regression results for gallon, fat-free and 1% milk relative quantities.

	Fat-free		1%	
Independent variable	1995–1997	1998–1999	1995–1997	1998–1999
constant	−6.59***	−7.04***	−4.05***	−3.51***
	(0.32)	(0.17)	(0.47)	(0.12)
marketsize	0.005***	0.006***	0.004***	0.001
	(0.002)	(0.0009)	(0.002)	(0.0009)
rBGHfreenonlabelled	2.01***	2.02***	−0.61	−1.51***
	(0.31)	(0.17)	(0.85)	(0.30)
rBGHfreelabelled	1.41***	1.45***	0.31	0.57**
	(0.47)	(0.28)	(0.42)	(0.23)
$P_{mi} - P_{mr}$	−3.54**	−3.43***	−2.87***	−3.93***
	(0.68)	(0.27)	(1.05)	(0.38)
Sample size	184	445	121	251
Degrees of freedom	179	440	116	246
F-statistic	74.93	144.39	9.21	34.78

Standard errors are corrected for heteroskedasticity and reported in parentheses.
** and *** denote coefficients that are statistically different from 0 at the 5% and 1% level, respectively.

Table 2.5. Regression results for gallon, 2% and whole milk relative quantities.

	2%		Whole	
Independent variable	1995–1997	1998–1999	1995–1997	1998–1999
constant	−5.35***	−5.05***	−4.78***	−4.47***
	(0.41)	(0.22)	(0.31)	(0.21)
marketsize	0.005***	0.006***	0.003***	0.003**
	(0.001)	(0.001)	(0.001)	(0.0007)
rBGHfreenonlabelled	0.78	0.39	−0.09	−1.04***
	(0.71)	(0.38)	(0.49)	(0.33)
rBGHfreelabelled	−0.13	1.16***	0.96***	1.57***
	(0.34)	(0.18)	(0.33)	(0.22)
$P_{mi} - P_{mr}$	−1.0*	−3.08***	−1.69***	−2.52***
	(0.52)	(0.27)	(0.54)	(0.29)
Sample size	183	353	164	376
Degrees of freedom	178	348	159	371
F-statistic	6.50	51.19	4.01	25.46

Standard errors are corrected for heteroskedasticity and reported in parentheses.
*, ** and *** denote coefficients that are statistically different from 0 at the 10%, 5% and 1% level, respectively.

theoretical model. They indicate that labelling, by reducing information search costs, improves the quality of information about product characteristics and increases the quantity demanded for rBGH-free milk. The qualitative differences in the coefficient estimates for the two time periods suggest that consumers may adjust their purchase decisions in response to changes in labelling policies over time, although these differences may well be associated with the change in the frequency with which the data were collected in 1998. There is no evidence, however, that consumer preferences for rBGH-free milk have declined over the period in response to information that milk containing rBGH has few, if any, harmful health effects.

Positive effects on quantities demanded were not consistently obtained for rBGH-free products that were not labelled as such. The estimated coefficients for *rBGHfreenonlabelled* are negative and significant for half-gallons with parameter estimates ranging from −2.88 to −0.72. Estimated coefficients were only positive and significant for fat-free milk in gallon containers, but the *rBGHfreenonlabelled* coefficient estimates were significantly greater in magnitude than the coefficient estimates for the *rBGHfreelabelled* variable. The coefficient estimates for the *rBGHfreenonlabelled* variable for fat-free gallons might be influenced by firm-specific factors because only one of the two milk processors that offered rBGH-free non-labelled milk sold fat-free milk over the analysed time period. In all other regressions for milk in gallon containers, the estimated coefficients for the *rBGHfreenonlabelled* variable were negative and significant, ranging from −1.51 to −0.09. These results indicate that the provision of relevant information on a product label is required if market segmentation is to take place between conventional and rBGH-free products.

These results also suggest that adding a label on the package enables consumers to make an improved product choice and, as a result, increases consumer surplus, *ceteris paribus*. Figure 2.2 illustrates these effects.

Starting from a demand curve for undifferentiated conventional milk products (D_{conv}) in Fig. 2.2, the provision of labelling information shifts the demand for differentiated rBGH-free milk products (D_{label}) to the right. The lighter shaded area illustrates the resulting increase in consumer surplus when both products have the same price, *P*.

The model specifications permit the computation of own price demand elasticity estimates for rBGH-free labelled and conventional fluid milk.[19] Table 2.6 presents

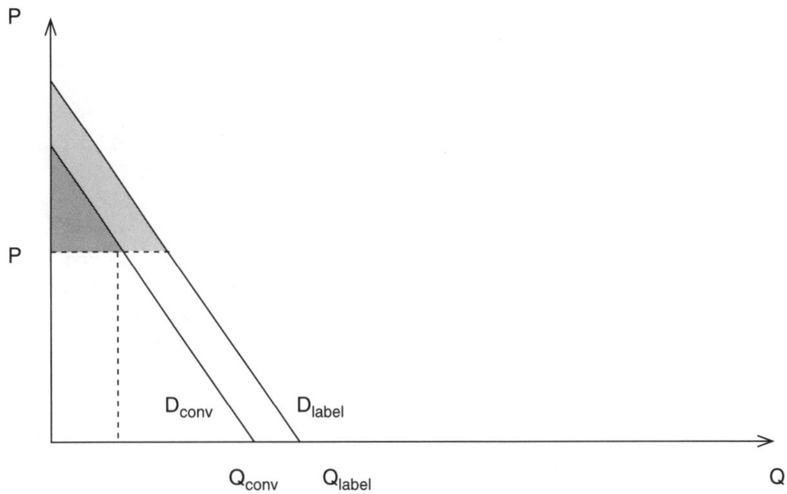

Fig. 2.2. Demand consumer surplus effects for rBGH-free labelled fluid milk.

[19] The following regression was estimated using an instrumental variables approach:

$$\ln\left(\frac{unitsales_{mi}}{unitsales_{mr}}\right) = \alpha_0 + \alpha_1 marketshare + \beta_1 rBGHfreelabeled + \beta_2(P_{mi} - P_{mr})rBGHfreelabeled + \beta_3(P_{mi} - P_{mr})conventional.$$

The price difference and rBGH-characteristic interaction terms were instrumented using the same exogenous variables as in the primary model. Carrying out appropriate transformations, price elasticities for rBGH-free labelled milk can be derived from: $\eta = unitsales_{mr} * e^{\beta_2 P_{mi}} * \beta_2 * \frac{P_{mi}}{unitsales_{mi}}$. For conventional milk, the same equation is used, substituting β_3 for β_2.

Table 2.6. Price elasticities computed at the sample means.

	1998–1999	
	rBGH-free labelled	Conventional
Half-gallon		
Fat-free	−0.28	−0.14
1%	−0.002	−0.68
2%	−0.14	–
Whole	−0.05	−2.35
Gallon		
Fat-free	−0.95	−0.00002
1%	−0.36	−0.0002
2%	−0.56	−0.0004
Whole	−0.12	−1.57

Estimated price elasticities reported in the table are significant at the 1% significance level.

the price elasticity estimates for rBGH-free labelled and conventional milk for the second time period.[20] Both of these categories only include branded milk products and the elasticity estimates reported here are specific to fat levels and container sizes. While some of these elasticity estimates are lower in absolute magnitude than previously reported price elasticities, in the aggregate, the results do not differ markedly from previous studies (e.g. Heien and Wessells, 1988; Gould, 1996; Green and Park, 1998, Glaser and Thompson, 2000). The elasticity estimates as a whole suggest no clear pattern in response to price changes between rBGH-free labelled and conventional milk. If only the estimated price elasticities for gallon milk products are considered, consumers appear more responsive to price changes in rBGH-free labelled milk than in conventional milk, but for half-gallon products there is no evidence of similar effects. It is important to note, however, that different fluid milk products do have different price elasticities of demand.

Conclusion

This study is innovative in several respects. A household production model of the effects of labelling has been developed that accounts for search costs and uncertainty about product attributes and the quality of information within a random utility framework. The model implies that the provision of more reliable information about product quality will, *ceteris paribus*, increase consumption of the commodity with a desirable but costly to observe characteristic and reduce consumption of a competing commodity with an undesirable characteristic. The predictions of the model were tested utilizing a new data set on actual purchases of fluid milk produced using rBGH and rBGH-free milk. The econometric results of the study indicate that the provision of labelling information increases the quantity demanded of rBGH-free milk, a result consistent with the predictions of the theoretical model. This result also confirms the findings of previous studies based on surveys of consumer attitudes (but not consumer

[20] Price elasticity measures were only computed if β_2 and/or (β_3 were statistically significant in the regression specification. For the first time period a number of the coefficients were insignificant, making comparisons between the elasticities impossible. In addition, using this criterion precluded estimating elasticities for unlabelled rBGH-free products.

behaviour in the marketplace) that indicate some consumers have a preference for milk and other foods that are not produced with biotechnology.

Another finding of interest in this study is that there is no evidence that consumer preferences for rBGH-free milk products have diminished since the introduction of rBGH milk products in 1994. The positive effects of labelling on rBGH-free fluid milk demand appear, if anything, to have increased in the period 1998–1999 as compared to the period 1995–1997. This suggests that food processors that postulate that consumer concerns over GMO products will diminish over time in response to a lack of evidence about adverse health effects may be wrong.

Finally, the results of the study are also important in that the findings indicate that own-price elasticities of demand for different categories of fluid milk are substantially different. While the elasticity estimates reported here are similar to those reported for aggregated fluid milk demand in other studies, they do differ between commodity milk and milk labelled as rBGH-free.

References

Aldrich, L. and Blisard, N. (1998) *Consumer Acceptance of Biotechnology. Lessons from the rBST Experience.* Agriculture Information Bulletin No. 747–01, Economic Research Service, US Department .of Agriculture, Washington, DC.

Becker, G.S. (1965) A theory of allocation of time. *The Economic Journal* September, 493–517.

Breusch, T.S. and Pagan, A.R. (1979) A simple test for heteroskedasticity and random coefficient variation. *Econometrica* 47, 1487–1494.

Burton, M.P., Metcalfe, J.S. and Smith, V.H. (2001) Innovation and the demand for food and drug labeling regulation in an evolutionary model of industry dynamics. *Structural Change and Economic Dynamics* 12, 457–477.

Caswell, J.A. and Padberg, D.I. (1992) Toward a more comprehensive theory of food labels. *American Journal of Agricultural Economics* 74, 460–468.

Chow, G.C. (1960) Tests of equality between sets of coefficients in two linear regressions. *Econometrica* 28, 591–605.

Glaser, L.W. and Thompson, G.D. (2000) Demand for organic and conventional beverage milk.' Paper presented at the Western Agricultural Economics Association Annual Meetings, Vancouver, British Columbia, 29 June–1 July .

Golan, E., Kuchler, F. and Mitchell, L. (2000) *Economics of Food Labeling.* Agricultural Economic Report No. 793. Economic Research Service, US Department of Agriculture, Washington, DC.

Gould, B.W. (1996) Factors affecting US demand for reduced-fat fluid milk. *Journal of Agricultural and Resource Economics* 21, 68–81.

Green, G.M. and Park, J.L. (1998) New insights into supermarket promotions via scanner data analysis: the case of milk. *Journal of Food Distribution Research* 29, 44–53.

Grobe, D. and Douthitt, R. (1995) Consumer acceptance of recombinant bovine growth hormone: interplay between beliefs and perceived risks. *The Journal of Consumer Affairs* 25, 128–143.

Grossmann, S.J. (1981) The informational role of warranties and private disclosure about product quality. *Journal of Law and Economics* 24, 461–483.

Heien, D.M. and Wessells, C.R. (1988) The demand for dairy products: structure, prediction, and decomposition. *American Journal of Agricultural Economics* 70, 219–228.

Ippolito, P.M. and Mathios, A.D. (1990) Information, advertising and health choices: a study of the cereal market. *RAND Journal of Economics* 21, 459–480.

Mathios, A.D. (2000) The impact of mandatory disclosure laws on product choices: an analysis of the salad dressing market. *Journal of Law and Economics* 43, 651–675.

McGuirk, A.M., Preston, W.P. and Jones, G.M. (1992) Introducing foods produced using biotechnology: the case of bovine somatotropin. *Southern Journal of Agricultural Economics* July, 209–223.

McFadden, D. (1973) Conditional logit analysis of qualitative choice behaviour. In: Zarembka, P. (ed.) *Frontiers in Econometrics*. Academia Press, New York.

McFadden, D. (1974) The measurement of urban travel demand. *Journal of Public Economics* 3, 303–328.

Misra, S.K. and Kyle, D.C. (1998) 'Demand for milk produced with and without recombinant bovine somatotropin. *Journal of Agribusiness* 16, 129–140.

Mojduszka, E.M. and Caswell, J.A. (2000) A test of nutritional quality signaling in food markets prior to implementation of mandatory labeling.' *American Journal of Agricultural Economics* 82, 298–309.

Monsanto (2000) *Status Update: Posilac Bovine Somatotropin*. Available at http://www.monsanto dairy.com/updates/Bovine.htm#general (accessed March 2002).

Rosen, S. (1974) Hedonic prices and implicit markets: product differentiation in pure competition. *Journal of Political Economy* 82, 34–55.

Smallwood, D.M. and Blaylock, J.R. (1991) Consumer demand for food and food safety: models and applications. In: Caswell, J.A. (ed.) *Economics of Food Safety*. Elsevier Science Publishing Co., New York, 4–27.

Stigler, G.J. (1961) The economics of information. *Journal of Political Economy* 69, 213–225.

Stigler, G.J. and Becker, G.S. (1977) De gustibus non est disputandum. *American Economic Review* 67, 76–90.

Teisl, M.F. and Roe, B. (1998) The economics of labeling: an overview of issues for health and environmental disclosure. *Agricultural and Resource Economics Review* October, 140–150.

Teisl, M.F., Roe, B. and Hicks, R.L. (2000) Can eco-labels tune a market? Evidence from dolphin safe labeling. Working paper, University of Maine.

Thompson, G.D. and Kidwell, J. (1998) Explaining the choice of organic produce, cosmetic defects prices and consumer preferences. *American Journal of Agricultural Economics* 80, 277–287.

Tweeten, L. (2000) Coexisting with alternative agriculture advocates (guest editorial). *Choices* Second Quarter, 3.

US Census Bureau (1999) State population estimates; annual time series, July 1, 1990 to July 1999. Available at http://www.census.gov/population/estimates/state/st.-99–3.txt (accessed September 2001).

White, H. (1980) A heteroskedasticity-consistent covariance matrix estimator and a direct test for heteroskedasticity. *Econometrica* 48, 817–838.

3 Consumer Purchasing Behaviour Towards GM Foods in The Netherlands

Leonie A. Marks,[1*] Nicholas G. Kalaitzandonakes[1] and Steven S. Vickner[2]
[1] *The Economics and Management of Agrobiotechnology Center (EMAC), University of Missouri-Columbia, Columbia, MO 65211, USA;* [2] *Department of Agricultural Economics, 400 Charles E. Barnhart Bldg, University of Kentucky, Lexington, KY 40546-0276, USA*

Introduction

It is June 1997 and Johanna van Buren[1] is at her local supermarket in the outskirts of Amsterdam. It's Wednesday afternoon and she is doing the weekly shop for her family of four. As usual, she is short on time so she is hastily walking up and down the aisles checking items off her list. When she reaches the aisle with the canned soups she is faced with an array of brands to choose from. Johanna already knows what she is looking for. She quickly picks up her favourite can of tomato soup; but as she puts the can in her shopping cart she notices the words *genetisch gemodificeerd* (genetically modified) on the ingredients list. She hesitates as she considers the new information. What will Johanna do next? Will she leave the soup in her cart or put it back on the shelf and look for another one that does not contain genetically modified (GM) ingredients?

The question of how consumers in Europe would respond to foods with labels indicating the presence of GM ingredients has been on the minds of biotechnology and food firms, researchers and policy makers alike. A large volume of research has been generated to gauge how European consumers might respond if faced with such a decision. Multiple opinion polls (e.g. European Commission (EC), 1997, 2000; Durant *et al.* 1998; Gaskell *et al.* 1999) have consistently indicated that a majority of Europeans would avoid GM foods in the market place. Yet few European consumers have ever been confronted with such a decision. Since 1999, food manufacturers and retailers in Europe have imposed voluntary bans against GM ingredients in their branded foods and have avoided mandated GM labelling altogether (Kalaitzandonakes, 2000).

Under these circumstances, no study has ever answered the question of how European consumers would *actually* behave in the presence of positively labelled GM foods. Instead, much of what is known today about consumer purchasing intentions towards GM foods in Europe is derived from attitude surveys. Divergence between attitudes and purchasing behaviour, however, is not uncommon and in the case of GM foods it has been observed in the past. As bovine somatotropin (rBST) was being introduced in the US market place in 1995, a survey conducted by Douthitt *et al.* (1996) found that 74% of consumers expressed concern about possible long-term health effects from

* Senior authorship is not assigned.

[1] Johanna is purely fictional, used in this context to represent a typical Dutch consumer.

consuming milk from rBST-treated cows and reluctance to purchase it. Hindsight being 20–20, we now know that such attitudes did not translate into significant changes in purchasing behaviour – or avoidance – on the part of US consumers (Aldrich and Blisard, 1998). The vast majority of US consumers purchased milk from rBST-treated cattle even when it was offered side-by-side with 'non-rBST' milk (Runge and Jackson, 2000).

In this study we depart from the previous literature and focus on revealed rather than stated consumer preferences towards GM foods in Europe. Specifically, we examine how consumers *actually behaved* when they could choose between GM-labelled and unlabelled food products in supermarkets across The Netherlands. We analyse national weekly point-of-purchase data for four separate food product categories that include products with GM labels. Our observations began in mid-1997. Three months after our initial observation, GM labels were introduced on a number of products that contained GM ingredients while the rest remained unlabelled. Almost 3 years later, GM labels were removed as manufacturers sourced non-GM ingredients. Hence, our data set allows us to investigate the behaviour of consumers when GM labels were introduced and then when they were removed from selected products.

Gauging Consumer Purchasing Intentions Towards GM Foods in Europe: What Do We Know?

There is a growing literature that is investigating consumer attitudes and stated preferences towards GM foods in Europe. What follows is a brief discussion of the main approaches used to date and their potential advantages and pitfalls in anticipating consumer preferences and behaviour in the market place.

Attitudinal surveys

Opinion and attitude surveys about GM foods range from the in-depth studies carried out by various academics, government agencies and organizations (e.g. Hoban, 1996, 1998; EC, 1997, 2000; Durant *et al.*, 1998; Gaskell *et al.*, 1999), to up-to-the-minute polls carried out by the media. A range of questions has been asked at different locations and points in time. Responses vary considerably depending on how the questions are framed, the kind of sample used (e.g. size, demographics, location), and over time.

Universally, Europeans have been more negative about GM foods than other applications of biotechnology, e.g. medical or environmental ones (Durant *et al.*, 1998; Gaskell *et al.*, 1999; EC, 2000). Likewise, public sentiment towards GM foods has been more negative in Europe than in North America and elsewhere (Hoban, 1996, 1998; Gaskell *et al.*, 1999). Even within Europe, however, attitudes have varied by country and over time. For instance, southern European countries were more accepting of GM foods than northern countries up until 1997 (Zechendorf, 1998). By 1999, Italy and Greece exhibited the most negative attitudes towards GM foods in Europe (EC, 2000).

Attitude surveys can capture public sentiment towards GM foods and biotechnologies, and investigate the sources of such sentiment and other associations. They are relatively easy to execute, cost effective and can be generalized to a population. Attitude surveys are constrained, however, by their hypothetical structure, especially since they do not account for price and income effects on consumer-stated preferences. Attitude surveys may also engage their subjects as citizens rather than strictly as consumers (Noussair *et al.*, 2002). Subjects may use information and beliefs to respond to the survey that they would not use while making narrower purchasing decisions (Noussair *et al.*, 2002). Accordingly, attitudes may or may not be effective proxies of consumer market behaviour.

Importantly, attitude surveys can be subject to significant biases. How questions are framed, the order in which information is presented, the degree of knowledge and understanding of the respondent, are just some of the potential sources of bias and error (Tolley and Randall, 1983; Kahneman and Tversky, 1984; Sterngold *et al.*, 1994). Biases, can be minimized but not eliminated through careful design and statistical analysis.

Willingness-to-pay studies

While the bulk of current research has focused on attitudinal surveys, some researchers have utilized the concept of 'willingness to pay' in order to capture how consumers might respond to GM foods if faced with realistic food choices. Consumers may be willing to pay more for GM foods exhibiting desirable attributes (e.g. Boccaletti and Moro, 2000; Burton and Pearse, 2002). Alternatively, consumers may be willing to pay more to avoid them altogether (Moon and Balasubramanian, 2001; Lusk *et al.*, 2003).

Contingent valuation is the best-known, most frequently used 'willingness to pay' method. Willingness-to-pay measures are estimated from direct consumer responses to a set of hypothetical questions. Consumers are given a hypothetical scenario and are asked to make a hypothetical choice. Surveys usually also gather socioeconomic data about the respondents, ask additional attitudinal questions, and use follow-up questions to evaluate whether the respondents understood the scenario presented.

Using willingness-to-pay surveys, Lusk *et al.* (2003) and Moon and Balasubramanian (2001) concluded that many European consumers would need a discount to consume foods with GM ingredients. Specifically, Lusk *et al.* (2003) compared consumer valuations of beef steaks from cattle fed with GM maize across four countries – France, Germany, the UK and the USA. They found that European consumers placed a higher value than US consumers on beef raised without GM maize. Moon and Balasubramanian (2001) compared US and UK consumer valuations of GM and non-GM foods with specific attributes. They concluded that UK consumers were more willing to pay price premiums to avoid GM foods than US consumers.

Willingness-to-pay surveys can yield important information about individual consumer-stated preferences. They can also usefully combine such information with demographic data and consumer attitudes. However, they are limited by their hypothetical design, and by the constraining complexity of GM food choices. This criticism is often stated as follows – if you ask a hypothetical question you get a hypothetical answer. Responses on hypothetical scenarios run the risk of giving unreliable results particularly when respondents are not well informed. This criticism is pertinent to GM food studies where consumers can exhibit a high level of unawareness about the technology.

The approach is also susceptible to two other important types of bias – strategic bias and hypothetical bias. Strategic bias can occur when consumers deliberately understate or overstate the true value they place on an attribute – for example, if they believe that by so doing they might influence a policy outcome. Hypothetical bias, on the other hand, typically occurs when consumers are unable to accurately assess their willingness to pay. Hypothetical bias is possible even in well-designed surveys, particularly when consumers have limited prior experience with the attribute (in this case GM versus non-GM). Lack of actual purchasing choices can make it very difficult for consumers to become aware of their own preferences so that they can place a value on changes in price, quantity and quality. Finally, some studies have shown that even the order in which questions are asked can affect willingness-to-pay measures by a significant magnitude (Tolley and Randall, 1983).

Experimental auction studies

A handful of recent studies have used experimental auction markets to directly elicit consumer preferences towards GM foods (Buhr *et al.*, 1993; Huffman *et al.*, 2001, 2002; Lusk *et al.*, 2001; Noussair *et al.*, 2004). Such experimental approaches can produce more detailed results on consumer-stated preferences than attitude surveys and willingness-to-pay studies.

Noussair *et al.* (2004) investigated European consumer response to GM-versus non-GM-labelled foods in an experimental laboratory setting in Grenoble, France. The study consisted of a representative sample of 97 consumers. The consumers compared various biscuit brands, some of which carried a GM label, while others carried an organic label or no label at all. Noussair *et al.* (2004) found that 35% of consumers boycotted GM-labelled foods, 40% were willing to purchase products containing GM ingredients if they were sufficiently inexpensive, and 25% of the participants were indifferent and would purchase them regardless.[2]

Experimental auction markets provide a more realistic environment than surveys for eliciting consumer preferences. Indeed, economists have developed approaches that offer incentives for consumers to bid their true reservation value (Menkhaus *et al.*, 1992). Likewise, experimental auction markets allow researchers to ask attitudinal questions of consumers and, in turn, relate them to stated preferences. Finally, experimental markets allow researchers to introduce different types of information shocks, and to observe changes in participants' behaviour (Shogren *et al.*, 1999).

Of course, experimental auction market analysis is not without limitations. Though elaborate, experimental auction markets are still artificial environments. The range of items for purchase is much more limited than in an actual retail store. Furthermore, participants are typically asked to bid, paying 'real' money for the purchase of 'real' goods at the end of the experiment. However, participants still know that they are being observed and can deviate from normal behaviour. For instance, participants may fall foul of what is known as the 'Hawthorne effect' where they can overstate their bids to please the monitor of the experiment (Shogren *et al.*, 1999, p.1192).

Accordingly, consumer-stated preferences elicited in experimental auction markets can be different from normal purchasing behaviour exhibited in a store environment. For example, Shogren *et al.* (1999) examined consumer response to irradiated chicken cuts in three different settings – an actual retail store where the irradiated food item was clearly labelled as such, an experimental auction market and a hypothetical market survey. They found that the acceptability of the product was greatest in the hypothetical setting and least in the retail outlets. Shogren *et al.* (1999) hypothesized that differences in acceptance were due to the importance of information provided in a leaflet accompanying the irradiated food in the experimental setting. In the store, consumers did not read the leaflet and were more negative towards food irradiation.

Generalization from experimental auction market research can also be problematic. Given that most such studies are carried out in specific locations and/or with small samples of consumers, demonstrating the representativeness and generality of the results can often be challenging.

Measuring Consumer Behaviour towards GM Labels in the Netherlands: Approach and Procedures

The literature review in the previous section suggests that substantial research effort has been focused on eliciting consumer attitudes and stated preferences towards GM foods in Europe and elsewhere. Yet attitudes and stated preferences are only proxies to actual consumer behaviour towards GM foods. Such behaviour has not been directly analysed in

[2] A few additional studies have used experimental auction markets to investigate stated consumer preferences outside of Europe. Buhr *et al.* (1993) used a split-valuation experimental auction to evaluate consumer response to porcine somatotropin (PST)-treated pork products in the USA. Despite early opinion surveys that indicated that consumers would avoid such products (Hoban and Burkhardt, 1991), Buhr *et al.* found that consumers were willing to pay a premium for leaner meat with fewer calories produced using PST. More recently, Huffman *et al.* (2002) used experimental auction markets to investigate consumer willingness to pay for both positively and negatively labelled GM foods in the USA. The experiments were conducted in two mid-western US cities across a random sample of 142 residents. They chose three different types of products, vegetable oil, tortilla chips, and a bag of potatoes, in order to capture how consumers might react to GM content in different types of foods. They found that consumers were able to distinguish between GM and non-GM labels and responded to such information (Huffman *et al.*, 2002, p. 21). Lusk *et al.* (2001) used experimental auction markets to examine willingness to pay for non-GM maize chips among college students in a particular US location. They found that 20% were willing to pay some premium for such chips.

Europe.[3] In this study, we examine revealed instead of stated consumer preferences by directly analysing consumer behaviour towards GM foods in one European country – The Netherlands.

The Netherlands provides an excellent case study for examining consumer response to GM food labels. First, a meaningful number of food products carried positive GM labels for an extended period of time allowing detailed analysis of actual consumer behaviour. Second, Dutch consumers have consistently exhibited higher overall levels of awareness of biotechnology and GM foods in opinion surveys (Hoban, 1998; Hamstra, 1993). Third, The Netherlands has a long history of active consumer involvement in food and nutrition policy, with over 10% of Netherlands households belonging to consumer organizations (Hillers and Lowik, 1998).

Model development

For our analysis, we use a statistical model of consumer response, consistent with the contemporary time series demand literature (Swartz and Strand, 1981; Smith *et al.*, 1988; Teisl *et al.*, 2002). The underlying framework is the well-known AIDS model (Deaton and Muellbauer, 1980). Other studies have used the logit-based, random utility framework (e.g. Mathios, 2000). This latter approach employs an arbitrary Gumbel error distribution and often violates the Independence of Irrelevant Alternatives (IIA) assumption (McFadden, 1973).

A national-level, syndicated point-of-purchase scanner data set is used for the analysis. Four product categories are considered: canned soup, frozen processed meat, frozen pizza and frozen processed fish. In each product category some products contained GM ingredients and carried a GM label while others did not and remained unlabelled. The data set spans 247 consecutive weeks, from the Sunday ending 13 April 1997 to the Sunday ending 30 December 2001. Within this time frame we capture two important natural experiments of consumer response to GM labelling. Across those relevant products containing GM ingredients, the labels were introduced in the 11th week of our data set, the week ending on Sunday 22 June 1997. The labels remained on relevant products for 151 consecutive weeks and were removed the week ending Sunday 14 May 2000 as non-GM ingredients were sourced.

The products in each product category were aggregated into two groups: GM-labelled and unlabelled. Our analysis therefore compares consumer behaviour towards GM-labelled products and unlabelled products sold side-by-side. Our non-linear conditional expenditure share model accounts for the separate influences of own price, the price of substitutes, per capita real expenditure for the product category, holiday effects, and the addition and removal of GM labels. Consumer behaviour could also change over time in response to new information about GM foods and biotechnology. To control for such effects, we include a variable that measures the amount of relevant information that could have been received by consumers from media sources. Accordingly our AIDS model is specified as follows:

$$\begin{aligned} w_{it} &= \alpha_i + \sum_{j=1}^{2} \gamma_{ij} \log p_{jt} + \beta_i \log(x_i / P_i) + \\ &\eta_i z_t + \phi_{i1} h_{1t} + \phi_{i2} h_{2t} + \lambda_{i1} on4_t + \lambda_{i2} on24_t + \\ &\lambda_{i3} off4_t + \lambda_{i4} off24_t + \theta_{i1} m_t \cdot on4_t + \\ &\theta_{i2} m_t \cdot on24_t + \theta_{i3} m_t \cdot off4_t + \theta_{i4} m_t + \\ &\cdot off24_t + \varepsilon_{it} \end{aligned} \quad (1)$$

where

$$\log P_t = \alpha_0 + \sum_{k=1}^{2} \alpha_k \log p_{kt} + \frac{1}{2} \sum_{j=1}^{2} \sum_{k=1}^{2} \gamma_{jk} \log p_{jt} \log p_{kt}. \quad (2)$$

i = 1 indicates GM-labelled products, i = 2 represents all unlabelled products and t indicates time measured in weeks.

[3] Outside Europe, two studies have focused on revealed consumer preferences towards GM foods. James *et al.* (2002) set up a limited market experiment and observed consumer purchasing patterns towards GM and non-GM sweetcorn they placed in a few grocery stores in a college town in Pennsylvania, USA. Kiesel *et al.* (2002) examined a national dataset of actual consumer purchases of fluid milk produced with rBST and rBST-free milk. Thus, they examined consumer response to negative ('does not contain') labels. Their results indicate that a small segment of consumers responded positively to such labels.

To clarify the specification above, consider the GM-labelled soup market share equation (w_{1t}) in week t. It is a function of own price ($\log p_{1t}$) expressed as an index (Moschini, 1995), the price of unlabelled soup ($\log p_{2t}$) expressed as an index, per capita real expenditure for the entire soup product category ($\log(x_t/P_t)$), a linear time trend (z_t), holiday effects (h_{1t},h_{2t}), product labelling variables ($on4_t$,$on24_t$,off4t, $off24_t$), external information variables ($m_t \cdot on4_t$,$m_t \cdot on24_t$,$m_t \cdot off4_t$,$m_t \cdot off24_t$) and a stochastic error. Equation (2) delineates the construction of the unobservable non-linear price index.

The holiday effects (h_{1t},h_{2t}) are set to one in the week of the calendar holiday and zero otherwise. The Queen's birthday (observed on 30 April), Remembrance Day (observed on 4 May) and Liberation Day (observed on 5 May) are combined into one holiday effect (h_{1t}) given the closeness in dates. Sinterklaas (observed the eve of 5 December and the following day) is given by (h_{2t}). The linear time trend (z_t) increments by one unit for each week in the database. The external information and GM labelling variables are discussed in more detail in the following sub-sections.

The conditional expenditure shares sum to one when the following 'adding up' conditions hold $\alpha_1 + \alpha_2 = 1$, $\gamma_{11} = \gamma_{21} = 0$, $\gamma_{12} + \gamma_{22} = 0$, $\beta_1 + \beta_2 = 0$, $\eta_1 + \eta_2 = 0$, $\phi_{11} + \phi_{21} = 0$, $\phi_{12} + \phi_{22} = 0$, $\lambda_{11} + \lambda_{21} = 0$, $\lambda_{12} + \lambda_{22} = 0$, $\lambda_{13} + \lambda_{23} = 0$, $\lambda_{14} + \lambda_{24} = 0$, $\theta_{11} + \theta_{21} = 0$, $\theta_{12} + \theta_{22} = 0$, $\theta_{13} + \theta_{23} = 0$ and $\theta_{14} + \theta_{24} = 0$. The homogeneity conditions are given by $\gamma_{11} + \gamma_{12} = 0$ and $\gamma_{21} + \gamma_{22} = 0$, and the symmetry conditions imply that $\gamma_{12} = \gamma_{21}$. These represent statistically testable hypotheses regarding the theoretical consistency of the empirical non-linear conditional expenditure share system. In the estimation of each product category, one share equation is estimated with non-linear least squares (Greene, 2000). The parameter estimates for the other share equation are then recovered using the adding up conditions.

Testing consumer response to labelling information

The main empirical question addressed in this paper is the degree to which Dutch consumers avoided labelled GM food products when they had the opportunity to do so. The product labelling variables ($on4_t$,$on24_t$,$off4_t$, $off24_t$) in equation (1) are used to measure consumer response to on-package labelling information related to GM ingredients.

A number of functional forms have been proposed in the literature to model the impact of labelling information on consumer purchases over time, including dummy variables, linear trends, non-linear trends and S-shaped response functions. After some experimentation with all of these functional forms, we constructed 4-week and 24-week linear trends to model potential immediate and/or gradual impacts from the introduction of GM labels.[4] Similarly, we constructed 4-week and 24-week linear trends to model the immediate and gradual potential response to the removal of GM labels from those same products. The weeks before and after each linear time trend were set to zero in each of the four series.

Our linear response functions allow for possible delays in consumer response over 4-week and 24-week intervals, both when the labels were introduced and when they were removed. It is expected that labelling information diffuses among consumers over some period of time. It might take some time for consumers to recognize that such labels have been placed on specific food products. Their avoidance might take even longer to manifest itself in the market place. Our specification accounts for such effects and allows for an immediate (1 month) response and a more gradual (6 months) consumer response, both when the labels were placed on the products and when they were removed.

For each of our four food product categories, if consumers adversely react to the introduction of the label immediately we expect $\lambda_{11} < 0$ in w_{1t} and $\lambda_{21} > 0$ in the w_{2t} *ceteris paribus*. If consumers do not change

[4] It should be noted that the choice of response variable affected the overall model fit but did not affect the qualitative results reported in this study.

their consumption patterns immediately after the introduction of the label, we expect $\lambda_{11} = \lambda_{21} = 0$ in both share equations *ceteris paribus*. Similarly, if consumers adversely react to the introduction of the label in a gradual fashion, then we expect $\lambda_{12} < 0$ in the w_{1t} and $\lambda_{22} > 0$ in the w_{2t} *ceteris paribus*. If consumers do not change their consumption patterns in response to the introduction of the GM labels in the long run, we expect $\lambda_{12} = \lambda_{22} = 0$ in both share equations *ceteris paribus*.

In a similar line of reasoning, if consumers favourably react to the removal of the label within a month's period, then, *a priori*, we expect $\lambda_{13} > 0$ in the w_{1t} and $\lambda_{23} < 0$ in the w_{2t} *ceteris paribus*. If consumers do not react to the removal of the label in the short run, we would expect $\lambda_{13} = \lambda_{23} = 0$ in both share equations *ceteris paribus*. Similarly, if consumers favourably react to the removal of the label in the long run, then we expect $\lambda_{14} > 0$ in the w_{1t} and $\lambda_{24} < 0$ in the w_{2t} *ceteris paribus*. If consumers do not react to the removal of the label in the long run, we expect $\lambda_{14} = \lambda_{24} = 0$ in both share equations *ceteris paribus*.

Modelling information on biotechnology through media coverage

It is possible that over a long period of time, consumers could change their purchasing behaviour towards GM foods in response to relevant new information. Since our study examines consumer behaviour over a 4-year period, influences from external information must be accounted for.

Over 90% of consumers receive information about food and biotechnology primarily through the popular press and television (Hoban and Kendall, 1993). Previous studies have found that media information on food risks (such as food contaminants) can affect food demand in the short run (Wessels *et al.*, 1995; Smith *et al.*, 1998; Dahlgran and Fairchild, 2002). Cumulative, long run effects have also been observed – particularly when health studies are diffused over long time periods (e.g. food additives, the link between cholesterol in meat/eggs and heart disease) (Brown and Schrader, 1990; van Ravenswaay and Hoehn, 1991; Wessels *et al.*, 1995). Therefore, both short run and long run effects are accounted for in our model specification.

We expect consumer reaction to be amplified during information-augmenting events, which raise awareness about biotechnology. Gaskell *et al.* (1999) and Durant *et al.* (1998) have found that heightened media coverage increases awareness of biotechnology. We therefore expect that shifts in consumer demand for specific product ingredients (e.g. GM versus unlabelled products), if they do exist, to be more distinguishable around heightened media coverage (Marks *et al.*, 2002). Accordingly, a media variable (m_t), which measures article frequency in key newspapers, is used interactively with the labelling variables ($on4_t$,$on24_t$,$off4_t$,$off24_t$) to capture the potential role of information received by consumers from global media sources.[5] That is, m_t was multiplied by the four label variables to create four media–label interaction effects that serve as additional shift variables in the AIDS model.

Similar to the label hypothesis tests above, if consumers decrease their purchases of GM-labelled foods in response to increased media coverage in the short run, then *a priori* we expect $\theta_{11} < 0$ in the w_{1t} and $\theta_{21} > 0$ in the w_{2t} or unlabelled share equation *ceteris paribus*. If consumers do not react to the label given media coverage, we expect $\theta_{11} = \theta_{21} = 0$ in both share equations *ceteris paribus*. Similarly, if over the long run consumers react to the label in response to increased media coverage by decreasing their purchases of labelled products we expect $\theta_{12} < 0$ in the w_{1t} and $\theta_{22} > 0$ in the w_{2t} *ceteris paribus*. If consumers do not react to information from the media in the long run, we would expect $\theta_{12} = \theta_{22} = 0$ in both share equations *ceteris paribus*. The same line of reasoning applies once the GM label is removed and details on relevant parameters need not be presented here.

[5] We include such an interaction term to account for the fact that the two effects are not independent. In other words, consumers can act (if they choose to do so) on external information provided by the media only when products are labelled.

Empirical results

Table 3.1 reports basic statistics on the series used in the analysis. The average expenditure shares for GM-labelled and unlabelled canned soup was 3.93% and 96.07%, respectively. Average expenditure shares for GM-labelled and unlabelled frozen processed meat was 39.03% and 60.97%, respectively. Average expenditure shares for GM-labelled and unlabelled frozen pizza was 6.05% and 93.95%, respectively. Average expenditure shares for GM-labelled and unlabelled frozen processed fish was 2.68% and 97.32%, respectively.

Prices, measured in Dutch guilders, were nearly identical between GM-labelled and unlabelled products across three of the four food categories. In the case of frozen pizza, the GM-labelled products enjoyed almost 40% higher prices over non-labelled products, indicating premium brands. Per capita total category expenditure was also measured in Dutch guilders. The population series used in the per capita series varied annually and was obtained from Statistics Netherlands.

Tables 3.2a–d catalogue the empirical parameter estimates and standard errors of α, γ, β, η, ϕ, λ and θ from equations (1) and (2). As with any singular system, $(n - 1)$ of the n system equations are estimated and the parameter estimates for the remaining equation are recovered using the adding up conditions and other relevant parameter restrictions of the model. The GM-labelled equation in each of the four product categories is estimated using non-linear least squares since equation (2), the unobservable price index, is itself a highly non-linear function. The parameter estimates for the labelled share equation in the four product categories are recovered from the adding up conditions. For each of the four models, Linear Approximate AIDS (LA/AIDS) model parameter estimates were employed as an initial feasible solution (i.e. starting values for the unknown parameter estimates) to the non-linear conditional AIDS model (Alston *et al.*, 1994). Moreover, consistent with the AIDS literature (Buse, 1994), the parameter α_0 in equation (2) was restricted to zero in all estimations. In the specification

Table 3.1. Descriptive statistics of selected demand system variables.[a]

	Mean	SD	Minimum	Maximum
Expenditure shares				
GM-labelled soup	0.0393	0.0101	0.0208	0.0840
Unlabelled soup	0.9607	0.0101	0.9160	0.9792
GM-labelled meat	0.3903	0.0831	0.1503	0.5424
Unlabelled meat	0.6097	0.0831	0.4576	0.8497
GM-labelled pizza	0.0605	0.0183	0.0321	0.1417
Unlabelled pizza	0.9395	0.0183	0.8583	0.9679
GM-labelled fish	0.0268	0.0090	0.0143	0.1361
Unlabelled fish	0.9732	0.0090	0.8639	0.9857
Prices				
GM-labelled soup	3.8530	0.2279	2.8101	4.2559
Unlabelled soup	3.8809	0.3420	2.1747	4.5619
GM-labelled meat	10.8219	0.5128	9.8323	12.0234
Unlabelled meat	10.0623	0.4623	8.3132	11.2424
GM-labelled pizza	14.9805	0.9747	9.5624	16.4487
Unlabelled pizza	10.9417	0.6106	8.5907	12.2628
GM-labelled fish	12.9019	0.6083	10.9048	13.9418
Unlabelled fish	12.6820	1.0975	10.4326	15.1471
Per capita total category expenditure				
Soup	0.1847	0.0351	0.1168	0.3044
Meat	0.0080	0.0013	0.0058	0.0117
Pizza	0.1995	0.0459	0.0971	0.3263
Fish	0.1717	0.0239	0.1168	0.2842

[a]Based on 247 weeks of data.

Table 3.2a. Estimated conditional expenditure share equations: canned soup.

	GM-labelled		Unlabelled[a]
	Parameter estimate	SE	Parameter estimate
Intercept (α)	0.0088***	0.0047	0.9912
Log of price (γ)			
GM-labelled	−0.1061*	0.0058	0.1061
Unlabelled	0.0193*	0.0045	−0.0193
Per capita			
Real expenditure (β)	−0.0208*	0.0022	0.0208
Linear time trend (η)	−0.0001*	0.00002	0.0001
Holidays (ϕ)			
Queen's Day	−0.0002	0.0013	0.0002
Sinterklaas	0.0029**	0.0014	−0.0029
Labels (λ)			
*on*4	0.0044	0.0044	−0.0044
*on*24	−0.0001	0.0002	0.0001
*off*4	0.0009	0.0015	−0.0009
*off*24	−0.0001	0.0001	0.0001
Media–label interaction (θ)			
*m·on*4	0.0067	0.0053	−0.0067
*m·on*24	−0.00004	0.0002	0.00004
*m·off*4	−0.0005	0.0004	0.0005
*m·off*24	−0.00004	0.00005	0.00004
Adjusted R^2	0.8589	–	–

[a]The parameter estimates in the unlabelled soup share equation are recovered using the adding up restrictions.
*, ** and ***denote signifance at the 0.01, 0.05 and 0.10 level, respectively.

testing phase, the two homogeneity restrictions and the symmetry condition were rejected and so were not imposed on the two-equation system. However, when imposed, the qualitative findings of this study remained unchanged. Finally, in each estimated equation, an *AR*(1) term was used to successfully purge first-order autocorrelation from the empirical residual series.

In the GM-labelled equation (w_{1t}) in the canned soup model (Table 3.2a), the intercept parameter (α_1) was found to be 0.0088 and statistically significant ($P < 0.10$). From the adding up restriction $\alpha_1 + \alpha_2 = 1$, we find α_2 equals 0.9912. Although these deviate slightly from the average soup expenditure shares found in Table 3.1, which is very common in applied demand analysis, the model fits the underlying data generation process well with an adjusted R^2 value of 0.8589. The parameter estimates on the natural logarithm of GM-labelled soup price and unlabelled soup price (i.e. γ_{11} and γ_{12}) were both statistically significant ($P < 0.01$). From the two adding up restrictions $\gamma_{11} + \gamma_{21} = 0$ and $\gamma_{12} + \gamma_{22} = 0$, we found γ_{21} and γ_{22} to be 0.1061 and −0.0193, respectively. The parameter estimate for per capita real expenditure (β_1) and the linear time trend (η_1) were also statistically significant ($P < 0.01$). The latter indicated an exogenous downward trend in w_{1t} not accounted for by the other observable effects controlled for in the AIDS model. Finally, one of the holiday effects, Sinterklaas, was statistically significant in the model ($P < 0.05$).

Since the dependent variables in an AIDS model are expenditure shares, not quantities demanded, the γ parameters do not have a direct interpretation as an own- or cross-price elasticity. Similarly, the β parameters do not have a direct interpretation as a conditional expenditure elasticity (Deaton and Muellbauer,

Table 3.2b. Estimated conditional expenditure share equations: frozen processed meat.

	GM-labelled		Unlabelled[a]
	Parameter estimate	SE	Parameter estimate
Intercept (α)	0.1970**	0.0792	0.8030
Log of price (γ)			
GM-labelled	−0.6770*	0.1787	0.6770
Unlabelled	0.2022*	0.0586	−0.2022
Per capita			
Real expenditure (β)	−0.0457*	0.0152	0.0457
Linear time trend (η)	−0.0001	0.0002	0.0001
Holidays (ϕ)			
Queen's Day	0.0128	0.0088	−0.0128
Sinterklaas	−0.0004	0.0092	0.0004
Labels (λ)			
*on*4	−0.0219	0.0286	0.0219
*on*24	−0.0008	0.0014	0.0008
*off*4	0.0011	0.0099	−0.0011
*off*24	−0.0004	0.0011	0.0004
Media–label interaction (θ)			
*m·on*4	−0.0237	0.0349	0.0237
*m·on*24	0.0008	0.0010	−0.0008
*m·off*4	0.0001	0.0025	−0.0001
*m·off*24	−0.0003	0.0003	0.0003
Adjusted R^2	0.8969	–	–

[a]The parameter estimates in the unlabelled meat share equation are recovered using the adding up restrictions.
* and ** denote significance at the 0.01 level and 0.05 level, respectively.

1980). Own- and cross-price elasticities are a somewhat complex function of the estimated and recovered α, γ and β parameters as well as the average expenditure shares and average natural logarithm of prices. In the case of Marshallian or uncompensated price elasticities (E_{ij}^U), we use the expression:

$$E_{ij}^U = -\delta_{ij} + \frac{1}{w_i}\left(\gamma_{ij} - \beta_i\left(\alpha_j + \sum_{k=1}^{2}\gamma_{kj}\log p_k\right)\right) \quad (3)$$

where α, γ and β are defined in equations (1) and (2), expenditure shares and the natural logarithm of prices are taken at their sample means and δ_{ij} is the Kronecker delta which equals 1 when $i = j$ and zero otherwise. In the case of Hicksian or compensated price elasticities (E_{ij}^C), we use the expression:

$$E_{ij}^C = -\delta_{ij} + \frac{1}{w_i}\left(\gamma_{ij} - \beta_i\left(\alpha_j + \sum_{k=1}^{2}\gamma_{kj}\log p_k - w_j\right)\right) + w_j \quad (4)$$

where α, γ and β are defined in equations (1) and (2), expenditure shares and the natural logarithm of prices are taken at their sample means and δ_{ij} is the Kronecker delta. Finally, the conditional expenditure elasticity ($E_{i,X}$) is given by:

$$E_{i,X} = 1 + \frac{\beta_i}{w_i} \quad (5)$$

where β is defined in equations (1) and (2) and the expenditure shares are taken at their sample means.

In Table 3.3a, the respective elasticities from equations (3), (4) and (5) may be found for canned soup. Uncompensated and compensated own-price elasticities are, as expected *a priori*, negative indicating usual downward sloping canned soup demand equations. For example, a 1% increase in own price results, on average, in a 3.6915% decrease in quantity demanded of GM-labelled soup. Also noteworthy, the GM-labelled soup demand equation is more elastic than the unlabelled soup demand equation regardless of which elasticity measure was

Table 3.3c. Estimated price and expenditure elasticities for frozen pizza.

	GM-labelled pizza	Unlabelled pizza
Uncompensated		
GM-labelled pizza	−2.7675	1.4108
Unlabelled pizza	0.1138	−1.0909
Compensated		
GM-labelled pizza	−2.8214	0.5728
Unlabelled pizza	0.0529	−2.0369
Expenditure	1.1080	0.9930

Table 3.3d. Estimated price and expenditure elasticities for frozen processed fish.

	GM-labelled fish	Unlabelled fish
Uncompensated		
GM-labelled fish	−5.1773	−0.8478
Unlabelled fish	0.1150	−0.9767
Compensated		
GM-labelled fish	−5.2033	−1.7915
Unlabelled fish	0.0882	−1.9507
Expenditure	1.0303	0.9992

expenditure share in that category, and in 1 week 54% of the market share – far in excess of the other three food categories. The price and expenditure elasticities for frozen processed meat were reasonable and consistent with theory.

In the frozen pizza category (Table 3.2c), the parameters α_1, γ_{11}, γ_{12}, and η_1 were found to be statistically significant ($P < 0.01$). The empirical model fits the data adequately with an adjusted R^2 of 0.68. The own- and cross-price elasticities, both uncompensated and compensated, as well as expenditure elasticities appear reasonable.

In the frozen processed fish category (Table 3.2d), the parameters α_1 and γ_{11} were found to be statistically significant at the 0.01 level, and η_1 was found to be significant at the 0.05 level. Since η_1 was positive it indicates an exogenous upward trend in w_{1t} not accounted for by the other observable effects controlled for in the AIDS model. The empirical model fits the data rather poorly with an adjusted R^2 of 0.28. However, most of the own- and cross-price elasticities, both uncompensated and compensated, as well as expenditure elasticities are reasonable.

Consumer response to GM labels

The most significant result in our empirical model is the lack of any statistically significant change in consumer response towards foods with GM labels. Indeed, we cannot detect any immediate or gradual consumer response to the introduction of GM labels or to their removal. Specifically, none of the other parameter estimates in the model, such as the label and media–label interaction variables, were statistically significant ($P > 0.10$), except for the parameter estimate on the $m_t \cdot off4_t$ variable (θ_{13}). This parameter was found to be −0.0025 and statistically significant at the 0.05 level. However, the result is indubitably spurious as the economic relationship it indicates – namely that the GM-labelled frozen pizza product is penalized in the first 4 weeks after the label is removed – is not sensible.

This result is in stark contrast with the bulk of the existing literature that anticipates Dutch – and most other European – consumers to actively discriminate against foods with GM labels in the market place, when given the opportunity. Indeed, a February 2002 consumer survey in The Netherlands directly asked consumers how they would respond if they noticed GM labels on food products they regularly purchase. It concluded that a large majority would stop purchasing them (Environics International, 2002). Our empirical results suggest that Dutch consumer preferences revealed in the market place were drastically different from stated preferences elicited through surveys.[6]

External information on GM foods and agrifood biotechnology did not have any significant impact on consumer response either. Despite substantial media coverage, no significant influences could be empirically identified.

Discussion and Concluding Comments

Our results indicate that, in aggregate, Dutch consumers did not significantly alter their purchasing behaviour in the presence of foods positively labelled as containing GM ingredients. Nor did they alter their purchasing behaviour towards such foods after the labels were removed nearly 3 years later.

We do not know *why* Dutch consumers did not alter their purchasing patterns in the presence of positive GM labels. Our data do not allow such insight. It could be that a majority of Dutch consumers are more accepting of the technology (Hamstra and Smink, 1996; Hoban, 1997; Zechendorf, 1998). Or it could be that Dutch consumers have a high level of trust in their food supply (Hamstra, 1993) and were therefore less concerned about purchasing GM foods. It could also be that Dutch consumers were reassured by the brand identity of each of the labelled products (Noussair *et al.*, 2004). In other words, it is possible that a majority of consumers read the labels, understood them and kept on purchasing them regardless. Alternatively, it could be that Dutch consumers did not see or read the labels.

Irrespectively of motives, however, the key result is that Dutch consumers, in aggregate, did not alter their behaviour towards positively labelled foods with GM ingredients. If the results obtained here could be generalized across products and markets in Europe, they would call into question the current European policy of mandatory labelling of GM foods and ingredients.

Protecting consumers' 'right to know' and the 'right to choose' is advanced as the main reason for the current European policy stance. In principle, there can be little objection to the argument that consumers should be able to exercise such rights. Market transparency is the linchpin of well-functioning markets. However, mandatory labelling is not the only option that would allow consumers a choice. Indeed, given that mandatory labelling systems are costly to implement (Kalaitzandonakes *et al.*, 2001) costs and benefits associated with such labelling regimes must be carefully weighted in order to decide their optimality (Giannakas and Fulton, 2002). In this context, the proportion of the consumers that would effectively discriminate between GM and conventional foods in the market place is a key parameter (Giannakas and Fulton, 2002). Indeed, Caswell (1998, 2000) and Giannakas and Fulton (2002) have argued that a voluntary labelling programme may better serve a country where only a minority of the population is interested in separating GM from non-GM foods. Mandatory systems, on the other hand, may better serve countries where a sizable percentage of the population would differentiate between genetically modified and conventional foods in the market place.

Understanding whether a majority of consumers, irrespective of motives, would use GM labels to discriminate against relevant products in the market is essential for effective policy decisions. Our results are national in scope, cover multiple product categories, and

[6] While our empirical results differ from results obtained through surveys and experimental auctions, they are consistent with the limited empirical evidence that is available on revealed consumer preferences and market behaviour. The vast majority of US consumers purchased milk from rBST-treated cows despite stated preferences for the opposite (Aldrich and Blisard, 1998). Similarly, anecdotal evidence from the UK indicates that a GM tomato paste offered in UK stores up until 1999 apparently outsold competing non-GM brands (Nunn, 2000).

suggest that at least in The Netherlands such majority response could not be detected despite the fact that consumers could 'vote with their wallets' against GM foods in the presence of alternatives. It is unclear to what extent these results could be generalized across Europe. At a minimum, however, they raise questions about the existing conventional wisdom on potential consumer behaviour, which is based on stated preferences, and beg for further research that focuses on revealed consumer preferences and actual behaviour.

References

Aldrich, L. and Blisard, N. (1998) *Consumer Acceptance of Biotechnology. Lessons from the rBST Experience* Agricultural Information Bulletin No. 7417–01, Economic Research Service, US Department of Agriculture, Washington, DC.

Alston, J.M., Foster, K.A. and Green, R.D. (1994) Estimating elasticities with the linear approximate almost ideal demand system: some Monte Carlo results. *Review of Economics and Statistics* 76, 351–356.

Boccaletti S. and Moro, D. (2000) Consumer willingness to pay for GM food products in Italy. *AgBioForum* 3(4), 259–267.

Brown, D.J. and Schrader, L.F. (1990) Cholesterol information and shell egg consumption. *American Journal of Agricultural Economics* 72(3), 548–555.

Burton, M. and Pearse, D. (2002) Consumer attitudes towards genetic modification, functional foods and microorganisms: a choice modeling experiment for beer. *AgBioForum* 5, 51–58.

Buhr, B.L., Hayes, D.J., Shogren, J.F. and Kliebenstein, J.B. (1993) Valuing ambiguity: the case of genetically engineered growth enhancers. *Journal of Agricultural and Resource Economics* 18, 175–184.

Buse, A. (1994) Evaluating the linearized Almost Ideal Demand System. *American Journal of Agricultural Economics* 76, 781–793.

Caswell, J.A. (1998) Should use of genetically modified organisms be labeled? *AgBioForum* 1(1), 22–24.

Caswell, J.A. (2000) Labelling policy for GMOs: to each his own? *AgBioForum* 3(3), 53–57.

Cotterill, R.W. (1994) Scanner data: new opportunities for demand and competitive strategy analysis. *Agricultural and Resource Economics Review* 23, 125–139.

Dahlgran, R.A. and Fairchild, D.G. (2002) The demand impacts of chicken contamination publicity – a case study. *Agribusiness* 18(4), 459–474.

Deaton, A. and Muellbauer, J. (1980) An Almost Ideal Demand System. *American Economic Review* 70, 312–326.

Douthitt, R., Zepeda, L. and Grobe, D. (1996) *Comparison of National and Poor Households: Results of a Survey of Consumer Knowledge and Risk Perception of Food-related Biotechnologies.* Special Report No. 68, Institute for Research on Poverty, University of Wisconsin-Madison.

Durant, J., Bauer, M.W. and Gaskell, G. (eds) (1998) *Biotechnology in the Public Sphere: a European Source Book.* Science Museum Press, London.

Environics International Ltd. (2002) *CIAA European Food Survey.* Available on the World Wide Web: http://www.ciaa.be/ciaa_summit/pages/hetherington,pdf.

European Commission (EC) (1997) The Europeans and modern biotechnology. *Eurobarometer* 46(1). European Commission, Brussels.

European Commission (EC) (2000) The Europeans and modern biotechnology. *Eurobarometer 52*(1). European Commission, Brussels.

Gaskell, G., Bauer, M.W., Durant, J. and Allum, N.C. (1999) Worlds apart? The reception of genetically modified foods in Europe and the U.S. *Science* 285, 384–387.

Giannakas, K. and Fulton, M. (2002) Consumption effects of genetic modification. What if consumers are right? *Agricultural Economics* 27, 97–109.

Greene, W.H. (2000) *Econometric Analysis*, 4th edn. Prentice Hall, Upper Saddle River, New Jersey.

Hamstra, A.M. (1993) *Consumer Acceptance of Food Biotechnology: The Relation Between Product Evaluation and Acceptance.* Research Report 137. SWOKA, Leiden.

Hamstra, A.M. and Smink, C. (1996) Consumers and biotechnology in the Netherlands. *British Food Journal* 98(4), 34–38.

Hillers, V.N. and Lowik, M.R.H. (1998) Incorporation of consumer interests in regulation of novel foods produced with biotechnology: what can be learned from the Netherlands? *Journal of Nutrition Education* January/February, 2–7.

Hoban, T.J. (1996) Trends in consumer attitudes about biotechnology. *Journal of Food Distribution Research* 27, 1–10.
Hoban, T.J. (1997) Consumer acceptance of biotechnology: an International perspective. *Nature Biotechnology* 15, 232–234.
Hoban, T.J. (1998). Trends in consumer attitudes about agricultural biotechnology. *AgBioForum* 1(1), 3–7.
Hoban, T.J. and Burkhardt, J. (1991) Biotechnology control of growth and product quality in meat production: implications and acceptability. In: Van der Wal, P. (ed.) *Proceedings of Determinants of Public Acceptance in Meat and Milk Production: North America Conference*. Wageningen Agricultural University, The Netherlands.
Hoban, T.J. and Kendall, P.A. (1993) *Consumer Attitudes about Food Biotechnology*. North Carolina Cooperative Extension Service, Raleigh, North Carolina.
Huffman, W.E., Shogren, J.F., Rousu, M. and Tegene, A. (2001) The value to consumers of GM food labels in a market with asymmetric information: evidence from experimental auctions. Paper presented at the *5th International Consortium on Agricultural Biotechnology Research (ICABR) Meetings*, Ravello, Italy.
Huffman, W.E., Rousu, M., Shogren, J.F., and Tegene, A. (2002) Should the United States initiate a mandatory labeling policy for genetically modified foods? Paper presented at the *6th International Consortium on Agricultural Biotechnology Research (ICABR) Meetings*, Ravello, Italy.
James, J., Parker, T., Fleischer, S. and Orzolek, M. (2002) Consumer acceptance of GMOs revealed: a market experiment with Bt sweet corn. Paper presented at the Northeastern Agricultural and Resource Economics Association Meetings, Camp Hill, Pennsylvania, USA.
Kalaitzandonakes, N. (2000) Agrobiotechnology and competitiveness. *American Journal of Agricultural Economics* 82, 1224–1233.
Kalaitzandonakes, N., Maltsbarger, R. and Barnes, J. (2001) Global identity preservation costs in agricultural supply chains. *Canadian Journal of Agricultural Economics* 49, 605–615.
Kahneman, D. and Tversky, A. (1984) Choices, values, and frames. *American Psychologist* 39, 341–350.
Keisel, K., Buschena, D. and Smith, V. (2002) Consumer acceptance and labeling of biotech in food products: a study of fluid milk demand. Paper presented at the *6th International Consortium on Agricultural Biotechnology Research (ICABR) Meetings*, Ravello, Italy.
Lusk, J.L., Daniel, S., Mark, D. and Lusk, C. (2001) Alternative calibration and auction institutions for predicting consumer willingness to pay for non-genetically modified corn chips. *Journal of Agricultural and Resource Economics* 26, 40–57.
Lusk, J.L., Roosen, J., and Fox, J.A. (2003) Demand of beef from cattle administered growth hormones or fed genetically modified corn: a comparison of consumers in France, Germany, the United Kingdom, and the United States. *American Journal of Agricultural Economics* 85, 16–29.
Marks, L.A., Kalaitzandonakes, N. and Zakharova, L. (2002) On the media roller-coaster will GM foods finish the ride? *Choices* Spring, 6–10.
Mathios, A.D. (2000) The impact of mandatory disclosure laws on product choices: an analysis of the salad dressing market. *Journal of Law and Economics* 43, 651–675.
McFadden, D. (1973) Conditional logit analysis of qualitative choice behavior. In: Zarembka, P. (ed.) *Frontiers in Econometrics*. Academic Press, New York, pp. 105–142.
Menkhaus, D.J., Borden, G.W., Whipple, G.D., Hoffman, E. and Field, R.A. (1992) An experimental application of laboratory experimental auctions in marketing research. *Journal of Agricultural and Resource Economics* 17, 44–55.
Moon, W. and Balasubramanian, S.K. (2001) Public perceptions and willingness-to-pay a premium for nonGMO foods in the US and UK. *AgBioForum* 4(3–4), 221–231.
Moschini, G. (1995) Units of measurement and the stone price index in demand system estimation. *American Journal of Agricultural Economics* 77, 63–68.
Noussair, C., Robin, S. and Ruffieux, B. (2004) Do consumers really refuse to buy genetically modified foods? *The Economic Journal* 114, 102–120.
Nunn, J. (2000) What lies behind the GM label on UK foods. *AgBioForum* 3, 250–254.
Runge, C.F. and Jackson, L.A. (2000) Negative labeling of genetically modified organisms (GMOs): the experience of rBST. *AgBioForum* 3, 58–62.
Shogren, J.F., Fox, J.A., Hayes, D.J. and Roosen, J. (1999) Observed choices for food safety in retail, survey, and auction markets. *American Journal of Agricultural Economics* 81, 1192–1199.

Smith, M.E., van Ravenswaay, E.O. and Thompson, S.R. (1988) Sales loss determination in food contamination incidents: an application to milk bans in Hawaii. *American Journal of Agricultural Economics* 70, 513–520.

Sterngold, A., Warland, R. and Herrman, R. (1994) Do surveys overstate public concerns? *Public Opinion Quarterly* 58, 255–263.

Swartz, D.G. and Strand Jr., I.E. (1981) Avoidance costs associated with imperfect information: the case of Kepone. *Land Economics* 57(2), 139–150.

Teisl, M.F., Roe, B. and Hicks, R.L. (2002) Can eco-labels tune a market? Evidence from dolphin-safe labeling. *Journal of Environmental Economics and Management* 43, 339–359.

Tolley, G.S. and Randall, A. (1983) *Establishing and Valuing the Effects of Improved Visibility in the Eastern United States*. Report of the US Environmental Protection Agency, Washington, DC.

van Ravenswaay, E.O. and Hoehn, J.P. (1991) The impact of health risk information on food demand: a case study of alar in apples. In: Caswell, J.A. (ed.) *Economics of Food Safety*. Elsevier Science Publishing Co., New York, pp.155–174.

Vickner, S.S. and Davies, S.P. (1999) Estimating market power and pricing conduct in a product-differentiated oligopoly: the case of the domestic spaghetti sauce industry. *Journal of Agricultural and Applied Economics* 31, 1–13.

Wessels, C.R., Miller, C.J. and Brooks, P.M. (1995) Toxic algae contamination and demand for shellfish: a case study of demand for mussels in Montreal. *Marine Resource Economics* 10, 143–159.

Zechendorf, B. (1998) Agricultural biotechnology: why do Europeans have difficulty accepting it? *AgBioForum* 1(1), 8–13.

4 The Welfare Effects of Implementing Mandatory GM Labelling in the USA[1]

Wallace E. Huffman,[1] Matthew Rousu,[2] Jason F. Shogren[3] and Abebayehu Tegene[4]

[1]*Department of Economics, Iowa State University, Ames, IA 50011, USA;* [2]*RTI International, 3040 Cornwallis Road, Research Triangle Park, NC 27709, USA;* [3]*Department of Economics and Finance, University of Wyoming, Laramie, WY 82070, USA;* [4]*Food and Rural Economics Division, Economic Research Service, US Department of Agriculture, Washington, DC 20036, USA*

Introduction

Although genetic engineering is a promising tool for crop varietal development, genetically modified foods continue to be controversial. Many groups that oppose these new goods are supporting a public policy of mandatory labelling for GM content. Debate, however, continues over whether the USA should impose a mandatory labelling policy for genetically modified (GM) foods. Groups that favour a mandatory labelling policy for GM foods include Greenpeace International (1997) Friends of the Earth (2001), and the Consumers Union (*Consumer Reports*, 1999). Groups opposing mandatory GM labels include the Council for Biotechnology Information (2001) and the US Food and Drug Administration (FDA) (2001). This contentious issue has engaged debate from all sides of the spectrum, yet only modest economic research has examined the merits and pitfalls of a new regulation that requires mandatory labelling for GM foods in the USA.

This chapter examines the potential welfare effects of imposing a mandatory GM-labelling policy in the USA. We first discuss when a mandatory labelling policy is likely to benefit consumers. We then describe an experimental auction designed to provide data that are needed to test whether consumers will benefit from a mandatory GM-labelling policy. For a sample of adult consumers living in two major Midwestern cities, our results do not contradict the hypothesis that consumers interpret voluntary and mandatory market signals identically. These findings suggest that it would be more efficient or welfare improving for the USA to continue its voluntary labelling policy and resist calls for new regulations that mandate labelling of GM foods.

[1] The authors gratefully acknowledge assistance from Daniel Monchuk and Terrance Hurley in conducting the auctions and assistance from Monsanto in providing some of the products used in the experiment. Authors received helpful comments on the chapter from participants at the ICABR Conference, Ravello, Italy, July 2002. This work was supported through a grant from the US Department of Agriculture Cooperative State Research, Education, and Extension Service, under Agreement 00-52100-9617 and from the US Department of Agriculture, Economic Research Service, under Agreement 43-3AEL-8-80125, and by the Iowa Agriculture and Home Economics Experiment Station. Views presented in this chapter are the authors and do not represent those of ERS or USDA.

Background on Labels

Caswell (1998, 2000) emphasizes that the list of potential GM labelling policies is large, and includes mandatory labelling of GM foods voluntary labelling of GM foods and bans on all labelling. An informed decision about labelling policies for GM foods should only be made after a careful benefit–cost analysis. Caswell points out that a voluntary labelling programme is likely to be a better policy option for a country that has only a small segment of the population that is concerned about GM foods, but a mandatory labelling system is likely to be the best policy option in countries where a large share of the population wants to know if their food is GM. A model by Kirchhoff and Zago (2001) reached a similar conclusion – voluntary GM-labelling policies may be better for a country that has more consumers who are concerned with cost savings, while mandatory GM-labelling policies may be better for more GM-averse consumers.

The USA does not require mandatory labelling for most GM foods. In January 2001, the FDA issued a 'Guidance for Industry' statement for labelling GM products, which stated that the only GM foods that need to be labelled are ones that have different characteristics from their non-GM versions. Labelling is not required for any other GM foods, but firms in the USA do have the option of voluntarily indicating whether their food is GM. Canada also has a similar voluntary labelling policy.

The European Union (EU) requires that all foods have the label 'genetically modified' if any ingredient in the food is at least 1% GM. The European Parliament voted for stricter regulations in early 2001.[2] The new regulations call for stricter labelling and monitoring of GM products and allow for the tracing of GM products through the food chain (CNN, 2001). The EU standards are the minimum standards that member countries must adhere to, although countries can have stricter standards. Several other countries around the world have mandatory labelling policies for GM foods, including Australia, Japan and New Zealand. For a detailed review of labelling policies, see Rousu and Huffman (2001) or Phillips and McNeill (2000).

Some groups think that mandating GM labels would improve a society's welfare. Many environmental and consumer advocacy groups call for mandatory labelling of GM foods, which they believe benefits consumers (Greenpeace International, 1997; Friends of the Earth, 2001; *Consumer Reports*, 1999). Greenpeace and Friends of the Earth both advocate labels on GM foods to give consumers the opportunity to choose whether to consume GM foods. Other benefits of labelling are that labels make it easier to find information on food products, can increase consumer information and can improve product design.

Relatively few estimates of the costs of GM food labelling exist. The accounting/consulting firm KPMG was commissioned for a study in Australia and New Zealand to examine the costs of complying with new labelling laws. They estimated that the costs of the labelling laws could mean an increase in consumer prices from 0.5 to 15%, and that firms could also receive lower profits (Phillips and Foster, 2000). Even though they commissioned the study, the Australia New Zealand Food Authority (2001) disregarded KPMG's study, citing two flaws. Whether this council had legitimate problems with the study or were responding to political pressure, we do not know. Smyth and Phillips (2002) estimated that a voluntary identity-preserved production and marketing system in Canada cost from 13 to 15% during 1995–1996. In a related study, Wilson and Dahl (2002) estimate that identity preservation costs would be $3.50 per bushel for GM wheat, assuming a 1% tolerance level. The Philippine Chamber of Food Manufacturers warned that mandatory GM food labels would increase production costs

[2] New regulations passed in 2002 will move this threshold down to 0.5% before the product must be labelled as GM.

by 15%, and that the increased costs would be passed on to consumers (*AgBiotech Reporter*, 2001). One issue seems apparent: implementing a labelling policy for GM foods is costly, even if the exact magnitude of the costs is unknown.

When Would Consumers Benefit from a Mandatory Labelling Policy?

With asymmetric information between food suppliers and consumers, consumers regularly must read signals about product quality. For brand-name products, consumer purchases are higher for some brands than others, and consumers are frequently willing to try a product based on external signals about the product (e.g. packaging, labelling, advertisements, etc.). The question we examine is how would mandatory labelling of GM foods help consumers in purchasing food products?[3]

The key benefit that a mandatory labelling policy could have is if it were to help consumers distinguish genetically modified from non-genetically modified food products. This is why many groups call for mandatory labelling of GM foods (e.g. see Greenpeace International, 1997). But without mandatory labelling there are still both GM and non-GM food products sold – so an ideal test of whether consumer welfare would improve from a mandatory labelling policy is to test if a market that has mandatory GM labelling makes it easier for consumers to distinguish GM food products from non-GM food products (relative to voluntary GM labelling). The next section outlines how we test this using experimental auctions.

Experimental Design

We designed two experimental auction markets, one emulating a market that has mandatory labelling in place and another which emulates a market that has voluntary labelling in place. We then test for similarity of consumer bids for three different food products – vegetable oil, tortilla chips and Russet potatoes. Experimental units are randomly assigned labelling treatments. Some consumers bid on foods with *positive GM labels* – the labels that would arise in a *mandatory* labelling regime; others bid on food with *negative GM labels* – the labels that would arise in a *voluntary* labelling regime.[4] If bidding behaviour for GM and GM-free food differs across the two markets, this would indicate that the mandatory market presents different signals than a voluntary market. If bidding behaviour does not differ, then we would find no evidence that a mandatory labelling policy assists consumers in accomplishing their main objective – informing consumers. However, finding a change in bidding behaviour is a necessary but not a sufficient condition for a mandatory labelling policy to be welfare-improving.

The experimental design consisted of four biotech information labelling treatments with each treatment replicated at least four times. The treatments were randomly assigned to ten experimental units, each consisting of 13–16 consumers drawn from the households of two major urban areas and who were paid to participate.

We now describe the four elements in our GM-labelling experiments – the GM food, the auction mechanism, the experimental units and the specific steps in the experiment, which includes the detailed information on the labels.

3 In this chapter we present the intuition behind when a mandatory labelling would benefit consumers, and that a voluntary labelling policy is more efficient than a mandatory labelling policy if consumers can distinguish between GM and non-GM foods identically in either market. A model which derives this formally is presented in our technical counterpart to this chapter (Huffman *et al.*, 2002).

4 These experimental markets were chosen to emulate the mandatory and voluntary GM-labelling regimes currently in place throughout the world. Our mandatory regime reflects the labels consumers might find in Europe, where foods that are GM must be labelled as such. Our voluntary labelling regime captures the labels that consumers might see in the USA, where food manufacturers can label their products as non-genetically engineered if they choose. We do not examine several other potential but currently non-implemented labelling policies, including a mandatory labelling policy that requires all non-GM foods to be labelled or a policy that requires every food product in a market to be labelled as GM or non-GM.

The food products and auction mechanism

Participants in our auction bid on three unrelated food items: a 32-ounce bottle of vegetable oil,[5] a 16-ounce bag of tortilla chips (made from yellow maize) and a 5-pound bag of Russet potatoes. They bid on these items using the random *n*th-price auction.

We chose the random *n*th-price auction for our GM food experiments because it is designed to engage both the on- and off-the-margin bidders (see Shogren *et al.*, 2001).[6] This is an aid in identifying the whole demand curve (rather than a short segment) for a new good. The random *n*th-price works as follows. Each of *k* bidders submits a bid for one unit of a good; then each of the bids is rank-ordered from highest to lowest. The auction monitor then selects a random number – the *n* in the *n*th-price auction, which is drawn from a uniform distribution between 2 and *k*, and the auction monitor sells one unit of the good to each of the $n - 1$ highest bidders at the *n*th-price. For instance, if the monitor randomly selects $n = 4$, the three highest bidders each purchase one unit of the good priced at the fourth highest bid. *Ex ante*, bidders who have low or moderate valuations now have a nontrivial chance to buy the good because the price is determined randomly. This auction increases the probability that insincere bidding will be costly.

Auctions were planned and conducted in two Midwestern US cities: Des Moines, Iowa, and St Paul, Minnesota, in 2001. Consumers were contacted through a random digit dialling method and were asked if they would participate in a group session that related 'to how people select food and household products', and they were informed that the session would last about 90 minutes.[7] They were also told that at the end of the session each participant would receive $40 in cash for his/her time. From the initial sample of usable randomly selected numbers, the percentage of people who accepted the offer to participate and then showed up at the auction was approximately 19%.

Our total sample size of participants is 142, and Table 4.1 summarizes the characteristics of the auction participants: 60% are female, mean age is 51 years and mean household income is $51,600.

Sequence of steps in the experiments

Figure 4.1 shows the ten steps in each experimental unit. In Step 1 when participants arrived at the experiment, they signed a consent form agreeing to participate in the auction. After they signed this form, they were given $40 for participating and an ID number to preserve their anonymity. The treatments were randomly assigned to each experimental unit, so the observed and unobserved characteristics of observations are uncorrelated with the treatments. The participants then read brief instructions and filled out a questionnaire. The questionnaire was purposefully given to consumers before the experiment to elicit demographic information and to capture consumers' prior perception of GM foods before bidding, which allowed us to compare their prior beliefs to their posterior beliefs after the experiment.

5 For the oil, soybean oil was used for the mandatory labelling trials, and canola oil was used for the voluntary/labelling trials. The soybean oil was initially used in the April experiments. We then tried to purchase non-GM soybean oil in 32-ounce bottles and were unsuccessful. The bids for the vegetable oil follow the same trend as the other products and are discussed in the results section. The other products (and packaging) were absolutely identical, except for the presence or absence of genetic modification.

6 The auction combines elements of two classic demand-revealing mechanisms: the Vickrey (1961) auction and the Becker–DeGroot–Marschak (1964) random pricing mechanism. The key characteristic of the random *n*th price auction is *a random but endogenously determined* market-clearing price. Randomness is used to give all participants a positive probability of being a purchaser of the auctioned good; the endogenous price ensures that the market-clearing price is related to the bidders' private values.

7 We considered the possibility that the demand for GM foods may change over the 8 months between auctions, so we replicated two experimental units, using the exact same procedures. We found no evidence that willingness to pay for GM-labelled foods had changed over time.

Table 4.1. Characteristics of the auction participants.

Variable	Definition	Mean	SD
Gender	1 if female	0.60	0.49
Age	The participant's age	51.40	18.1
Married	1 if the individual is married	0.65	0.48
Education	Years of schooling	14.74	2.36
Household	Number of people in participant's household	2.56	1.49
Income	The household's income level (in thousands)	51.60	33.40
White	1 if participant is white	0.92	0.27
Read_L	1 if never reads labels before a new food purchase	0.02	0.14
	1 if rarely reads labels before a new food purchase	0.11	0.32
	1 if sometimes reads labels before a new food purchase	0.32	0.47
	1 if often reads labels before a new food purchase	0.36	0.48
	1 if always reads labels before a new food purchase	0.18	0.39

In Step 2, participants were given detailed instructions (both oral and written) about how the random *n*th-price auction works, including an example written on the blackboard. After the participants learned about the auction, a short quiz was given to them to ensure that everyone understood how the auction worked. All experimental instructions are available from the authors on request.

Step 3 was the first practice round of bidding, in which participants bid (in a real auction) on a brand-name candy bar. The participants were all asked to examine the product and then to place a bid on the candy bar. The bids were collected and the first round of practice bidding was over. Throughout the auctions, when the participants were bidding on items in a round, they had no indication of what other items they may be bidding on in future rounds.

Step 4 was the second practice round of bidding, and in this round the participants bid separately on three different items. The products were the same brand-name candy bar, a deck of playing cards and a box of pens. Participants knew that only one of the two rounds would be chosen at random to be binding, which prevented anyone from taking home more than one unit of any product. By using only one binding round, we avoided problems of demand reduction that can occur in multi-unit auctions (List and Lucking-Reiley, 2000). The consumers first examined the three products and then submitted their bids.

After the two practice auction rounds were completed, the binding round and the binding *n*th-prices were revealed in Step 5. All bid prices were written on the blackboard, and the *n*th-price was circled for each of the three products. Participants could see immediately what items they won, and the price they would pay. The participants were told that the exchange of money for goods was in another room nearby and would take place after the entire experiment was completed.

In Step 6, participants received one of two potential info-packets that provided non-food label information about biotechnology (for a detailed look at how information affected the demand for foods labelled as GM, see Rousu *et al.*, 2002). These info-packets were produced as follows. We created three information sources: (i) the *industry perspective* – a collection of statements and information on genetic modification provided by a group of leading biotechnology companies, including Monsanto and Syngenta; (ii) the *environmental group perspective* – a collection of statements and information on genetic modification from Greenpeace, a leading environmental group; and (iii) the *independent third-party perspective* – a statement on genetic modification approved by a third-party group, consisting of a variety of people knowledgeable about GM goods, including scientists, professionals, religious leaders and academics, who do not have a financial stake in GM foods. We limited each information source to one full page, organized into five categories: general infor-

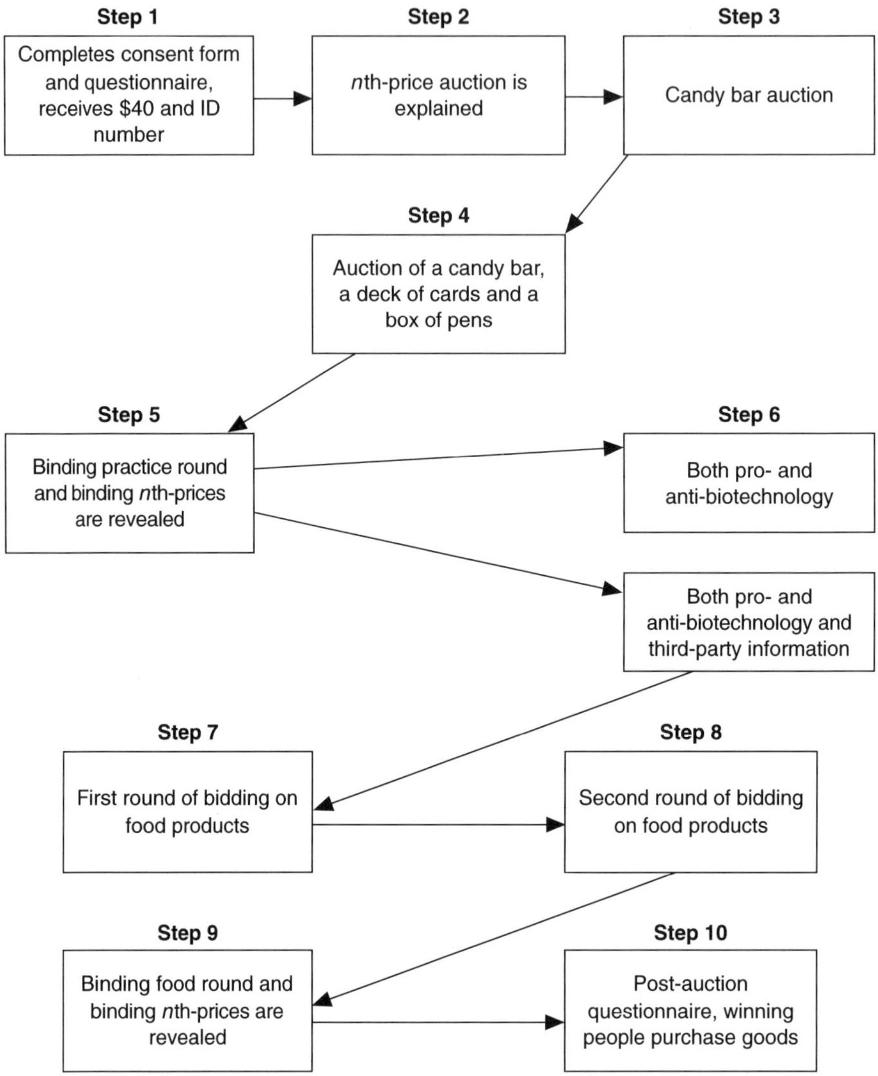

Fig. 4.1. Steps in the experiment.

mation, scientific impact, human impact, financial impact and environmental impact. The information sheets are available in Rousu *et al.* (2002) or by request from the authors.

These information sources were then randomized to create the two info-packets: (i) both pro- and anti-biotechnology and (ii) pro-biotechnology, anti-biotechnology, and independently verifiable.[8] These info-packets were then randomized among all ten experimental units, with each info-packet going to four experimental units. By giving all participants both positive and negative information on GM foods, and by giving some participants a third-

[8] The order of the positive information and negative information was, also, randomized across consumers. Participants who received the third-party, verifiable information always received it after the other information sources.

party perspective on GM foods, we could determine the willingness to pay for individuals who received all perspectives on the GM food debate.

Once we distributed the appropriate info-packet to the participants in a given unit, we then conducted two auction rounds. The rounds were differentiated by the food label – either the food had a standard food label or a label that indicated the status of genetic modification (e.g. see Fig. 4.2).[9] In one round (which could be Round 1 or 2 depending on the experimental unit), participants were bidding on the three food products each with the standard food label. We made these labels as plain as possible to avoid any influence on the bids from the label design. In the other round, participants were bidding on the same three food products with either a GM label or a non-GM label. The GM and non-GM labels differed from the standard label only by the inclusion of one extra sentence. The GM label said 'This product is made using genetic modification (GM)', while the non-GM label said 'This product is made without genetic modification'. For each experimental unit, participants knew that only one round would be chosen as the binding round that determined auction winners.

In Step 7, participants bid on three different food products: a bottle of vegetable oil, a bag of tortilla chips and a bag of potatoes, either with the standard label or the label indicating the product's GM status. Six groups bid on foods with plain labels and foods with labels saying 'made using genetic modification (GM)'. Four groups bid on foods with plain labels and foods with labels saying 'made without using genetic modification'. The participants were instructed to examine the three products and then to write down their sealed bid for each of the three goods. Participants bid on each good separately. The monitor then collected the bids from the people and told them they were next going to look at another group of food items. Table 4.2 summarizes the four treatments.

Step 8 had participants examine the same three food products, each with a different label from Round 1.[10] Again the participants examined the products and bid on the three products separately. The bids were then collected from all of the individuals. In contrast to early experimental auction work using repeated

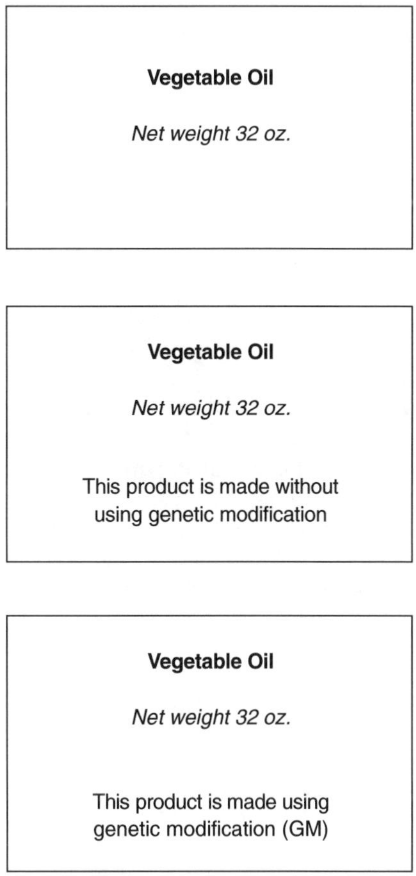

Fig. 4.2. The three types of labels used for the vegetable oil.

[9] Note that our labels are clearly displayed on the front of the package, where consumers would see them. See Noussair *et al.* (2002) for evidence of how consumers are not always likely to read food labels that are on the back of packages.

[10] We randomize the order the participants were presented the food products across groups. The null hypothesis that the round the consumer bid on foods led to the same bids could not be rejected at a 5% level for any of the three goods under both a *t*-test and a Wilcoxon rank-sum test. Hence, the order in which consumers saw the items did not appear to matter.

Table 4.2. Information and labelling given to the four treatments.

Treatment	Labelling regime type	Third party	Number or trials per treatment
1	Voluntary regime	No	2
2	Voluntary regime	Yes	2
3	Mandatory regime	No	4
4	Mandatory regime	Yes	2

trials, this auction used only two rounds to avoid any chance of affiliation of values and changes in willingness to pay due to the posted-market behaviour of other bidders (see List and Shogren, 1999; Knetsch *et al.*, 2001).

Step 9 selected the binding round, and the binding random *n*th-prices for the three goods. The winners were notified. In Step 10, each participant was asked to complete a brief post-auction questionnaire, and then the monitors dismissed the participants who did not win. The monitors and the winners then exchanged money for goods, and the auction winners were also dismissed.

Data and Results

The statistical analysis of our experimental data supports the hypothesis that consumers read similar signals in the two markets. The mean and median bids in the two markets are reported in Table 4.3. Eighty-six participants were in treatments that bid on the plain-labelled and GM-labelled food products (the mandatory GM-label market), and 56 participants were in treatments that bid on the plain-labelled and non-GM-labelled food products (the voluntary GM-label market). For the participants who bid on the GM-labelled and plain-labelled foods, consumers discounted the GM-labelled oil by an average of 11 cents, the GM-labelled tortilla chips by 8 cents and the GM-labelled potatoes by 8 cents. The participants who bid on the plain-labelled food and the non-GM-labelled food discounted the plain-labelled oil by an average of 4 cents, the plain-labelled tortilla chips by 7 cents and the plain-labelled potatoes by 9 cents.

Our main goal is to determine whether consumers can accurately decipher which

Table 4.3. Mean bids: markets with mandatory and voluntary labels.

	n	Mean bid	SD	Median	Minimum	Maximum
Mean bids for the mandatory GM-labelling market						
GM oil	86	0.63	0.65	0.50	0	2.75
Oil	86	0.74	0.75	0.50	0	3.29
GM tortilla chips	86	0.61	0.70	0.43	0	3.25
Tortilla chips	86	0.69	0.72	0.50	0	2.89
GM potatoes	86	0.59	0.54	0.50	0	2.00
Potatoes	86	0.67	0.54	0.50	0	2.25
Mean bids for the voluntary GM-labelling market						
Non-GM oil	56	0.80	0.80	0.50	0	4.75
Oil	56	0.76	0.68	0.50	0	3.00
Non-GM tortilla chips	56	0.75	0.81	0.50	0	4.00
Tortilla chips	56	0.68	0.77	0.50	0	4.00
Non-GM potatoes	56	0.84	0.75	0.75	0	4.00
Potatoes	56	0.75	0.70	0.68	0	4.00

food is GM irrespective of the labelling treatment. The size of the discount for the perceived GM food provides evidence about consumers' perception of the signals from the two labelling regimes. We tested null hypotheses that consumers did not discount the perceived GM food in the two markets differently. Table 4.4 provides these results. The first column shows the difference in bids in the mandatory labelling trials; the second column shows the difference in bids in the voluntary labelling trials. The third column is the difference between these columns. The absolute difference is an average of 7 cents for vegetable oil, 1 cent for the tortilla chips and 1 cent for the potatoes. At the 10% significance level, the tests show that one cannot reject the null hypothesis that the difference in bids is zero for any of the three food products.[11] Although none of the differences are statistically significant, at first glance it is curious that the mean discount under mandatory and voluntary labelling regimes is virtually identical for the tortilla chips and potatoes, yet it is considerably larger for the vegetable oil. A possible explanation for the vegetable oil having an average of 7 cents difference is the fact that we used two different types of vegetable oil.[12]

Consumers discounted perceived GM food the same, irrespective of whether the market had mandatory or voluntary GM labelling. *This result provides evidence that consumers receive the same signals under either regime*. By not rejecting the thesis that consumers know GM from non-GM food regardless of the labelling regimes, we have no evidence that the necessary condition of consumers reading signals differently in a mandatory GM-labelling policy than in a voluntary GM-labelling policy is met. Without speculating beyond the reach of the laboratory, this finding supports those who believe the USA has been prudent in avoiding calls to initiate a mandatory GM-labelling policy.

Conclusion

GM food labelling remains a controversial and an important issue in the USA. Some groups have called for mandatory labelling of GM foods, but others want to keep labelling voluntary. The benefit of mandatory labelling that is cited by its supporters is that it will help consumers to choose between GM and non-GM food products. We designed two experimental auction markets, one emulated a market that had a mandatory labelling pol-

Table 4.4. *t*-Test of null hypothesis that differences in bid differences are equal under two labelling regimes.

	Difference in bids for GM and plain labelled – mandatory regime ($n = 86$)	Difference in bids for the plain-labelled and non-GM – voluntary regime ($n = 56$)	Difference	*t*-Test statistic
Oil	0.11	0.04	0.07	0.90
Tortilla chips	0.08	0.07	0.01	0.03
Potatoes	0.09	0.08	0.01	0.20

[11] Regression models were also fitting to test whether demographic characteristics made a difference on the discount for the perceived GM food. No demographic characteristic affected significantly the discount for the perceived GM food. Also, one could not reject the null hypothesis that third-party information did not affect the difference in the discount for the perceived GM food.

[12] We also ran Wilcoxon rank-sum tests to see if one could reject that consumers had different behaviour for the different label types. The results for the Wilcoxon rank-sum tests are similar to those of the *t*-test results, showing that one cannot reject the null hypothesis that consumers perceive the signals from the two labelling policies the same.

icy in place, the other emulated a market that had a voluntary labelling policy in place. We found no evidence that consumers could more easily distinguish which product was GM or non-GM in the mandatory labelling market. This provides evidence that the voluntary labelling policy in the USA is the best policy.

One further avenue for research would be to examine the international dimension to GM food labels, say in Europe or Australia. For example, do consumers in those countries read the same signals of genetic modification in voluntary labelling markets as in mandatory labelling markets? If people can read the signals for which food is GM accurately in either mandatory or voluntary GM-labelling regimes, this calls into question the relevance and usefulness of the mandatory labelling policies throughout the world.

References

AgBiotech Reporter (2001) 18(8).

Australia New Zealand Food Authority (2001) Genetically modified foods. Available at http://www.anzfa.gov.au/GMO/.

Becker, G., DeGroot, M. and Marschak, J. (1964) Measuring utility by a single response sequential method. *Behavioral Science* 9, 226–236.

Caswell, J.A. (1998) Should use of genetically modified organisms be labeled? *AgBioForum* 1, 22–24.

Caswell, J.A. (2000) Labeling policy for GMOs: to each his own? *AgBioForum* 3, 53–57.

CNN (2001) Europe approves tough GM food rules. Available at http://www.cnn.com/2001/WORLD/europe/02/14/eu.gm/index.html.

Consumer Reports (1999) Genetically altered foods, recommendations. Available at http://www.consumerreports.org/main/detail.jsp?CONTENT%3C%3Ecnt_id=19339andFOLDER%3C%3Efolder_id=18151andbmUID=1011290002896.

Council for Biotechnology Information. Frequently asked questions. Available at http://www.whybiotech.com/en/faq/default.asp?MID=10 (accessed October 2001).

Friends of the Earth. (2001) The need for labelling genetically engineered foods. Available at http://www.foe.org/safefood/factshtgelabel.htm.

Greenpeace International (1997) Greenpeace launches genetech labelling policy as European Commission fails to do so. Available at http://www.greenpeace.org/pressreleases/geneng/1997nov3.html.

Huffman, W.E., Rousu, M., Shogren, J.F. and Tegene, A. (2002) Should the United States regulate a mandatory labelling policy for GM foods?' Working paper, Iowa State University, Ames, Iowa.

Knetsch, J.L., Tang, F.-F. and Thaler, R.H. (2001) The endowment effect and repeated trial auctions: is the Vickrey auction demand revealing?' *Experimental Economics* 4, 257–269.

Kirchhoff, S. and Zago, A. (2001) A simple model of mandatory vs. voluntary labeling of GMOs. Working paper, Istituto Nazionale di Economia Agraria.

List, J.A. and Lucking-Reiley, D. (2000) Demand reduction in multiunit auctions: evidence from a sportscard field experiment. *American Economic Review* 90, 961–972.

List, J.A. and Shogren, J.F. (1999) Price information and bidding behavior in repeated second-price auctions. *American Journal of Agricultural Economics* 81, 942–949.

Noussair, C., Robin, S. and Ruffieux, B. (2002) 'Do consumers not care about biotech foods or do they just not read the labels? *Economics Letters* 75, 47–53.

Phillips, P.W.B. and Foster, H. (2000) Labelling for GM foods: theory and practice. Paper presented at the International Consortium on Agricultural Biotechnology Research (ICABR) on 'The Economics of Agricultural Biotechnology', Ravello, Italy, 24–28 August.

Phillips, P.W.B. and McNeill, H. (2000) A survey of national labeling policies for GM foods.' *Agbioforum* 3, 219–224.

Rousu, M. and Huffman, W.E. (2001) GM food labeling policies of the U.S. and its trading partners. Staff paper No. 344, Iowa State University, Department of Economics, Ames, Iowa.

Rousu, M., Huffman, W.E., Shogren, J.F. and Tegene, A. (2002) The value of verifiable information in a controversial market: evidence from lab auctions of genetically modified foods.' Staff working paper No. 3, Iowa State University, Department of Economics, Ames, Iowa.

Shogren, J.F., Margolis, M., Koo, C. and List, J.A. (2001) A Random *n*th-price auction. *Journal of Economic Behavior and Organization* 46, 409–421.
Smyth, S. and Phillips, P. (2002) Competitors co-operating: establishing a supply chain to manage genetically modified canola.' *International Food and Agribusiness Management Review* 4, 51–66.
US Food and Drug Administration (2001) Guidance for industry: voluntary labeling indicating whether foods have or have not been developed using bioengineering. Available at http://vm.cfsan.fda.gov/~dms/biolabgu.html.
Vickrey, W. (1961) Counterspeculation, auctions, and competitive sealed tenders. *Journal of Finance* 16, 8–37.
Wilson, W.W. and Dahl, B.L. (2002) *Costs and Risks of Testing and Segregating GM Wheat*. Agribusiness and Applied Economics Report No. 501, North Dakota State University, Fargo.

5 Using Simulated Test Marketing to Examine Purchase Interest in Food Products that are Positioned as GMO-free

Marianne McGarry Wolf, Angela Stephens and Nicci Pedrazzi
Agribusiness Department, California Polytechnic State University, San Luis Obispo, CA 93407, USA

Introduction

Research has been conducted to examine consumer attitudes toward genetically modified food. However, attitude research does not address the specific question: 'Will purchase interest in a specific product be impacted by the consumer's knowledge that a product is a genetically modified food product?' According to the *Wall Street Journal* on 5 April, 2001, there is a segment of the consumer market that wants to know of the presence of genetically modified ingredients and wants to avoid them. In response to the consumers' desire to avoid genetically modified ingredients in food products, numerous products have appeared on the grocery shelves in the USA that bear the label 'non-GMO'. The *Wall Street Journal* indicated that industry executives believe that the non-GMO segment is growing approximately as fast as organic food products. The organic market is a $7.8 billion market that is growing at eight times the rate of the packaged foods market (Callahan and Kilman, 2001).

Methodology

This research uses simulated test marketing methodology to examine the impact of a GMO-free label on purchase interest for two snack products. Simulated test marketing research is a valid methodology that has been used by the marketing community since the 1960s to forecast purchase interest in new products and new positionings for existing products. For example, microwaveable, fat-free, low calorie and organic have been examined through the use of simulated test marketing as characteristics of food products that impact purchase interest. Simulated test marketing is a combination of mathematical modelling and a laboratory experiment. The laboratory experiment is used to simulate the purchase environment for consumers. It may be hypothesized that consumer reaction to a GMO-free positioning will be related to the type of food product that has such a positioning. Therefore, this research examines consumer response to the GMO-free positioning to two types of branded convenience food products: a salty snack food and a fresh vegetable snack food. This research uses the laboratory experiment component of simulated test marketing in a four-cell study design. One cell examines the newly positioned GMO-free salty snack product. The second cell examines the same product without the GMO-free positioning at the same price. Cell three examines the newly positioned GMO-free vegetable snack product. The fourth cell examines the same product without the GMO-free positioning at the same price. A print advertisement is the stimulus used to represent the concept in the laboratory experiment.

The research uses a survey instrument and a concept exposure that were administered through the use of a personal interview of 558 randomly selected respondents at food stores in February 2001 in San Luis Obispo County, California. San Luis Obispo County was designated the best test market in the USA by *Demographics Daily* (Thomas, 2001). San Luis Obispo was found to be the best of 3141 counties to represent a microcosm of the USA based on 33 statistical indicators.

Simulated Test Marketing

Simulated test marketing technology has evolved over time through a combination of methodologies for generating market response and mathematical models that simulate the marketing environment. The result of this combination is a reliable and valid methodology for forecasting awareness, penetration, share, and volume for new and repositioned products and services.

There are a number of simulated test marketing systems that are used by marketing research companies. Bases, Litmus^R, Designator, and FYI are such systems (Clancy *et al.*, 1994). They are branded research methodologies. The specific types of forecasts provided by simulated test marketing research vary between systems. However, all simulated test marketing systems provide the option of generating volumetric forecasts for year 1 sales. Year 1 sales are a function of the awareness of the new product generated by the marketing plan, the distribution of the product in the market place and consumer response to the new product. The Litmus^R system provides both awareness and volume forecasts. In particular, the Litmus^R system forecasts total brand awareness for new products, campaign awareness in the case of the repositioning of an established brand, penetration (percentage of consumers purchasing), repeat buyers, unit sales and share for each month of the introductory year for new and repositioned products.

The validation history for year 1 projections is very strong for the forecasting systems using simulated test marketing methodology. For 250 cases reported by Bases, in-market sales have been within 10% of predicted sales (Clancy *et al.*, 1994). In-market sales generated by products tested using the Designator system have been within 9% (Clancy *et al.*, 1994).

The Litmus^R system uses two types of laboratory experiment. One uses behaviourally generated market response and the other uses attitudinally generated market response. Over 2600 cases have been tested using behaviourally generated purchase interest data. More than 700 of these cases have been available for validation and in 92% of the cases the forecasts were within 10% of the actuals (Clancy *et al.*, 1994). Among the 287 cases that have been tested using attitudinally generated trial, 49 were available for validation. In 85% of the cases the forecasts were within 14% of the actuals.

To forecast accurately the launch of a new product or the repositioning of an existing brand through a simulated test market methodology, the actual marketing environment that will exist at the time of the launch/restage is modelled. The Litmus^R system models the entire marketing mix. It uses three categories of model input: category data, market response and marketing plans (Clancy *et al.*, 1994). Market response is the consumers' response to the test product generated by laboratory experiment. The purchase probability, the repurchase probability, and the purchase cycle for the new or repositioned product are the key components of consumer response. The category data and marketing plans provide the competitive environment that will exist during the launch or restage.

The interview process for most simulated test marketing methodologies occurs in three stages. The initial interview stage generates purchase interest. The second phase occurs after the consumer has taken the product home to use. The second phase generates the first repeat purchase probability and the purchase cycle. The third phase generates loyalty and refines the purchase cycle. The repeat purchase probability, the purchase cycle and loyalty are generated attitudinally.

The initial phase of research is a one-to-one interview. This interview screens the consumers to confirm qualifications and collects background information concerning attitudes and practices in the category. This information is used for two purposes: to simulate the shopping experience by having the consumer think about the category; and to collect information that is used to profile the consumer. For example, the number of units purchased in the test product's category identifies heavy, medium and light purchasers (Clancy *et al.*, 1994). The brand share of requirements for the category identifies brand purchasers.

After the background information is collected, the respondent is exposed to the concept in a competitive context. The concept exposure is done through a simulated store environment and an advertisement. In the absence of a finished product, a printed concept is used as the exposure. In the Litmus[R] system the concept is always priced, accompanied by a competitive array with the key competitors priced appropriately for the market. The respondent purchases the product in the simulated store environment or evaluates the concept for potential trial through a model in the absence of a simulated store.

Market Response

This research is conducted through the use of a concept exposure phase only. Purchase interest in this research uses an 11-point purchase intent scale. Each of the 11 points is coupled with a verbal anchor from: 'Certain will buy – 99 chances in 100' to 'No chance will buy – zero chances in 100': The following question is used after concept exposure in the competitive context:

> If you find this new product in a store where you shop, how likely would you be to purchase this new product in the next 12 months?

Certain will buy	(99 chances in 100)
Almost sure will buy	(90 chances in 100)
Very probably will buy	(80 chances in 100)
Probably will buy	(70 chances in 100)
Good possibility will buy	(60 chances in 100)
Fairly good possibility will buy	(50 chances in 100)
Fair possibility will buy	(40 chances in 100)
Some possibility will buy	(30 chances in 100)
Slight possibility will buy	(20 chances in 100)
Very slight possibility will buy	(10 chances in 100)
No chance will buy	(0 chances in 100)

The 11-point behaviour probability scale was developed by Dr Thomas Juster and published in different forms during the 1960s (Clancy *et al.*, 1994). The 11-point scale couples word meanings with probability estimates to enhance serious thinking. It is more discriminating than the traditional five-point or seven-point scale that is used by forecasting systems such as Bases. It has been employed by marketing researchers such as Yankelovich Partners since the early 1960s in numerous consumer behaviour studies including packaged goods, durables, financial products and services, and other categories.

Yankelovich Clancy Shulman found that, like all self-reported measures of consumer behaviour, the 11-point scale overstates buying (Clancy *et al.*, 1994). Much of the overstatement is a result of the 100% awareness and distribution in the research environment which is never realized in the real world. Even taking this into account by factoring responses by forecasted awareness and distribution, people are more likely to say they will buy than in fact do buy. This was found to be true in all product categories examined (Clancy *et al.*, 1994).

Concept Exposure

This research uses simulated test marketing methodology to examine the impact of a GMO-free label on purchase interest for two snack products. It may be hypothesized that consumer reaction to a GMO-free positioning will be related to the type of food product that has such a positioning. Therefore, this research examines consumer response to the GMO-free positioning to two types of branded convenience food products: a salty snack food, Ruffles Snack Pack, and a fresh vegetable snack food, Cool Cuts. Ruffles Snack Pack is a 5$\frac{1}{2}$ oz container of Ruffles brand potato crisps and onion dip in a convenience package. Ruffles is the number two brand of potato crisps in the USA (Gale Research Inc.,

2000). The Ruffles Snack Pack is sold in the crisps aisle of the supermarket. Cool Cuts is a three-pack of carrots and ranch dip with a picture of Bugs Bunny on the package to attract the attention of consumers. Cool Cuts is sold in the value-added section of the produce department.

This research uses the laboratory experiment component of simulated test marketing in a four-cell study design. One cell examines the newly positioned GMO-free salty snack product. The second cell examines the same product without the GMO-free positioning at the same price. Cell three examines the newly positioned GMO-free vegetable snack product. The fourth cell examines the same product without the GMO-free positioning at the same price. A print advertisement is the stimulus used to represent the concept in the laboratory experiment. The retail price for all four cells is $1.99.

Consumers in all cells were shown the same competitive board. The competitive board displayed pictures of crisps, vegetable snacks, snack mixes and both of the new products. Each product was priced based on current market conditions. Ritz Snack Mixers were $1.99 for a 7 oz. container. Pringles were $1.99 for a 6 oz. container. Carrot Dippers were $1.99 for a three pack. After the respondent had time to review the concept and the competitive boards, purchase interest was evaluated using the 11-point scale.

Consumer Purchase Interest in the GMO-free Positioning

The 11-point purchase interest scale was used to determine purchase interest in the GMO-free positioning compared to the positioning that did not discuss GMOs. Respondents who indicated a 90% chance or higher probability are designated likely purchasers of the product. Respondents with lower than a 90% probability of purchasing the product are designated non-purchasers. There is no difference in the purchase interest between the products or positionings. Therefore, the GMO-free positioning did not have an impact on purchase interest (Table 5.1).

Positioning of a New or Existing Product

A successful product positioning is based on the factors that motivate consumers to purchase one product versus other products. The products that are examined here are branded snack food products. In order to develop a successful positioning, the characteristics that are desirable to consumers when they shop for a snack food must be identified. The characteristics that consumers want when they purchase snack foods are examined by desirability ratings (Clancy *et al.*, 1994). The most desirable characteristics should be used in the development of a product positioning since those are the most important to consumers when they purchase a new product. The product positioning should also stress the characteristics that the consumers perceive the product to have relative to the competition.

In order to understand how consumers perceive the products in the competitive array, each product is rated on the characteristics that were evaluated for desirability. It is important to note that consumers develop perceptions about products, in this case snack foods, from the concept exposure, experience, seeing the products in the store, advertisements, word of mouth, public relations and the media. The perceptions about

Table 5.1.

	Ruffles GMO-free (n = 129)	Ruffles not GMO-free (n = 129)	Cool Cuts GMO-free (n = 155)	Cool Cuts not GMO-free (n = 141)	Chi squared
Likely purchasers	14.0%	11.6%	8.4%	13.5%	2.70
Non-purchasers	96.0%	88.4%	91.6%	86.5%	

a product provide the consumer with the information they use to decide to purchase a product. It is the responsibility of the promotional campaign for a product to communicate the appropriate information to consumers who have not had experience with the product. The promotional campaign also reinforces the perceptions of the consumers that have had experience with the product.

Desirability Ratings of New Product Characteristics

Before the respondents were exposed to the new products, GMO-free Ruffles Snack Pack, Ruffles Snack Pack, GMO-free Cool Cuts and Cool Cuts, they rated eight characteristics that describe the products on a five-point desirability scale. The purpose of the rating is to identify the characteristics of the products that impact a consumer's purchase decision. Characteristics of the products concerning taste, natural product, contains vegetables, good for children, satisfying, free of genetically engineered ingredients, healthy and flavourful were rated. Consumers were asked the following question:

> Please rate the following characteristics you look for when shopping for snack foods where: 5 = Extremely desirable; 4 = Very desirable; 3 = Somewhat desirable; 2 = Slightly desirable; 1 = Not at all desirable.

Analysis of the mean ratings of the interval data indicates that the characteristics are divided into two groups: very to extremely desirable characteristics and somewhat to very desirable characteristics (Table 5.2). The attributes that are *very to extremely desirable* to consumers are tasty, flavourful, satisfying and healthy. The *somewhat to very desirable* characteristics to consumers are natural product, contains vegetables, good for children and free of genetically engineered ingredients. Therefore, when marketing these products, the characteristics concerning tasty, flavourful, satisfying and healthy are important to the consumer. The characteristic, free of genetically engineered ingredients, is only somewhat desirable. Therefore, it is not an important positioning characteristic for the typical snack consumer.

Product Ratings

In order to understand how consumers perceived the new products and the conventional products, the products were rated on the characteristics that were also rated for desirability. Respondents answered the following question:

> Based on your perceptions, please use the following scale to describe how these characteristics describe GMO-free Ruffles, GMO-free Cool Cuts, Ruffles, and Cool Cuts where: 5 = Describes completely; 4 = Describes very well; 3 = Describes somewhat; 2 = Describes slightly; 1 = Does not describe at all.

The mean rating of 1.95 generated for the characteristic, the product is free of genetically modified ingredients, for the GMO-free Ruffles indicates that the respondents that

Table 5.2. Desirability of characteristics ($n = 558$).

Characteristic	Mean	SE
Very to extremely desirable		
Tasty	4.62	0.03
Flavourful	4.56	0.03
Satisfying	4.32	0.04
Healthy	4.06	0.04
Somewhat to very desirable		
Natural product	3.48	0.05
Contains vegetables	3.27	0.05
Good for children	3.25	0.06
Free of genetically engineered ingredients	3.06	0.06

were exposed to the GMO-free Ruffles did not perceive the product as being free of genetically modified ingredients (Table 5.3). Similarly, the mean rating 2.85, generated for the GMO-free Cool Cuts indicates that respondents only somewhat agreed that the product is free of genetically modified ingredients. The term, GMO-free, does not appear to mean free of genetically modified ingredients to the average consumer.

Free of Genetically Engineered Ingredients Segmentation

Consumers were segmented based on the desirability of snack products to be free of genetically engineered ingredients. Customers are segmented into those that find a snack product that is free of genetically engineered ingredients either extremely or very desirable and consumers that find a snack product that is free of genetically engineered ingredients to be either somewhat, slightly or not at all desirable. Not surprisingly, those that think being free of genetically engineered ingredients is an important attribute are less likely to purchase a genetically modified food (Table 5.4).

However, there is no difference in purchase interest in the GMO-free positioning for the snack products examined here between the segment that indicated that free of genetically engineered ingredients is an important attribute and those that do not think it is an important attribute (Table 5.5). This is further evidence that the term, GMO-free, does not appear to mean free of genetically modified ingredients to the average consumer.

Table 5.3. Mean product ratings.

Free of genetically engineered ingredients	Mean	SE
GMO-free Ruffles (*n* = 130)	1.95	0.10
GMO-free Cool Cuts (*n* = 140)	2.85	0.13

Table 5.4. Purchase interest in a genetically modified food product.

	GMO-free (*n* = 197)	Not GMO-free (*n* = 357)	Chi squared
Definitely	4.1%	3.7%	
Probably	10.7%	26.1%	
Maybe	31.6%	44.4%	
Probably not	25.5%	19.7%	
Definitely not	28.1%	6.2%	65.1*

*Significance at the 0.10 level.

Table 5.5. Purchase interest in GMO-free products.

	GMO-free	Not GMO-free	Chi squared
Both GMO-free products	*n* = 90	*n* = 178	
Likely purchasers	13.3%	11.8%	0.13
Non-purchasers	86.7%	88.2%	
GMO-free Ruffles	*n* = 41	*n* = 87	
Likely purchasers	12.2%	10.3%	0.75
Non-purchasers	87.8%	89.7%	
GMO-free Cool Cuts	*n* = 49	*n* = 91	
Likely purchasers	14.3%	13.2%	0.86
Non-purchasers	85.7%	86.8%	

Conclusions

The GMO-free positioning had no impact on purchase interest for either of the products examined, a salty snack food and a fresh vegetable snack food product. Consumers who indicated that *free of genetically modified ingredients* was extremely or very desirable in their purchase decision for a snack product did not have a higher purchase interest for the products labelled GMO-free. The same consumers indicated that they were less likely to purchase genetically modified food products.

Positioning research was conducted to determine the characteristics of snack food products that are important to consumers when purchasing snack products. Eight characteristics were examined. The least important of the eight characteristics was free *of genetically modified ingredients*. Each product was also rated on how well each of the eight characteristics described the product. Consumers who rated the salty snack that was positioned GMO-free indicated that the phrase *free of genetically modified ingredients* described the product slightly. Consumers who rated the salty snack that was positioned GMO-free indicated that the phrase *free of genetically modified ingredients* described the product somewhat. Therefore, it appears that many consumers do not understand the term GMO-free.

The GMO-free positioning for a snack food does not impact the purchase interest for consumers. This may indicate that while consumers report in attitudinal research a likelihood of not purchasing genetically modified food products, when deciding to purchase branded and priced products in a competitive context, the presence of genetically modified ingredients is a factor of low importance. In addition, many consumers are not familiar with the meaning of GMO-free.

References

Callahan, P. and Kilman, S. (2001) Seeds of doubt. *Wall Street Journal*, 5 April. Section A, p. 1.

Center for Food Safety (2000) Why voluntary labelling of genetically engineered foods won't help consumers. Available at http://www.centerforfoodsafety.org/facts&issues/VoluntaryLabelingMemo.html.

Clancy, K.J. Shulman, R.S. and Wolf, M.M. (1994) *Simulated Test Marketing: Technology for Launching Successful New Products*. Lexington Books, New York, 1994.

Consumer International Briefing Paper. Genetically modified foods: magic solution or hidden menace? Available at www.consumerinternational.org/campigns/biotech/breifing.html.

Deis, R.C. (2000) Tortilla chip. *Food Product Design*. Available at http://www.foodproductdesign.com/srchive/2000/100ap.html.

Gale Research Inc. (2000) *Market Share Reporter*. Gale Research, Detroit, Michigan.

Nuffield Council on Bioethics. Genetically modified crops: the ethical and social issues. Available at www.nuffield.org/bioethics/publication/modifiedcrops/rep0000000082.html.

Thomas, G.S. (2001) Playing in San Luis Obispo, *Demographics Daily*, 6 February. Available at wysiwyg://44/http://bizjournals.bcentral.com/journals/demographics/.

Whitman, D.B. (2000) Genetically modified foods: harmful or helpful? *Cambridge Scientific Abstract*. Available at www.csa.com/hottopics/gmfood/oview.html.

6 Measuring the Value of GM Traits: The Theory and Practice of Willingness-to-pay Analysis

Simbo Olubobokun and Peter W.B. Phillips
Department of Agricultural Economics, University of Saskatchewan, 51 Campus Drive, Saskatoon, Canada S7W 5AB

Introduction

As scientists continue to do research and develop new crops, along with studies on the food safety of such crops, it is also important to conduct economic studies on how consumers perceive the benefit–cost of the potential traits.

Presently, there is growing attention towards genetically modified (GM) foods. While there are a wide range of citizen and consumer surveys that show varying degrees of support or antagonism to GM foods, it is not clear how these views would or could influence the operation of the market in the absence of regulation.

Consumer welfare is measured by the amount of money that must be paid (or received) by the consumer facing a change in some parameter in order to keep him or her at the particular level of utility. Welfare of an agent is usually measured based on the fact that value is derived from the utility that individuals derive from satisfying their wants (according to their preferences). This value is usually measured in terms of what people are willing to pay. Thus, an individual's willingness to pay (WTP) for an item or product can be used as a measure of the utility they derive from the product or as a measure of the benefit (the value) of the product to the individual. WTP studies have generally been conducted in the area of environmental economics using tools such as option price, option demand and option value to measure consumers' willingness to pay for certain resources. This chapter reviews and assesses six theoretical approaches used to measure the willingness of a consumer to pay for the perceived benefits of a good.

Background

The introduction of GM crops has economic significance. The 17 crops that have been genetically modified and approved for production somewhere in the world are reputed to be potentially used in 70% or more of the processed foods sold in developed countries. Consumers have expressed a number of concerns about this development, ranging from fears of new or enhanced health and environmental safety risks to questions about the economic, commercial, ethical and social impacts of the technology. Governments and regulators are seeking ways to address consumer concerns, using partial or total bans on the technology, voluntary or mandatory labelling rules, or trade embargoes. Each of these has significant potential to complicate the already difficult international trading relationships around food.

Many surveys show that consumers are beginning to differentiate their views about different GM foods. The majority of consumers in most markets are not keen on GM traits that simply change agronomic practices and increase yields, such as herbicide tolerance. There is somewhat greater support for input traits that reduce the use of chemicals, such the *Bacillus thuringiensis* (Bt)-resistant and viral-resistant crops. Support rises for those products that are perceived to have greater consumer benefits, such as those that have been modified to improve health, nutrition or increase shelf-life. A survey conducted by Environics International in 2000 showed, for example, that 86% of respondents were strongly in favour of biotechnology applications in new medicines, 69% for more nutritious crops, 66% for pest-resistant crops and 33% for farm animal production. These survey results are borne out in a similar survey conducted by Yann Campbell Hoare Wheeler (1999) and in consumer studies where GM and GM-free produce in the same food category are actually offered for sale to consumers (Powell, 2000).

Generally, the scientific research process involves scientists identifying a problem such as variability in income of farmers due to a short growing season and then developing a crop that is cold-tolerant. The gap so far is that scientists and investors have invested substantial resources to develop new agricultural crops that address certain agronomic issues with very little assessment of consumer acceptance of such crops.

This rest of this chapter is organized into three sections. The next section reviews the theoretical background and empirical measurements of WTP, examines theoretical models used to analyse willingness to pay for a product and assesses the appropriateness of the various modelling options for measuring willingness to pay for a product. The following section provides a short discussion of the applications of the theoretical model and the final section provides some concluding comments and policy options.

Modelling Consumer Preferences

Theoretical background

Using the framework of utility theory, this chapter will review six theoretical approaches used to measure the willingness of a consumer to pay for the perceived benefits of a good. We will show that while there are different methods of evaluating the benefits to a consumer, some of these methods overlap. The price the individual is willing to pay for a good represents the expected utility of the good[1]. This value could be greater than, the same as or less than the market price for the good. The theoretical model shows that if a consumer perceives a health benefit from the consumption of a particular product, it will be demonstrated by a willingness to pay a premium above the market price of the product.

Consumer information and knowledge are very important in a period when consumers appear to be changing their attitude towards quality, especially their definition of quality in the area of GM foods. An attitude is an overall evaluation that expresses how much we like or dislike an object, issue or action. Attitudes reflect the consumer's overall evaluation of something based on the set of associations linked to it. Attitudes are learned, and they tend to persist over time. Attitudes are important because they guide the consumer's thoughts, feelings and behaviour and ultimately influence the consumer's buying behaviour during the acquisition, consumption and disposition of an offering (Hoyer and MacInnis, 2001).

Perception of quality could be based on a number of factors such as taste, knowledge of the fact that the product was genetically modified, the level of pesticide used in production and brand name. Usually consumers define a set of determining attributes for a particular product class by acquiring information and learning about the available alternatives. After comparing available products based on the determining attributes, consumers eliminate some alternatives and develop final 'choice sets' of products from which to choose. Options facing the consumer are: purchase, delay purchase or not to purchase (Louviere, 1988).

[1] The authors acknowledge Richard Gray for his suggestions.

With the increased awareness of GM products, various surveys have been conducted and certain issues have been identified to be important to consumers. Four broad groups of concerns are apparent: specific food safety and quality concerns, environmental concerns, fear of the unknown, and ethical objections (Hobbs and Plunkett, 1999). By investigating the willingness to pay for specific traits in GM products, this paper will apply some economic analysis to some of these specific issues of consumer perception.

While consumers cannot directly purchase units of food safety or quality, they can choose to avoid the goods that they perceive to be unsafe or those that have lower quality. They can also choose to pay a higher price for the goods that they perceive to be less risky or have a higher quality (Kuperis *et al.*, 1999). Consumers typically purchase the attributes that are embodied in a product, rather than purchase the product for itself. The product, *per se*, does not give utility to the consumer – the characteristics of the product do (Lancaster, 1966).

van Ravenswaay (1995) identified three classes of consumer activities involved in household production of a healthy state that influence utility levels and have implications for welfare analysis and marketing research. Health maintenance, rehabilitation and protection affect current expenditures and future levels of utility. A consumer who is protecting his/her health might not consume a good because of the perceived health cost that could result from the consumption of such a good. A consumer who is maintaining or rehabilitating his/her health might consume a good because of the perceived health benefit that could result from the consumption of such a good.

Goering (1985) allowed the expectation about product quality rather than taste to vary among consumers. The results indicated that the higher an individual's expectation about product quality, the higher the price the individual will be willing to pay for the product. The distribution of consumer expectations was made to determine the demand for the product. From this analysis, we can assume that at a given quantity, a consumer having a perceived health benefit will be willing to pay a higher price relative to a consumer who has a perceived health cost.

Welfare from the stated WTP values can be measured by the consumer surplus, compensating variation, equivalent variation, option price or demand and option value. The approach used in most cost–benefit studies when measuring consumer welfare is to estimate the future demand for the goods under consideration and the expected consumer surplus value. It should be noted that surveys which elicit the WTP of consumers and estimate future demand obtain results that are classified as stated preferences which may not give the same result as actual purchase preferences. In the rest of this section, we will discuss six economic tools used to measure the willingness of a consumer to pay for the perceived benefits of a good.

Consumer surplus

Consumer surplus (CS) is the measure of welfare change with a given price change. Consumer surplus is usually defined as how much money a consumer would pay for the right to continue to buy at the current price something that he/she is now buying or intends with certainty to buy in the future. Expected consumer surplus, on the other hand, will apply to the good whose consumption is uncertain. Byerlee (1981) modified the definition by adding that consumer surplus is how much money a consumer would pay for the right to buy at the current price something that he/she is now buying or may buy in the future. This definition includes both certain as well as uncertain situations.

Suppose good X is being consumed because of its perceived health benefits. For simplicity of analysis, one can consider the demand for good X to be equal to the demand for the quality of good X[2]. Assuming a given quantity, the WTP values can be interpreted as the WTP for the health benefits in good X. To determine the welfare gain from a price decrease, one can determine how much the consumer will be willing to pay for the price change from P_0 to P_1. The welfare change has two components. First, the consumer will be willing to pay $(P_0 - P_1)X_0$ which is the savings on total expenditure on the health benefits (in good X) at the original

[2] The authors acknowledge Mobinul Huq for his suggestions.

amount X_0. Second, the lower price makes it possible for the consumer to afford to purchase more health benefits of good X. See Varian (1978, pp. 207–209) for the derivations.

If we assume that the marginal utility of income is constant, then the consumer surplus is measured by

$$\int_{P_1}^{P_0} X(P,y)dP_1 = [V(P_1,y) - V(P_0,y)] / [\partial V(P,y)/\partial y]$$

where the right-hand side of the above expression is the money equivalent of the indicated utility change (given the price change).

Compensating and equivalent variation

Compensating variation (CV) in income is defined as the amount of money that the consumer would need to be paid at P_1 so as to maintain the same level of utility (at P_0) before the price change (Varian, 1978). Assume a consumer with an initial price vector P_0 (with demand $X(P_0)$) and a final price vector P_1 (with demand $X(P_1)$), has a constant money income W_0 in each situation.

$$V(P_1, W_0 + CV) = V(P_0, W_0)$$
$$CV = e(P_1, V_0) - e(P_0, V_0)$$
$$V_0 = V(P_0, W_0).$$

Equivalent variation (EV) is the amount of income that could be taken away from the consumer at P_0 to make him as well off as he would be at P_1

$$V(P_0, W_0 - EV) = V(P_1, W_0)$$
$$EV = e(P_1, V_1) - e(P_0, V_1)$$
$$V_1 = V(P_1, W_0).$$

CV or EV can be represented by the amount usually referred to as a premium that the consumer is willing to pay for the health benefits above the market price of goods. In measuring the welfare benefits to different agents, the Hicksian surplus measure is commonly used. Compensating variation (CV) and equivalent variation (EV) are usually estimated off the Hicksian demand curve. Since Hicksian demand curves are unobservable, Slutsky's equation can be used to relate the slope of the Hicksian to the Marshallian demand curve, which is the observed demand curve.

$$[\partial X_1(P, W_0)/ \partial P_1 = \partial h_1(P, V(P, W_0))/\partial P_1 - \partial X_1 (P, W_0)/ \partial W] \cdot X_1.$$

The consumer surplus, CV and the EV are measured off the observed demand curve. As such, they are *ex post* assessments in the sense that welfare is evaluated after the true state of the world is known. The *ex post* analysis, while useful, is not directly measurable for potential GM traits in non-market goods. For such goods, estimation of CS, CV and EV can be done *ex ante* by using an expected demand function obtained through a contingency valuation method or in an experimental setting.

Uncertainty in the demand for a good could be due to factors such as uncertainty concerning income, uncertainty concerning complements or substitutes, or uncertainty concerning personal preferences. Most research studying potential products are evaluated *ex ante* in the sense that the compensation paid or received is measured while the consumer is uncertain about the future demand for the good. This compensation is sometimes referred to as 'willingness to pay' or the option price of an item. Willingness to pay for the perceived health benefits in a product represents the full value to the individual of the health benefits in the product. What a consumer is willing to pay above the market price for a product with perceived health benefits could be a representation of the CV or EV of the consumer, but probably is not equal to it due to uncertainty.

Certainty equivalence

Based on expected utility theory, a rational agent's preferences over risky alternatives can be ordered by the mathematical expectations of utilities for the possible outcomes of the alternatives (Varian, 1978). In addition, a rational agent will choose among risky alternatives so as to maximize expected utility. Let L be a lottery which results in a known state (W_0) with a probability of P and a known state (W_1) with a probability of 1 – P. The expected value of the lottery $E(L) = PW_0 + (1 - P)W_1 = W^*$.

Certainty equivalence (CE) of a lottery L is usually defined as the amount of 'sure' wealth, which gives the same utility as the expected utility of a lottery (L). The utility of the certainty equivalence is equal to the expected utility (EU) derived from the consumption of the good: $U(CE(L)) = EU(L)$.

Risk aversion means that the individual is willing to accept a lower amount than the expected value of the lottery in order to avoid risk. Consequently, a risk-averse individual derives less utility from the expected value of a given lottery ($E(L)$) than having the same amount of sure wealth (W^*). The difference between the utility of a sure wealth and the expected utility of a lottery is the utility loss due to uncertainty ($U(W) - EU(L)$). If $U(W) = EU(L)$, the individual is said to be risk neutral with a linear utility function.

Based on expected utility theory, the utility of the certainty equivalence is equal to the expected utility derived from the consumption of the good. By assuming that the consumer is rational with an objective of maximizing expected utility, the amount that the consumer is willing to pay for a good can be considered a representation of the certainty equivalence.

Option price and value

The option price (OP) is usually defined as the maximum willingness to pay in the present period for the option of demanding a resource in the future. The option price is an *ex ante* value in the sense that it is payable in the present and determined while the consumer is still uncertain about future preferences, income levels, resource availability or other economic parameters.

Since the individual is paying now for an expected future benefit, an option price could be considered equal to an *ex ante* willingness to pay for a product and a representation of the certainty equivalence. Byerlee (1981) defined option demand as the amount a consumer is willing to pay for the option of consuming a good in the future. This definition is the same as the term option price. Long (1967) and Byerlee (1981) showed that under certainty, option demand and consumer surplus are identical.

In a general case when there is uncertainty, option demand may be greater than, equal to, or less than the expected consumer surplus. Weisbrod (1964) and Lindsay (1969) showed that uncertainty provides an additional component (option value) to consumer surplus. The option value (OV) is the difference between what a consumer would be willing to pay for the option of using a resource and the expected consumer surplus of that usage.

One could consider the option value as a premium paid to protect from uncertainty. Zeckhauser (1969) and Cicchetti and Freeman (1981) interpreted option value as a risk premium for risk-averse consumers, the premium being the amount of money that a risk averse individual will pay so as to avoid a risk. Cicchetti and Freeman (1981) demonstrated that option values exist separately from consumer surplus when there is uncertainty and individuals are risk averse. They defined option value as the difference between maximum option price and the expected consumer surplus: $OV = OP - E(CS)$. Their definition of option value posits that an additional measure of benefit is equivalent to a willingness to pay in excess of consumer surplus.

Option value can be interpreted as an *ex ante* allowance for uncertainty that the consumer has about a future event taking place. An individual facing uncertainty will be willing to pay a little extra now above his/her expected value of consumer surplus to assure that the resource would be available in the future if he/she decides to demand the resource. Fisher and Hanemann (1986) applied the concept of option value to situations wherever a decision has the characteristics that one of the possible outcomes is irreversible and there is some future prospect of better information about the benefits and costs of these outcomes.

Links between the theoretical approaches

If probability values can be assigned to the future consumption of a product, then option value and option price (or option demand) can be used. However, if it is not possible to assign probability values then the elicited amount that the consumer is willing to pay for a product will represent the value of the product to the consumer.

Willingness to pay for the perceived health benefits in a product represents the full value to the individual of the health benefits in the product. Given the assumption that the consumer is rational with an objective of maximizing expected utility, the amount that the consumer is willing to pay for a good can be considered a representation of the certainty equivalence. Where the utility of the certainty equivalence is equal to the expected utility derived from the consumption of the good. What a consumer is willing to pay above the market price for a product with perceived health benefits could be a representation of the CV or EV of the consumer, but probably is not equal to it due to uncertainty. When the marginal utility of income is constant, the area under the demand curve (measured as consumer surplus) gives an exact measure of the willingness to pay for a price change (the extra consumption of good X).

Model

Economic theory usually assumes that people make choices based on their preferences with the objective of maximizing utility. This means that the marginal utility (benefit) derived from the consumption of a product is equal to the marginal cost of the product.

This section looks at the economic aspect of the concerns that consumers have for GM goods. It uses a two-period model to explain the utility maximizing conditions for individuals having different concerns for GM foods. In period $t = 1$, the consumer has incomplete information. Current consumption in time $t = 1$ is decided subject to uncertainty about future prices, future income, future health impact and future environmental impact. The uncertainty facing the consumer is assumed to be temporal such that at time $t = 2$ the consumer is able to select X_2 with certainty. Between $t = 1$ and $t = 2$, the consumer observes (at zero cost) the realized values of the random variables from period $t = 1$. This model reveals insight into four areas of consumer concerns: health, environmental, ethical issues and uncertainty about future preferences.

The decision to purchase or not to purchase a GM product is based on factors such as price, perception of health benefit or risk, ethical issues, environmental factors, budget constraint and individual characteristics. Based on the Lancaster (1966) approach, the demand for a good is based on the attributes of the good, consequently, the utility that the consumer derives from the consumption of the good is based on the utility derived from the different attributes of the good. This model utilizing an *ex ante* welfare measure was adapted from the example used in Epstein (1975) and Chavas *et al.* (1986).

Utility is derived from the amount of X consumed, and the wealth of the consumer. In this model, any health benefit derived from the consumption of X will result in an addition to wealth whereas any health cost experienced from the consumption of X will result in a reduction of wealth.

$$U = U(X, W)$$
$$W = Wo - Px + H(x) + B + e$$

where U is the utility obtained from the consumption of the particular good X; Wo is the initial wealth of the consumer; Px is the expenditure on good X; H, which is an increasing function of X, could be positive or negative depending on the perception of the consumer, $+H$ is the perceived health benefit while $-H$ is the perceived health cost that results from consuming good x. B is a measure that accounts for factors such as environmental impact and ethical issues, the value that B takes will depend on the type of GM food being considered. For a particular good, B will be zero if the consumer only has health concerns (perceives health benefits or costs from the consumption). However, if in addition to health concerns the consumer has other concerns such as ethical/religious and/or environmental then B can be modelled as follows:

$$B = B_t + B_n(x)$$

where B_n is an increasing function of X; $+B_n$ is the perceived benefit to the environment and $-B_n$ is the perceived cost to the environment that results from the direct use of the crop and/or the agronomic practices used in the production of good X; and B_t, which is fixed, is the measure of ethical concern. When there is a possible concern regarding the ethics surrounding the production process of a product, a consumer will either have or not have some concern. Consequently, B_t is modelled fixed and could be measured using dummy variables. e is the measure of uncertainty about future preferences. Like B_t, e is assumed to be fixed. In the first order condition (FOC) conditions, B_t and e do not show up because they are not modelled as increasing functions of X (they are fixed values).

Perceived health impact

For some goods, the expected health impact from the consumption of such goods could increase or reduce the utility derived by the consumer. In such cases, the expected utility in t = 1 incorporates the perceived health benefits (costs) into the optimization condition. The utility function U is assumed to be a Von Neumann–Morgenstern utility function that is twice continuously differentiable and strictly quasi-concave in X.

$$V_1 = \text{Max } E_1U(X_1, X_2), \text{Max } E_1[V_2]$$
$$\text{subject to} \quad W_1 = P_1X_1 \quad \text{Period 1 } (t = 1).$$
$$V_2 = \text{Max } U(X_1, X_2)$$
$$\text{subject to} \quad W_2 + Hx = P_2X_2 \quad \text{Period 2 } (t = 2).$$

Hx is positive in cases of perceived health benefits and negative for perceived health costs. E_1 is the expectation operator that is conditional on the information available at time $t = 1$. At the current time ($t = 1$), an amount of X is consumed which will give future health benefits. This future health benefit is modelled as an addition to wealth in the future at time $t = 2$ when the consumer chooses X_2.

Using backward induction, the FOC is first solved in $t = 2$ and then solved in $t = 1$.

FOC in $t = 2$:
$$UX_2 - U_{w2}[P_2 + H] = 0$$
$$UX_2/U_{w2} = P_2 + \text{H}.$$

FOC in $t = 1$:
$$E_1(U_{x1} - U_{w1}[P_1 + H]) = 0$$
$$E_1(U_{x1}/U_{w1}) = P_1 + H.$$

The expected utility in $t = 1$ incorporates the perceived health benefits into the optimization condition. A shift from the original utility level to the new utility level, *ceteris paribus*, will change utility by an amount + H (perceived health benefits).

For goods with perceived future health costs, Hx is modelled as a deduction from wealth in the future at time $t = 2$ when the consumer chooses X_2. The expected utility in $t = 1$ ($E_1(U_{x1}/U_{w1}) = P_1 - H$) incorporates the perceived health cost into the optimization condition. A shift from the original utility level to the new utility level, *ceteris paribus*, will change utility by an amount −H (perceived health cost).

Perceived environmental impact

Some consumers will consider for certain GM traits, the perceived positive impact on the environment as an additional benefit of consuming the good, while a negative impact on the environment will be considered a reduction in utility.

$$\text{Max } E\,\{U[X, Wo - Px + H(x) + B]\}.$$

$$\text{Max } E_1U(X_1, X_2), \text{Max } E_1[V_2]$$
$$\text{s.t} \quad W_1 = P_1X_1 \quad \text{Period 1 } (t = 1).$$

$$V_2 = \text{Max } U(X_1, X_2)$$
$$W_2 + H_x + B_{nx} = P_2X_2 \quad \text{Period 2 } (t = 2).$$

Using backward induction, the FOC is first solved in $t = 2$ and then solved in $t = 1$.

FOC in t = 2:
$$U_{x2} - U_{w2}\,[P + H_x + B_{nx}] = 0$$
$$U_{x2}/U_{w2} = P + H + B_n.$$

FOC in $t = 1$:
$$E_1(U_{x1} - U_{w1}[P_1 + H_x + B_{nx}]) = 0$$
$$E_1(U_{x1}/U_{w1}) = P_1 + H + B_n.$$

The expected utility in $t = 1$ incorporates the perceived health and environmental benefits into the optimization condition. A shift from the original utility level to the new utility level, *ceteris paribus*, will change utility by an amount $+H + B_{nx}$ [perceived health benefits and perceived (positive) environmental impact].

For goods with perceived (negative) environmental impact, Bx is modelled as a deduction from wealth in the future at time $t = 2$ when the consumer chooses X_2. The expected utility in $t = 1$ ($E_1(U_{x1}/U_{w1}) = P_1 - H - B_n$) incorporates the perceived environmental cost into the optimization condition. A shift from the original utility level to the new utility level, *ceteris paribus*, will change utility by an amount $-\text{H} - B_{nx}$ [perceived health cost and perceived (negative) environmental impact].

It should be noted that the perceived benefits may not be symmetrical to the perceived costs due to the fact that on average, perceptions of positive benefits in a good by some individuals may not be equal to perceptions of costs in the same good by other individuals. As a result, the optimizing values for perceived health (and environmental) benefits should be solved differently from perceived health (and environmental) costs.

Links between the model and the theoretical approaches

From the survey results, the elicited amount that the consumer is willing to pay for a product will represent the value of the product to the consumer. Willingness to pay for the perceived health benefits in a product represents the full value to the individual of the health benefits in the product. Given the assumption that the consumer is rational with an objective of maximizing expected utility, the amount that the consumer is willing to pay for a good can be considered a representation of the certainty equivalence, where the utility of the certainty equivalence is equal to the expected utility derived form the consumption of the good.

What a consumer is willing to pay above the market price for a product with perceived health benefits could be a representation of the CV or EV of the consumer, but likely is not equal to it due to uncertainty. When the marginal utility of income is constant, the area under the demand curve (measured as consumer surplus) gives an exact measure of the willingness to pay for a price change. If the consumer is willing to pay a premium for the perceived health benefit above the market price, this will show that the WTP value includes the CS and some measure of perceived benefit or cost (premium such as option value).

It should be noted that it is possible for an interaction to exist between the different elements of the model, for example, a consumer having personal health concerns or living with a sick person could have a lot of concern for the environment. Consequently, the presence of such interaction may cause the total WTP values for good X to be other than a straight summation of the individual WTP for each perceived benefit in the good.

Chavas *et al.* (1986) analysed the effects of consumer compensation (option price) on the optimum future consumption levels of the goods in the bundle. They decomposed the option price into two additive terms; the first term corresponds to the surplus measure commonly used in a risk-less situation and the second term (correction factor) involves the covariance between the compensated marginal utility of future income and the compensated future demand evaluated at the current time. They concluded that the correction factor is the amount by which the discounted expected value of the *ex post* welfare triangle must be adjusted to obtain an exact *ex ante* welfare measure. It reflects the influence of temporal uncertainty on *ex ante* welfare measurement. This interaction factor is beyond the scope of this chapter. However, it is a potential area of future research.

Given the economic analysis of what consumers are willing to pay for a good (with perceived health benefits), the next section will review three empirical methodologies used in WTP studies.

Empirical measurements of willingness to pay

Valuation research has been used by a lot of economists when evaluating policy options or the value of a product. In most instances, it is easier to measure the cost of a good or action but it is not always easy to measure the benefits (van Ravenswaay, 1995). In the case of market goods, when consumers are behaving in an optimizing manner (maximizing utility) the marginal benefit is equal to the marginal cost (the market price of such a good). For non-market goods however, estimated benefits are generally measured by what consumers are willing to pay for the hypothetical good.

Hedonic pricing

Hedonic pricing involves observing choices (revealed preference) made in an actual or constructed market, and inferring the value of the perceived benefits or risks (cost) of the product. It is based on the assumption that quality of a good is related to measurable specification variables. That is, quality determines price and such prices can be said to be dependent on the characteristics (quality) of the product. Empirical applications usually involve the regression of prices (or the logarithm of prices) of the different varieties of a type of good on the specification variables. Hedonic price functions are usually estimated in an attempt to reveal the behavioural information about the marginal values of the characteristics of a product. The price that people are willing to pay for a product can be interpreted as the price they are willing to pay for the quality attributes of the product. It is also possible to evaluate the trade-offs between price and the attributes of the product.

In the hedonic model, the increment in price due to increases in any characteristics will equal the buyers' WTP for the characteristics and the marginal cost of producing the characteristic for sellers. When buyers and sellers have time to adjust their responses, the marginal hedonic price equals the marginal value to consumers and the marginal cost to suppliers. For non-market goods, it is not possible to use actual hedonic pricing. However, hedonic pricing can be used for non-market goods in an experimental setting.

Experimental auctions

Experimental auctions involve the use of real products and real incentives along with some information on the different products being auctioned. By giving the participants repeated opportunities to participate in the auction market, learning is enhanced; hence the participants can show their real preferences for the products (Fox *et al.*, 1995). The objective of the auction is to elicit the value of the good to the consumer.

Experimental auctions can be used as a complement or alternative to elicitation methods such as contingency valuation (CV) surveys and hedonic pricing. By pre-testing the CV survey design, experimental auctions can be used *ex ante* to improve the design of contingency valuation surveys (Coursey and Schulze, 1986). The *ex ante* auction could be used to observe how bidding behaviour is affected by alternative incentives and by repeated market experience. Prior to running a survey instrument, an experimental auction could provide the researcher with the opportunity to design, test and replicate the preference revealing the incentives of the elicitation method (Fox *et al.*, 1995).

When experimental auctions are used *ex post*, they could be run independently as a valuation process or the results from such auctions, which revealed the learning and market experience from the experimental auction, could be used to adjust the bids of the respondents of the contingency valuation method.

If respondents have vague or undefined incentives when evaluating a (GM) product, results from non-market methods could be inconsistent. The laboratory experimental auctions provide participants with a well-defined incentive structure that enables the researcher to elicit more accurately the value of a non-market product.

Experimental auctions do not have the non-response bias that is common in survey techniques. Usually when participants are recruited, they do not have an indication of the nature of the experiment they will be involved in. As a result, their willingness to participate is unrelated to their attitude towards the product being studied (Fox *et al.*, 1995)

A general drawback is the fact that if respondents do not fully understand the questions posed to them, the responses may not conform to theoretical expectation. The evaluation and the subsequent survey response to a given product will vary among individuals due to the fact that individuals have different perceptions about different products in addition to varying levels of familiarity with a particular product.

Valuing non-market goods is difficult because there is no formal market to obtain price or other information relevant for economic analysis. While it is possible to simulate the actual purchase of certain products in an experimental setting, it is not possible to simulate a market experience for issues such as food safety or environmental concern.

A lot of non-market valuation research has been conducted in the area of food safety and environmental issues. For non-market goods, the absence of primary market data (e.g. demand curves) makes it impossible to estimate welfare changes. This has led to the use of methods such as contingency valuation to estimate the demand for non-market goods.

Contingency valuation

WTP has been empirically measured using the contingency valuation (CV) method. The CV method involves the use of surveys to elicit consumers' WTP for non-market goods contingent on the specified scenario. Stated WTP values can serve as prices and can be used to determine the relationship between the determinants of WTP.

A primary advantage of CV over hedonic pricing is the fact that CV is a flexible tool that can be tailored to analyse specific issues (Buzby *et al.*, 1995). Usually, CV results are comparable to other methods of valuing non-market goods and are generally less expensive than actual market experiments.

CV relies on the subjective responses of consumers, which could be a source of bias. Generally, elicited values inflate what consumers would actually pay because consumers take hypothetical situations less seriously than real-life situations.

The researcher should recognize the potential for bias when using CV. Bias could be minimized through careful phrasing and testing of scenario statements and WTP questions. When biases are present in CV studies, the WTP estimates may be overstated and the comparison of WTP values between different issues may be questionable.

In addition, care should be taken when interpreting and comparing the results from CV studies. A distinction should be made between measures of WTP above the market price (premium) and actual WTP values, which represent the price of the product.

Research Applications

Although many surveys are conducted on a regular basis, in order to measure effectively the economic analysis of consumers' perceived value of a GM trait the survey instrument has to be based on an economic model. An obvious application of this study is as an economic background for a consumer survey instrument. Results from such surveys can then be used to test the model. In surveying one has to be mindful of the welfare measure to be used. Generally, WTP applies to the perceived value of the benefit to the consumer. If the survey instrument is to elicit WTP then welfare measures such as option price, option demand and certainty equivalence will apply. On the other hand, if the survey instrument is to elicit WTP a premium above the market price, welfare measures such as option value, compensating variation and equivalent variation will apply.

Concluding Comments

The price the individual is willing to pay for a good represents the expected utility of the good. This value could be greater than, the same as or less than the market price for the good. If a consumer perceives a health benefit from the consumption of a particular product, this will be demonstrated by a willingness to pay a premium above the market price of the product subject to a budget constraint.

Attitudes of consumers are directly linked to the utility function. Generally, consumers will be willing to pay a positive premium for traits they perceive to have positive health benefits, which adds to their utility derived from consuming the product. Given information on the demand for the final product, farm supply of the agricultural product along with information on the type of the processing sector, the distribution of economic benefits can be predicted for the adoption of a new GM crop.

When consumer attitudes toward quality change, the demand for such goods is affected. Generally, consumers develop their attitudes towards the quality of a product based on perceptions of benefits and risks. These perceptions are usually based on prior knowledge and experience in addition to information such as brand name, retailer reputation and labelling.

Producers need adequate consumer information to be able to provide the quality assurance that consumers of GM foods need. If producers are able to provide the quality assurance that consumers need, then producers can have a high level of confidence that the goods being produced will eventually be purchased.

The differentiation of consumer preferences or the degree of perceived quality differences have implications in the separation of the markets of GM foods and non-GM foods through labelling. The superiority of a policy of labelled versus unlabelled GM foods will depend on consumer perceptions of GM foods in addition to the magnitude of the labelling and sorting costs. If the perceived quality difference is sufficiently large, obligatory labelling might be preferred. If the perceived quality difference between GM foods and non-GM foods is sufficiently small, the labelling and sorting costs of moving to a separating market situation could exceed the benefits, such that mandatory labelling might not be required. Consumer preference studies could be used to document and support both labelling requirements and import restrictions on GM foods when there are public concerns based on possible, but yet unknown, environmental and health risks. Given the growing awareness of GM foods, exporting countries will have to be aware of the different consumer segments in a region and understand that the veto of acceptance in different countries needs to be carefully considered when introducing a new GM food or product.

References

Buzby, J.C. (1995) Using contingency valuation to value food safety: a case study of grapefruit and pesticide residues. In: Caswell, J. (ed.) *Valuing Food Safety and Nutrition*. Westview Press, Boulder, Colorado.

Byerlee, D.R. (1981) Option demand and consumer surplus: comment. *Quarterly Journal of Economics 85*, 523–27.

Chavas, J., Bishop, R.C. and Segerson, K. (1986) Ex ante consumer welfare evaluation in cost–benefit analysis. *Journal of Environmental Economics and Management,* 13, 255–568.

Cicchetti, C.J. and Freeman III, A.M. (1981) Option demand and consumer surplus: further comment. *Quarterly Journal of Economics* 85, 528–539.

Coursey, D.L. and Schulze, W.D. (1986) The application of laboratory experimental economics to the contingent valuation of public goods. *Public Choice* 49, 47–58.

Environics International (2000) Global public perception of food biotechnology. Presented at the Convergence of Global Regulatory Affairs: Its Potential Impact on International Trade and Public Perception Conference. Ag-West Biotech Inc., Saskatoon, Canada.

Epstein, L.G. (1975) A disaggregated analysis of consumer choice under uncertainty. *Econometrica* 43, 877–891.

Fisher, A.C. and Hanemann, W.M. (1986) Option value and the extinction of species. *Advances in Applied Micro-economics* 4, 133–152.

Fox, J.A., Shogren, J.F., Hayes, D.I. and Kliebensein, J.B. (1995) Experimental auctions to measure willingness to pay for food safety. In: Caswell, J. (ed.) *Valuing Food Safety and Nutrition*. Westview Press, Boulder, Colorado.

Goering, P.A. (1985) Effects of product trial on consumer expectations, demand, and prices. *Journal of Consumer Research* 12, 74–82.

Hobbs, J.E. and Plunkett, M.D. (1999) Genetically modified foods: consumer issues and role of information asymmetry. *Canadian Journal of Agricultural Economics* 47, 445–455.

Hoyer, W.D. and MacInnis, D.J. (2001) *Consumer Behavior*. Houghton Mifflin Co., Boston, Massachusetts.

Kuperis, P., Veeman, M. and Adamowicz, W.L. (1999) Consumers' response to the potential use of bovine somatrophin in Canadian dairy production. *Canadian Journal of Agricultural Economics* 47, 151–163.

Lancaster, K.J. (1966) The new approach to consumer theory. *Journal of Political Economy* 74, 132–157.

Lindsay, C.M. (1969) Option demand and consumer's surplus. *Quarterly Journal of Economics* 83, 344–346.
Long, M.F. (1967) Collective consumption services of individual consumption goods: comment. *Quarterly Journal of Economics* 81, 351–352.
Louviere, J.J. (1988) *Analysing Decision Making: Metric Conjoint Analysis*. Quantitative Applications in the Social Sciences, Vol. 67. Sage Publications, Thousand Oaks, California.
Powell, D. (2000) Safe enough. Enhancing consumer confidence in food production technologies. Unpublished report, University of Guelph, Canada.
van Ravenswaay, E.O. (1995) Valuing food safety and nutrition: the research needs. In: Caswell, J. (ed.) *Valuing Food Safety and Nutrition*. Westview Press, Boulder, Colorado.
Varian, H.R. (1978) *Microeconomic Analysis*. W.W. Norton and Co., New York.
Weisbrod, B.A. (1964) Collective consumption services of individual-consumption goods. *Quarterly Journal of Economics* 78, 471–477.
Yann Campbell Hoare Wheeler, (1999) Public attitudes towards biotechnology. Biotechnology Australia. Available at http://www.biotechnology.gov.au/library/content_library/BA_pYCHW.pdf
Zeckhauser, R. (1969) Resource allocation with probabilistic individual preferences. *American Economic Review* 56, 546–552.

7 Willingness to Pay for GM Food Labelling in New Zealand

William Kaye-Blake, Kathryn Bicknell and Charles Lamb
Commerce Division, PO Box 84, Lincoln University, Canterbury 8150, New Zealand

Introduction

New Zealand is one of an estimated 28 countries, plus the European Union (EU), that have adopted or planned to adopt labelling for genetically modified (GM) food (Phillips and McNeill, 2000). The specific legislation governing labelling varies, so that each country's experience with labelling is likely to be unique. New EU regulations, for example, will require labels on food products containing GM soybean oil and glucose syrup from maize starch (European Commission, 2001), products which do not require labelling in New Zealand. In both places, products containing genetically modified organisms (GMOs) require labels, whereas the voluntary programme in the USA provides for labelling non-GM food (Phillips and McNeill, 2000). In New Zealand's case, the Australia New Zealand Food Authority (ANZFA) requires that food sold in supermarkets be labelled if it contains GM ingredients.

All labelling programmes, however, are likely to have something in common: added costs. For example, as a result of the ANZFA regulations, manufacturers conducted detailed surveys of the ingredients in their products (Robertson, 2002). The segregation or Identity Preservation (IP) necessary for a trustworthy label would also add to the cost of production. The Australia New Zealand Food Standards Council (ANZFSC) estimated that its labelling regime would entail start-up costs in New Zealand of NZ$43 million, with ongoing costs of $42 million annually (ANZFA, 2001). The European Commission estimated that IP would increase grain farmgate prices by 6–17% (European Commission, 2000). The US Department of Agriculture (USDA) estimated that segregating non-biotech maize would add $0.22 per bushel (Lin *et al.*, 2001–2002). Generally, the costs of segregation and IP depend on the definition of GM food: the more ingredients that need labelling and the lower the tolerances allowed, the higher the costs (OECD, 2000; Wright, 2000; Lin *et al.*, 2001–2002). An estimate of overall willingness to pay (WTP) for labelling whether food is GM would provide some idea of an economically efficient level of spending on labelling programmes, and thus on the nature of IP systems and regulations that might best suit consumers' demand.

There is evidence to suggest that the general level of support for mandatory labelling of GM food is high. Interestingly, opinion polls indicate that a larger proportion of the population desires labelling compared to the proportion that wishes to purchase non-GM food. The European Commission (2000) reported that nearly three-quarters of the European population favours a clear labelling of GM food, versus just over one-half who

(eds R.E. Evenson and V. Santaniello)

would pay more to avoid GM food. The Commission also reported the results of various opinion polls conducted in the USA since 1995. They indicate that the level of support for mandatory labelling amongst Americans is high and stable, ranging from 81% to 93%. However, only 43% are not likely to buy food enhanced through genetic engineering (GE) (Wirthlin Quorum Poll, 2000, quoted in Campbell *et al.*, 2000).

Economic efficiency, however, requires more than a general level of support. If mandatory labelling implies higher costs, it is desirable to determine whether the public is generally willing to pay for the information contained on the label. The objective of this study is to begin to understand the key factors that influence a consumer's WTP for labelling of GM food, and to draw some initial conclusions about the magnitude of that WTP.

Factors Affecting the Demand for Labelling

Demand for labelling the GM status of food has not been widely investigated. Discussions of labelling have tended to focus on policy alternatives, and demand for labelling is often conflated with demand for non-GM food. Therefore, guidance for the selection of variables in the current study was sought from literature on the demand for non-GM food.

Prior research is not unanimous on which economic or demographic characteristics influence attitudes towards GM food or WTP for non-GM food. Usually important is gender – women are more likely to have negative attitudes towards GM food than men (Sparks *et al.*, 1994; Anon., 2001; Cook, 2001; Gamble *et al.*, 2000). It has been argued that education also affects how people view GM food, but empirical work indicates that the effect is unclear (Couchman and Fink-Jensen, 1990). Occupation may be important, with those in higher-skilled occupations seeing more benefits in the technology (Sparks *et al.*, 1994). Although economic theory suggests that income and WTP for labelling are likely to be positively correlated, the results of empirical research on how income affects the demand for non-GM food are inconclusive (Gamble *et al.*, 2000). Age has also appeared in some research as an important factor affecting this demand (Sparks *et al.*, 1994; Gamble *et al.*, 2000). Finally, price sensitivity or the marginal utility of the food dollar is important (Bredahl *et al.*, 1998; Burton *et al.*, 2001).

Research into consumer acceptance of GM food has also focused on the importance of prior attitudes. This research, often based in sociological or psychological theory, gauges such factors as respondents' beliefs about nature or government, or feelings about chemicals in agriculture or food additives. In fact, these factors are often more important than socioeconomic variables with which economists are more comfortable (Bredahl *et al.*, 1998; Senauer, 2001). One behaviour that seems a good indicator for this 'lifestyle' factor is buying organically grown food. Burton *et al.* (2001), for example, successfully used respondents' preferences for organically grown food to segment their sample.

Discussions about labelling often raise the question of its usefulness (Valceschini, 1998; Huffman *et al.*, 2001). In particular, if consumers do not know what genetic engineering is, then the label is potentially meaningless. A closely related issue is whether consumers actually read food labels. Reading labels has been shown to affect food consumption (Nayga, 2001–2002), and so could be important in determining which consumers care about GM labelling.

There may be other factors affecting the demand for labelling aside from those that affect the demand for non-GM food. Presumably, consumers who wish to refuse GM food want the information to determine what food comes from GMOs and then make their refusal. As noted before, however, support for labelling exceeds rejection of GM food. This is consistent with findings that consumers want more information about GE (Bredahl *et al.*, 1998). Furthermore, consumers may be interested in purchasing an option regarding future consumption. They may not currently be worried about GMOs, but they may want to preserve their right to refuse them should they receive new information (Gollier *et al.*, 2000; Huffman *et al.*, 2001). The complexity of consumer reactions to GM food therefore supports the idea of examining the demand for labelling separately from the demand for non-GM food.

The Survey Data

As part of a larger study that sought to understand better how consumers perceive and attempt to manage food risks, a total of 450 households in Christchurch, New Zealand, were personally interviewed. The respondents were adults aged 18 or over who were identified as those individuals primarily responsible for their households' food purchasing and/or food preparation. The sample was drawn by stratifying suburbs into groups based on socioeconomic and property valuation data, then proportionately allocating the total sample size across the groups relative to the proportions of households in each group. Each interview took 30 to 60 minutes. After coding, editing and evaluation for completeness, there were 441 usable questionnaires.

As part of the survey enumeration process, respondents were asked the weekly food expenditure for the household. Interviewers added 2% to the figure given and asked the respondent if he or she would be willing to spend the resulting dollar amount on weekly food expenditure in order to know whether food had been genetically modified. The same procedure was followed for 5% and 10% of the food bill. The order in which the respondent was asked about the different levels of spending was randomized.

In addition, several attitudinal and behavioural questions were asked. At the beginning of the interview, respondents were given the opportunity to describe the most important food-related issues facing their household. Each respondent could provide up to three responses, which were used to explore attitudes such as price consciousness that might affect WTP for labelling. Attitudes were also assessed by asking respondents whether they agreed with a series of statements, for example: 'I believe that there are definite benefits to the consumer associated with GE/GMO food'. They responded using a five-point Likert scale from 'strongly disagree' to 'strongly agree'. In addition, respondents were questioned about food-related behaviour, such as whether they bought organically grown food. Responses were again recorded on a five-point scale, from 'never do' to 'always do'.

Finally, demographic and economic data were collected: age, income, occupation, educational attainment and gender.

Methodology

Respondents to the survey provided an indication of the strength of their demand for labelling by answering questions as to whether they were willing to pay an additional 2%, 5% or 10% more for their groceries in order to know the GM status of their food. The results represent the outcome of a decision between a finite set of alternatives, which generated a discrete dependent variable. As the underlying dependent variable is assumed to be continuous, but only a discrete response is observed, it is appropriate to analyse the data using a qualitative response model (Maddala, 1983; Greene, 1993).

In this application, a respondent's choice falls into one of four categories that are naturally ordered. This gives rise to the following latent regression:

$$y^* = \beta' x + \varepsilon \tag{1}$$

where y^* refers to some unobserved measure of the respondents' WTP for labelling, β is a vector of parameters that reflect the impact of changes in the independent variables (x), and ε is an unobserved error term. While y^* is not observed, the following four values of the dependent variable (y) are observed:

$$\begin{aligned} &y = 0 \text{ (WTP = 0\%)} && \text{if } y^* < \gamma_0 \\ &y = 1 \text{ (WTP = 2\%)} && \text{if } \gamma_0 \le y^* < \gamma_1 \\ &y = 2 \text{ (WTP = 5\%)} && \text{if } \gamma_1 \le y^* < \gamma_2 \\ &y = 3 \text{ (WTP = 10\%)} && \text{if } y^* \ge \gamma_2. \end{aligned} \tag{2}$$

Here γ is a vector of unknown parameters to be estimated along with β.

If ε is assumed to be distributed logistically, the parameters can be estimated using an ordered logit model. Under these circumstances, the probability that a respondent will answer 0% is given by the following:

$$\text{Prob (response} = 0\%) = \frac{e^{\gamma_0-\beta' x}}{1+e^{\gamma_0-\beta' x}} \tag{3}$$

Similarly,

$$\text{Prob (response} = 2\%) = \frac{e^{\gamma_1-\beta' x}}{1+e^{\gamma_1-\beta' x}} - \frac{e^{\gamma_0-\beta' x}}{1+e^{\gamma_0-\beta' x}} \tag{4}$$

$$\text{Prob (response} = 5\%) = \frac{e^{\gamma_2-\beta' x}}{1+e^{\gamma_2-\beta' x}} - \frac{e^{\gamma_1-\beta' x}}{1+e^{\gamma_1-\beta' x}} \tag{5}$$

$$\text{Prob (response} = 10\%) = 1 - \frac{e^{\gamma_2-\beta' x}}{1+e^{\gamma_2-\beta' x}} \tag{6}$$

In the current study, the observed responses (y) and the survey data (x) were used to estimate γ and β using a maximum likelihood procedure.

A further assumption of the ordered logit is that the β parameters do not change for each response category. The responses should represent a true order, drawn from a single underlying distribution. The effects of the explanatory variables should therefore be consistent across all categories. The test of parallel lines tests this assumption by comparing the results of the ordered logit with a general (i.e. non-ordered) multinomial logit model. If the data indicate that the β parameters do, in fact, vary across response categories, then it is likely that a multinomial logit would be a more appropriate modelling choice. This approach, however, loses the information contained in the ordered nature of the responses.

Tables 7.1 and 7.2 summarize the variables included in the preferred ordered logit model. Dummy variables were used to indicate price consciousness, label awareness and purchase practices regarding organically grown food. Price conscious respondents were identified from their 'top-of-the-mind' answers to the survey's open-ended question regarding the most important food issues facing their household. The label variable was used to indicate whether or not respondents used ingredient or nutritional labelling. If the respondent either mostly or always purchased organically grown food, then the variable *ORG* took on a value of 1.

Table 7.1. Definitions of variables.

Variable	Definition
L	Strength of demand for labelling: none = 0, 2% = 1, 5% = 2, 10% = 3.
PC	= 1 if the respondent is price conscious, or 2 otherwise (reference category).
RDL	= 1 if the respondent reads nutrition or processing information, or 2 otherwise (reference category).
ORG	= 1 if the respondent mostly or always buys organic food, or 2 otherwise (reference category).
GE	= 1 if the respondent had some knowledge of GE/GM, or 2 otherwise (reference category).
BEN	In response to the question 'I believe that there are definite benefits to the consumer associated with GE/GMO food', = 1 if the respondent agrees or strongly agrees, 2 if disagrees or strongly disagrees, or 3 if neutral or no response (reference category).
RSK	In response to the question 'I believe the risks to health from consuming GE/GMO food are low', = 1 if the respondent agrees or strongly agrees, 2 if disagrees or strongly disagrees, 3 if neutral or no response (reference category).
TRST	In response to the question 'I do not trust the large food manufacturing companies', = 1 if the respondent agrees or strongly agrees, 2 if disagrees or strongly disagrees, 3 if neutral or no response (reference category).
OCC	= 1 if the respondent's occupation is professional or managerial, = 2 if the respondent's occupation is clerical, sales or service, = 3 if the respondent's occupation is technical or engineering, = 4 if the respondent's occupation is agricultural or farming, = 5 if the respondent's occupation is student, = 6 if the respondent's occupation is unemployed, receiving a benefit or not given, = 7 if the respondent's occupation is tradesperson or labourer, = 8 if the respondent's occupation is retired or housewife (reference category).
UNI	= 1 if the respondent has a university degree or bursary, university entrance or scholarship, or 2 otherwise (reference category).
GEN	= 1 if the respondent is female, or 2 otherwise (reference category).

Table 7.2. Descriptive statistics for the sample ($n = 441$).

Variable	Category	Count	Percentage of sample
L	None	111	25.2
	2%	107	24.3
	5%	121	27.4
	10%	102	23.1
PC	Yes	130	29.5
	No	311	70.5
RDL	Mostly or always	237	53.7
	Other	204	46.3
ORG	Mostly or always	51	11.6
	Other	390	88.4
GE	Yes	360	81.6
	No	81	18.4
BEN	Agree	123	27.9
	Disagree	165	37.4
	Neutral/no response	153	34.7
RSK	Agree	327	74.1
	Disagree	37	8.4
	Neutral/no response	77	17.5
TRST	Agree	132	29.9
	Disagree	135	30.6
	Neutral/no response	174	39.5
OCC	Professional/managerial	104	23.6
	Clerical/sales/service	60	13.6
	Technical/engineering	33	7.5
	Agricultural/farming	6	1.4
	Student	46	10.4
	Unemployed/beneficiary/not given	29	6.6
	Trades/labourer	46	10.4
	Retired/housewife	117	26.5
UNI	Degree or UE	172	39.0
	No degree indicated	269	61.0
GEN	Female	285	64.6
	Male	156	35.4

In an attempt to measure familiarity with genetic modification, a dummy variable was included that indicated some degree of prior knowledge. Specifically, respondents who could provide a description of genetic engineering or genetic modification, e.g. 'gene transfer' or 'selective breeding', were considered 'knowledgeable'. This approach did not assess the quality of respondents' knowledge, but such an assessment was considered beyond the scope of this research. However, the results are similar to others' (Couchman and Fink-Jensen, 1990; Macer, 1992), in that 81.6% of respondents were judged to be knowledgeable.

Since trust in institutions and the food system has been highlighted as important in the demand for non-GM foods (Sparks *et al.*, 1994; Bredahl *et al.*, 1998; Campbell *et al.*, 2000; Bredahl, 2001; Cook, 2001), respondents were also asked whether they agreed with the statement, 'I do not trust the large food manufacturing companies'. In an attempt to capture attitudes towards GM food, respondents were asked whether they agreed with two statements: 'I believe that there are definite benefits to the consumer associated with GE/GMO food' and 'Risks from consuming GE/GMO food are unknown' (on GM food and risk perception, see Bredahl, 2001).

Demographic and economic data were also included in this analysis. A simplified educational indicator was developed: whether or not the respondent had a university degree (this also included those whose highest qualification was bursary, university entrance or scholarship, i.e. those who were on a university track). Respondents were also grouped into several occupational categories. Finally, gender was incorporated into the current analysis with a dummy variable.

The survey instrument included a question on income. Unfortunately, a large number of respondents refused to answer the question. Because preliminary analysis indicated that income was not a useful explanatory variable, it was omitted from the final analysis. It should be noted, however, that the variable on price consciousness may be important in reducing the impact of income, since lower-income consumers tend to be more price sensitive (e.g. Jones, 1997).

Results

The model specified above was estimated with the ordinal regression command (PLUM) in SPSS, using the logit link function. Table 7.3 contains the parameter estimates obtained, as well as several statistics evaluating the model. The significant chi-squared statistic indicates that the null hypothesis that all of the coefficients of the explanatory variables are zero can be rejected. However, the value of the pseudo R^2 is disappointingly low, suggesting that a large proportion of the variation in WTP for labelling is not captured by the independent variables.

Another measure of the model's overall strength is its predictive power. Table 7.4 compares the actual responses with the model's predicted responses. The model predicted the correct category about 40% of the time (compared to a naïve or random expectation of 25%), and the average error between

Table 7.3. Estimated model.

Variable	Category	Parameter	SE	Significance
γ_0		0.846	0.387	0.029
γ_1		2.109	0.397	0.000
γ_2		3.538	0.419	0.000
PC		−0.347	0.200	0.083
RDL		0.302	0.186	0.104
ORG		0.897	0.290	0.002
GE		0.603	0.245	0.014
BEN	Agree	−0.087	0.229	0.705
(ref = neutral)	Disagree	0.529	0.219	0.015
RSK	Agree	0.772	0.250	0.002
(ref = neutral)	Disagree	0.853	0.385	0.027
TRST	Agree	−0.162	0.222	0.465
(ref = neutral)	Disagree	−0.539	0.215	0.012
OCC	Professional/managerial	0.549	0.260	0.035
(ref = retired/housewife)	Clerical/sales/service	0.634	0.301	0.035
	Technical/engineering	0.839	0.368	0.023
	Agricultural/farming	−0.330	0.775	0.671
	Student	0.263	0.353	0.456
	Unemployed/beneficiary/not given	−0.599	0.401	0.135
	Trades/labourer	0.759	0.329	0.021
UNI		0.455	0.201	0.024
GEN		0.518	0.192	0.007
McFadden's pseudo R^2	0.080			
Models compared	**Purpose**	**Chi-squared**	**df**	**Significance**
Intercept-only vs. model	Assess model fit	98.113	19	0.000
Ordered vs. general logit	Test of parallel lines	30.447	38	0.803

Table 7.4. Predicted demand categories versus actual demand categories.

Predicted categories	Actual demand categories: None	2%	5%	10%	Total
None	62	43	26	16	147
2%	18	9	10	6	43
5%	25	42	57	35	159
10%	6	13	28	45	92
Total	111	107	121	102	441
Correct %	55.9%	8.4%	47.1%	44.1%	39.2%

the predicted and actual categories was less than one category. The model's predictive ability varied by WTP category, however. In particular, the model correctly predicted only 8.4% of the 2% WTP responses.

Finally, the test of parallel lines (Table 7.3) indicates that the parameters do not vary by category, which confirms the choice of an ordered logit regression.

The individual estimated parameters provide a basic level of information: the direction of the relationship between the independent and dependent variables. Parameters with negative signs indicate that the variable is associated with a lower WTP for labelling and positive signs indicate the opposite.

However, the raw parameter does not directly indicate the marginal effect of an explanatory variable. Whilst effects on the lowest and highest categories are clear (a positive sign indicates a shift out of the zero category and a shift into the 10%), the effects on the intermediate categories are uncertain (Greene, 1993). Greene suggests directly evaluating the magnitude of an explanatory variable's effect on the dependent variable by calculating the changes in predicted categories that result from a change in the explanatory variable. For a dummy variable, the model is evaluated twice, once with the variable in question set to 1 and once with it set to 0, with all other variables set to their mean values. Table 7.5 gives the effects of the different explanatory variables on the predicted categories.

Respondents least likely to pay more for labelling were unemployed or on public benefits. Other factors decreasing the demand for labelling were being price conscious about food, being involved in an agricultural occupation and agreeing that GE has definite benefits. Respondents who either agreed or disagreed that they trusted large companies, as opposed to those who were neutral or had no opinion, were also less willing to pay for labelling. Presumably, those who trust companies do not see a need for labelling, and those who distrust large companies would not find the label trustworthy. Retirees and housewives (the reference occupation category) were less willing to pay for labelling than other occupational groups.

At the other end of the scale, respondents were most likely to be willing to pay 10% more for food in order to have labelling if they were consumers of organically grown food. This variable was only a little more important than those who disagreed that the risks of GE are unknown. Respondents who were female and those who had a higher degree of education tended to be more willing to pay for labelling. Information about GE was important: those who had some knowledge of GE or GM and those who thought GE did not offer definite benefits were more willing to pay for labelling. Occupationally, those in paid employment (except in agriculture) were more willing to pay for labelling, as were students.

Discussion

Some of the results of this analysis support earlier research. Women, for example, consistently score lower on measures of the

Table 7.5. Marginal effects of explanatory variables on predicted categories.

		Demand for labelling			
Variable	Category	None	2%	5%	10%
PC		0.062	0.025	−0.035	−0.052
RDL		−0.052	−0.023	0.028	0.047
ORG		−0.125	−0.089	0.045	0.169
GE		−0.114	−0.034	0.065	0.084
BEN	Agree	0.015	0.007	−0.008	−0.013
(ref = neutral)	Disagree	−0.087	−0.044	0.045	0.086
RSK	Agree	−0.146	−0.044	0.082	0.108
(ref = neutral)	Disagree	−0.118	−0.086	0.042	0.161
TRST	Agree	0.028	0.012	−0.016	−0.025
(ref = neutral)	Disagree	0.098	0.036	−0.055	−0.079
OCC	Professional/managerial	−0.087	−0.049	0.042	0.093
(ref = retired/housewife)	Clerical/sales/service	−0.095	−0.060	0.042	0.113
	Technical/engineering	−0.116	−0.084	0.041	0.159
	Agricultural/farming	0.062	0.020	−0.035	−0.047
	Student	−0.042	−0.023	0.022	0.044
	Unemployed/beneficiary/not given	0.117	0.029	−0.067	−0.079
	Trades/labourer	−0.109	−0.075	0.043	0.140
UNI		−0.076	−0.037	0.040	0.073
GEN		−0.093	−0.036	0.051	0.077

acceptability of GM food; it is therefore unsurprising that they would be more willing to pay for labelling. Organic buyers, as Burton *et al.* (2001) have shown, are particularly likely to be willing to pay more for non-GM food, and the results presented above are consistent with this. In addition, these results confirm that attitudinal variables are important. This fact makes the job of predicting the demand for labelling more difficult, because attitudinal variables are difficult to collect and provide a less reliable basis for comparison with a wider population.

Although preliminary analysis indicated that there was no significant relationship between income and the demand for labelling, there seems to be a relationship between WTP and what might be termed an income constraint. Whether the respondent feels an income constraint, i.e. whether a respondent feels as though there is disposable income to spend on labelling, seems to affect the demand for labelling. In the present model, this is captured both by those who are price-conscious shoppers and by those who are retired, housewives, unemployed or beneficiaries, i.e. likely to be on a fixed income.

How GM products should be labelled has been widely discussed. This research indicates that respondents who read either the nutritional or ingredient labelling are more willing to pay for GM labelling. This finding suggests that the ANZFA labels, which indicate the GM status of ingredients on the ingredient label, make the information available to one group of interested consumers in a relatively subtle manner. However, label-readers had a small WTP compared to nearly every other characteristic examined. Thus, some sort of prominent labelling may be desirable from a welfare perspective. It may also help producers recoup more of the costs of segregation than a less visible label.

One of the interesting attitudinal questions was whether or not respondents thought that the risks of GE were unknown. Those who had an opinion on the risks seemed to be in opposition to those who were neutral or did not know. Those with opinions had approximately the same WTP regardless of whether they thought the risks were known or unknown. The results suggest that those who think the risks are known think they are bad. They also suggest that those who believe the risks are unknown are estimating the risks, anyway. Furthermore, it supports the idea that consumers are, in the absence of specific risk information, applying a sort of standard risk discounting (Lamb *et al.*, 2001).

The survey results suggest that there is sufficient overall WTP for labelling in New Zealand to warrant the type of labelling mandated by ANZFA. Consumption of food at home in New Zealand is approximately NZ$6.85 billion annually (Statistics New Zealand, 2001). If 23.1% of the population is willing to pay 10% more for food in order to have labelling, as was the proportion of the sample, then aggregate WTP for this group is $158 million (assuming no covariance between WTP for labelling and spending on food). If another 27.4% of the population is willing to pay 5% more for labelling, then this group's aggregate WTP is $93.8 million. Finally, if another 24.3% is willing to pay 2%, the aggregate figure is $33.3 million. The total WTP for the three categories is $285 million annually. As the ANZFA labelling is estimated to cost approximately $42 million annually, a potential Pareto improvement is likely.

Conclusion

Drawing firm conclusions from this research is problematic, given the low goodness of fit and strength of predictions from the model. By and large, the results tend to support other similar research. Specifically, attitudes are at least as important as economic or demographic factors in determining consumers' demand for labelling. Willingness to pay for labelling is spread across the economic spectrum, although consumers who feel their food budgets are constrained are not as willing to pay.

The results also suggest that more work should be done on identifying factors that affect demand for labelling. The current model accounts for some variation in WTP, but more variation remains to be explained. Moreover, the relationship between occupation, price consciousness and income should be explored further. That being said, the results of this preliminary analysis suggest that total WTP seems to be enough to warrant the labelling programme currently in place in New Zealand.

A future application of this methodology could involve specific GM food products or biotechnology processes. Although Bredahl (2001) found that consumers tend to reject the technology overall rather than evaluate products on a case-by-case basis, survey results indicate that the type of genetic engineering application affects its acceptability (Couchman and Fink-Jensen, 1990; Macer, 1994; Sparks *et al.*, 1994). In addition, ANZFA research found that the use of labels varies across food categories (Gamble *et al.*, 2000). The present research has examined GM food labelling generally, but the costs of labelling will vary by product and it is likely that the WTP for labelling will, too. Economic efficiency may turn out to be more complex than simply labelling all GM food.

References

Anonymous (2001) Women are concerned about genetically modified foods. *Marketing to Women: Addressing Women and Women's Sensibilities* 14, 6.

Australia New Zealand Food Authority (2001) Genetically modified foods. Available at http://www.anzfa.gov.au (accessed 29 April 2002).

Bredahl, L. (2001) Determinants of consumer attitudes and purchase intentions with regard to genetically modified foods – results of a cross-national survey. *Journal of Consumer Policy* 24, 23–61.

Bredahl, L., Grunert, K.G. and Frewer, L.J. (1998) Consumer attitudes and decision-making with regard to genetically engineered food products – a review of the literature and a presentation of models for future research. *Journal of Consumer Policy* 21, 251–277.

Burton, M., Rigby, D., Young, T. and James, S. (2001) Consumer attitudes to genetically modified organisms in food in the UK. *European Review of Agricultural Economics* 28, 479–498.

Campbell, H., Fitzgerald, R., Saunders, C. and Sivak, L. (2000) *Strategic issues for GMOs in primary production: key economic drivers and emerging issues.* CSAFE Discussion Paper No. 1. Centre for the Study of Agriculture, Food and Environment, University of Otago, Dunedin.

Cook, A. (2001) New Zealand consumer reactions to GM food: studies of beliefs, attitudes and intentions to purchase. In: *Proceedings of the Seventh Annual Conference of the New Zealand Agricultural and Resource Economics Society*. NZARES, Christchurch.

Couchman, P.K. and Fink-Jensen, K. (1990) *Public Attitudes to Genetic Engineering in New Zealand*. DSIR Crop Research Report No. 138. Department of Scientific and Industrial Research Crop Research, Christchurch.

European Commission (2000) *Economic Impacts of Genetically Modified Crops on the Agri-food Sector: A First Review*. Directorate-General for Agriculture, Brussels.

European Commission (2001) Commission improves rules on labelling and tracing of GMOs in Europe to enable freedom of choice and ensure environmental safety (press release). Available at http://europa.eu.int/comm/dgs/health_consumer/library/press/press172_en.pdf (accessed 25 July 2001).

Gamble, J., Muggleston, S., Hedderley, D., Parminter, T. and Richardson-Harman, N. (2000) *Genetic Engineering: The Public's Point of View*. Report to Stakeholders, HortResearch Client Report No. 2000/249, Mt Albert Research Centre, The Horticulture & Food Research Institute of New Zealand Ltd, February. HortResearch, Auckland.

Gollier, C., Jullien, B. and Treich, N. (2000) Scientific Progress and irreversibility: an economic interpretation of the 'Precautionary Principle'. *Journal of Public Economics* 75, 229–253.

Greene, W.H. (1993) *Econometric Analysis*, 2nd edn. Maxwell Macmillan International Publishing Group, Sydney.

Huffman, W.E., Shogren, J.F., Rousu, M. and Tegene, A. (2001) The value to consumers of GM food labels in a market with asymmetric information: evidence from experimental auctions. Paper presented at the annual meeting of the American Agricultural Economics Association, Chicago, 5–8 August.

Jones, E. (1997) An analysis of consumer food shopping behavior using supermarket scanner data: differences by income and location. *American Journal of Agricultural Economics* 79, 1437–1443.

Lamb, C., Mollenkof, D.A. and Ozanne, L.K. (2001) An exploratory look at New Zealand consumers' perceptions of food risks. In: Chetty, S., and Collins, B. (eds) *Bridging Marketing Theory and Practice*. Proceedings of the 5th Australia New Zealand Marketing Academy Conference, 3–5 December Auckland, New Zealand.

Lin, W., Price, G.K. and Allen, E. (2001–2002) StarLink™: where no Cry9C Corn should have gone before. *Choices* 16, 31–34.

Macer, D.R.J. (1992) *Attitudes to Genetic Engineerig: Japanese and International Comparisons*. Eubios Ethics Institute, Christchurch.

Macer, D.R.J. (1994) *Bioethics for the People by the People*. Eubios Ethics Institute, Christchurch.

Maddala, G.S. (1983) *Limited-dependent and Qualitative Variables in Econometrics*. Cambridge University Press, New York.

Nayga, R.M. Jr (2001–2002) Looking for the nutritional label: does it make a difference? *Choices* 16, 39–42.

OECD (2000) *Modern Biotechnology and Agricultural Markets: A Discussion of Selected Issues*. Directorate for Food, Agriculture and Fisheries, Committee for Agriculture, OECD.

Phillips, P.W.B. and McNeill, H. (2000) A survey of national labelling policies for GM foods. *AgBioForum* 3, 219–224.

Robertson, D. (2002) Marking time: Australian rules on genetically modified food labels aren't as tough as they're made out to be. *Far Eastern Economic Review* 165, 41.

Senauer, B. (2001) The food consumer in the 21st century: new research perspectives. Working paper 01–03, The Retail Food Industry Center, University of Minnesota. Available at http://trfic.umn.edu/.

Sparks, P., Shepherd, R. and Frewer, L.J. (1994) Gene technology, food production, and public opinion: a UK study. *Agriculture and Human Values* 11, 19–28.

Statistics New Zealand (2001) Household spending (year ended 30 June 2001) – standard tables. Wellington. Available at http://www.stats.govt.nz/ (accessed 31 May 2002).

Valceschini, E. (1998) L'étiquetage obligatoire des aliments est-il la meilleure solution pour les consommateurs? Éléments de théorie économique. In: *Organismes Génétiquement Modifiés à l'INRA: Environnement, Agriculture et Alimentation*. Institut National de Recherches Agronomiques, Paris, pp. 111–115.

Wright, J.C. (2000) The economics of genetic modification. Background paper for the (New Zealand) Royal Commission on Genetic Modification. Available at http://www.gmcommission.govt.nz/ (accessed 30 May 2002).

8 Contingent Valuation of Breakfast Cereals Made of Non-biotech Ingredients

Wanki Moon[1] and Siva K. Balasubramanian[2]

[1]*Department of Agribusiness Economics, Southern Illinois University, Carbondale, IL 62901, USA;* [2]*Department of Marketing, Southern Illinois University, Carbondale, IL 62901, USA*

Introduction

Controversy over biotech foods continues to be an issue of profound importance across the globe to stakeholders involved in the food supply chain including farmers, grain handlers, food processors and retailers, regulatory agencies as well as consumers. A number of major US and European food manufacturers and retailers announced that they would accept only non-biotech crops (Josling *et al.*, 1999; Economic Research Service (ERS), 2000a,b). This is coupled with the recent recall of taco shells made of Starlink™ maize which has stirred up a wave of turmoil in the domestic food supply chain as well as in export markets. The uncertain prospect of agrobiotechnology is in sharp contrast to the initial promise of agrobiotechnology as a major technological breakthrough that would revolutionize the way crops are produced while enhancing the nutritional value of food products. Growing public concerns appear to be altering the dynamic path of the progress of agricultural biotechnology, raising such intriguing issues as adoption of identity preservation, market segregation, and labelling as ways of segregating GMOs from non-GMOs throughout the food supply chain.

Identity preservation and market segregation are not without additional costs to the production and marketing system. For example, ERS (2000a,b) estimates that segregation could add about $0.22 per bushel for maize and $0.54 per bushel for soybean to marketing costs from country elevator to export elevator. Moreover, labelling incurs tangible and intangible costs to the society in the forms of regulatory requirements (e.g. testing, standardization, certification and enforcement). Hence, food supply chain participants would want to ensure that market demand for non-biotech foods is sizeable enough to guarantee that market prices cover these costs. The key question then is whether instituting segregated markets for non-biotech foods generates benefits greater than those segregation costs. Benefits associated with segregating and labelling non-biotech foods can be measured by estimating whether and how much consumers would be willing to pay more for non-biotech foods.

Despite the heated debate over the needs of instituting segregation and labelling, we have little knowledge about how consumers would behave if they are given the right to choose between biotech and non-biotech foods. This lack of knowledge is detrimental to shaping constructive dialogue among stakeholders involved in the food supply chain. In practice, the lack of information could critically disrupt the basic supply–demand rela-

 Consumer Acceptance of Genetically Modified Food (eds R.E. Evenson and V. Santaniello)

tionships for major crops. For example, a large demand for non-biotech crops relative to the supply is likely to bring about substantial premiums for non-GMOs and deep discount for GMOs (Babcock and Beghin, 1999). Theoretically, government can serve market-facilitating functions such as provision of information or regulations that would prevent such market disruptions from taking place. However, if government fails to adapt rules and regulations for biotech foods to evolving consumer preferences, consumers will increasingly lose confidence in the credibility of the regulatory agency, which would deepen instability further in the food marketing system. Consequently, the lack of information in regard to the demand for non-biotech foods could undercut governmental regulations as well as market itself.

The primary objectives of this chapter are: (i) to estimate whether and how much consumers would be willing to pay more to purchase non-biotech foods, and (ii) present empirical insights into the current debate concerning the need for redesigning the food supply chain to separate non-biotech from biotech crops, and (iii) evaluate how differences in individual characteristics including nationality and perceived attributes of agro-biotechnology affect individual valuation of the non-biotech attribute of foods. In recognition of the new product or non-market property of non-biotech foods, a stated preference approach (i.e. contingent valuation) is used to measure consumer willingness to pay a premium for non-GMO foods using cross-sectional data collected in December 2000 in the USA and UK.

Segregation Costs and Labelling

Institutional underpinnings fundamental for the market for non-biotech foods to emerge include segregation or identity preservation throughout the food supply chain and labelling of final food products at food manufacturing and retailing levels. Segregation of non-GMOs from GMOs is essentially an extension of the handling process for speciality grains and oilseeds, which has been in place for some time (ERS, 2000a,b). In the speciality grains case, segregation costs include the additional costs of storage, handling, risk management, analysis and testing, and marketing (i.e. expenses associated with negotiating contract terms).

At this time, several estimates are available regarding the additional costs associated with segregating, certifying and testing of non-GMO crops. ERS (2000a,b) estimates that segregation could add about $0.22 per bushel to marketing costs of non-GMO maize from country elevator to export elevator. Segregation of non-GMO soybeans at these elevators could add $0.54 per bushel. These are averaged over 84 surveyed elevators. According to Bullock *et al.* (2000), non-GMO soybeans average a 50 cent premium per bushel on the Tokyo market compared to GMO soybean. He found that most of the premium resulted from the additional cost of segregating GMOs from non-GMO crops within the grain-handling system but not from at the farmers' level.

A survey conducted in 1999 by Spark Companies disclosed that the non-GMO premiums were estimated by a number of sources at 10–15 cents per bushel for soybeans and 5–10 cents for maize (ERS, 2000a). The lower end of the premium ranges reflects less strict tolerance levels for GMO content. Good *et al.* (2000) conducted a survey to estimate additional marketing costs associated with speciality grains in Illinois. The survey reported an average additional handling cost of 17 cents per bushel for maize and 48 cent per bushel for soybeans in 1998.

In conjunction with the segregation throughout the food supply chain, an appropriate labelling system needs to be instituted to communicate the segregation and identity preservation to consumers, the final destination of the food supply chain. What type of labelling is used has important implications for all participants in the food supply chain. For example, while mandatory labelling of products that use GMOs has the advantage of giving consumers full information, it would be over-regulating if there is a significant segment of the population who do not have a preference for non-biotech foods. As a consequence, while mandatory labelling benefits consumers, it may incur a substantial cost to society. Besides, it can give the impression that biotech foods are not safe, potentially worsening the uncertain prospect of agrobiotechnology.

Voluntary labelling of products that do or do not use GMOs is desirable in the sense that it allows consumers to choose products consistent with their preferences. That is, it relies on market forces to determine the acceptance of new technologies. Voluntary labelling can be accompanied by a disclaimer that may be necessary to prevent consumers from being misled about safety differences (Caswell, 2000). For example, the Food and Drug Administration (FDA) chose this option in regard to the marketing of dairy products from cows treated with supplemental rBST, allowing the voluntary labels to include a statement: 'No significant difference has been shown between milk derived from rBST-treated and non-rBST-treated cows'. Caswell (2000) notes that this FDA approach facilitates consumers' right to informed choices while restricting the scope of the claims by the disclaimer.

Determination of Premium for Non-GMO Crops

If non-biotech crops are segregated from biotech crops throughout the food supply chain and a labelling system is instituted to offer consumers the opportunities to choose between biotech and non-biotech foods in grocery stores, then prices may or may not be differentiated between biotech and non-biotech crops depending upon segregation costs and the strength of the demand for non-biotech crops. Figures 8.1 and 8.2 show hypothetical situations concerning the supply of and demand for non-GMO crops and illustrate how premium is determined for non-GMO crops.

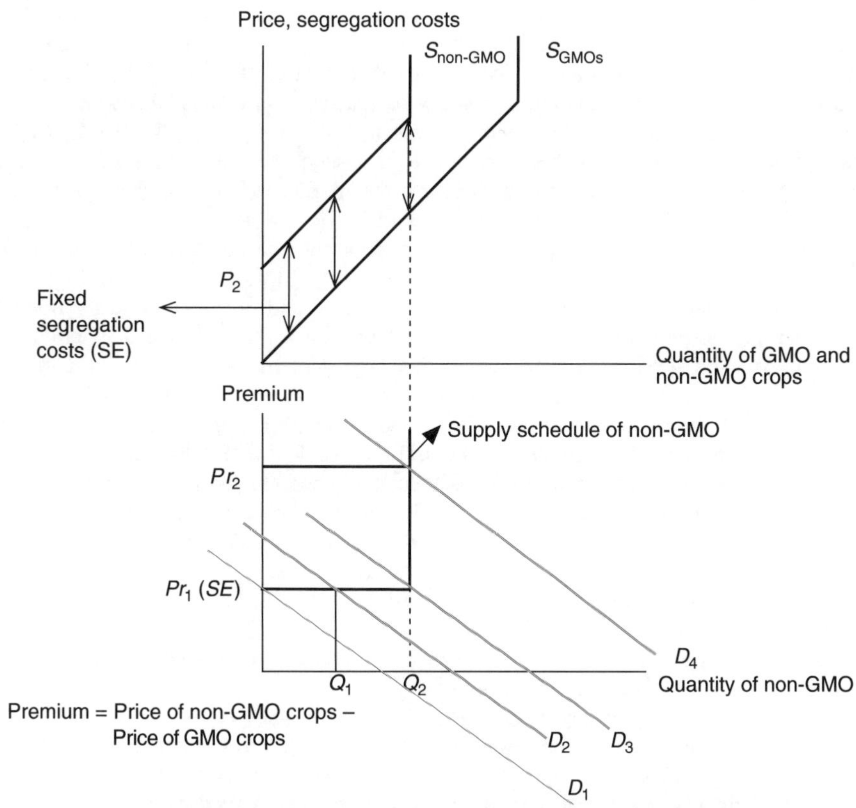

Fig. 8.1. Determination of premium for non-GMO crops when the costs of segregation are constant.

Fixed segregation costs

Figure 8.1 illustrates the important role that the strength of the demand plays in determining the size of segregation and premium for non-biotech crops when there are some costs of segregating non-biotech from biotech crops and such costs do not vary across differing levels of quantity.

The upper panel in Fig. 8.1 shows the supply schedules for biotech and non-biotech crops with the vertical section indicating that the supply for non-biotech crops is limited by total acreage planted (Q_2) in the short run (e.g. in a given year). Based on segregation costs in the upper panel, the lower panel derives the supply schedule of segregation as a function of segregation costs or premiums. In essence, the lower panel describes three possibilities concerning the role of the demand for non-biotech crops in determining the size of segregation and premium for non-biotech crops.

First, as long as the demand curve is positioned below D_1, the food supply chain would have no incentive to segregate non-biotech from biotech crops. No segregation would take place due to the insufficient strength of the demand for non-biotech crops. Second, only part of non-biotech crops planted would be segregated if the demand curve lies between D_1 and D_3. For example, if the demand for non-biotech crops is given by D_2, the difference between non-biotech crops planted and demanded as represented by $Q_1 - Q_2$ would be commingled with biotech crops instead of being segregated. Last, when the magnitude of the demand is larger than D_3, all non-biotech crops planted would be segregated and the size of premium would begin to diverge with segregation costs as illustrated by Pr_2 associated with D_4.

Increasing segregation costs

Figure 8.2 exhibits the determination of premium for non-biotech crops if segregating non-biotech from biotech crops incurs additional costs and such costs increase as the scale of segregation expands. Analogous to Fig. 8.1, the upper panel depicts the supply curves of biotech and non-biotech crops. The supply curve of non-biotechs is shown to be steeper than that of biotechs to account for the increasing segregation costs. The lower panel depicts the supply schedules for segregation as a function of segregation costs. As far as the demand for non-GMOs lies below the vertical line of the supply curve, the size of premium will be identical with the costs associated with segregation, testing, and certifying. For example, if the demand for non-GMO crops is represented by D_0, the premium would be Pr_1. Pr_1 coincides with the costs of segregation. Once the demand reaches the maximum the food supply chain can offer, the strength of the demand for non-GMOs will be the only determinant of the size of premiums. A shift in demand for non-biotech crops from D_1 to D_2 will increase the magnitude of premium from Pr_2 to Pr_3.

Discounts for biotech foods

While we analysed several potential outcomes in relation to the market for non-GMO crops, there could be parallel developments in the market for GMOs. On one hand, if the demand for non-GMOs lies within the vertical section of the supply curve such as D_2 in Fig. 8.2, it suggests that there would be a surplus in the market for biotech crops. Any markets with a surplus would need to offer discounts to clear the market. Consequently, discounts will emerge in the market for GMO foods. On the other hand, if the demand for non-GMOs is given by a relatively small magnitude such as D_0, then, the amount of non-GMO crops represented by $Q_3 - Q_2$ would be commingled with GMOs rather than being segregated. There would be no discounts for GMOs in such a case. In sum, discounts for biotech foods are likely to arise only if demand for non-GMO crops lies on the vertical section of the supply curve.

Survey Design and Data

Given such a key role of the demand for non-biotech crops in determining whether or not segregation is needed, a survey instrument was designed to shed light on the size of the demand for non-biotech foods and to evaluate public attitudes toward agrobiotechnology. The survey instrument is composed of two sections: (i) measuring attitudes and perceptions as related to agrobiotechnology issues, and (ii) measuring behavioural intentions with a focus on willingness to pay for breakfast cereals made of non-biotech ingredients. The instrument was tested and question wording was refined in light of the results of three focus group studies. A mail survey was conducted in the USA and an online survey in the UK in December 2000 using household panels maintained by the National Panel Diary (NPD) group (a marketing consulting firm specializing in research on consumer behaviour and food marketing).

The survey method using such an established panel is called 'permission-based survey' and is increasingly used in exploring various respects of consumer behaviour for academic or commercial purposes. Advantages associated with the permission-based survey include: (i) the response rate is higher than other regular surveys, and (ii) demographic information is disclosed for non-returners as well as returners, which would permit researchers to assess potential non-response bias. Questionnaires were distributed to 5200 households selected across the USA by random sampling: about 3000 households returned completed questionnaires, yielding a response rate of nearly 58%.

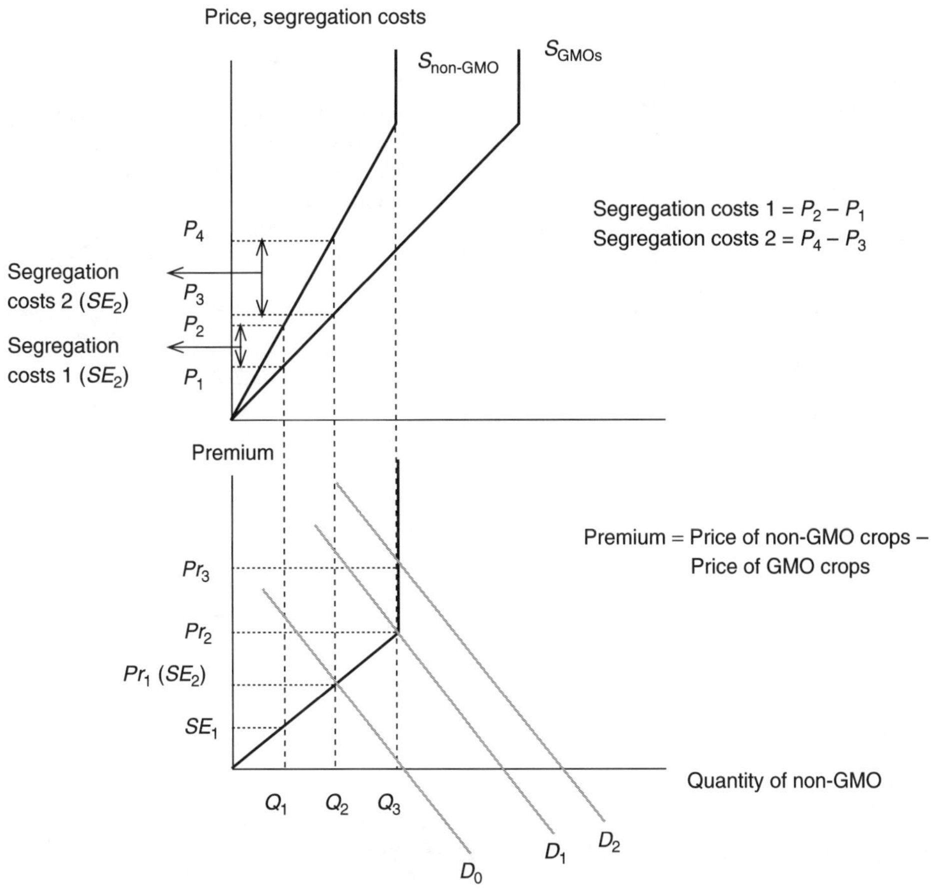

Fig. 8.2. Determination of premium for non-GMO crops when the costs of segregation increase across the scale of segregation.

The sample is drawn stratified by geographic regions, market size, household head age, education and income to balance with the US census for adults. Table 8.1 compares the sample with the US census based on socioeconomic and demographic profiles. The comparison suggests that the survey sample is remarkably well representative of the US census in most of the demographic categories. In addition, Table 8.1 shows that the sample is drawn from four geographic regions (Northeast, Midwest/North Central, South and West) quite in proportion with the US census. The only noticeable discrepancy is the moderate under-representation of the 'Not Employed' category in the sample as compared with the US census.

Table 8.1. Comparison of the survey sample with the US census for adults.

	Survey sample (%)	US census for adults (%)
Census region		
Northeast	21.2	18.9
Midwest	28.0	23.6
South	34.1	35.8
West	16.8	21.8
Market size		
<1,000,000	31.0	28.0
1,000,000–2,499,999	22.6	24.6
2,500,000	22.5	28.4
Non-MSA	23.8	19.0
Household head age		
18–24 years	3.7	6.4
25–34 years	17.9	21.4
35–44 years	27.5	27.3
45–54 years	24.2	22.9
55–64 years	19.0	15.5
65–69 years	7.7	6.6
Household head education		
Grade	1.9	4.7
Some High	9.3	8.8
Graduated High	36.8	35.6
Some College	23.4	23.8
Graduated	18.9	18.2
Post College	9.6	9.2
Household head occupation		
White collar	41.3	44.7
Blue collar	31.9	30.4
Student	1.0	0.9
Retired	20.5	23.5
Other	5.2	0.5
Household income		
<$15,000	14.9	14.7
$15,000–$24,999	11.6	12.2
$25,000–$34,999	13.6	13.1
$35,000–$49,999	18.2	16.9
$50,000–$74,999	21.3	20.5
$75,000 +	20.4	22.7

Source: NPD Genetically Modified Organism (GMOs) Consumer Survey, 2000.

The same instrument was administered to consumers in the UK. The survey in the UK used an online method instead of mail. The online method uses the Internet as a data collection tool and provides the following advantages over the more traditional methods (e.g. mail, mall, telephone): (i) faster response time; (ii) lower costs versus mall or mail/telephone; (iii) ability to follow up with panelists quickly and inexpensively; and (iv) more current samples.

The online method could result in a bias due to the fact that the sampling was restricted to those who have access to a computer and the Internet. A possible conjecture about the direction of the bias is that consumers with an access to the Internet would be more technologically oriented and more likely to accept new technologies than those without access, leading to an upwardly biased acceptance rate. Accordingly, the results from the UK survey should be interpreted in light of the potential biases that may be associated with this particular sampling method.

About 9000 consumers voluntarily registered to participate in the NPD online survey panel in the UK. For this survey, the NPD group sent all of the 9000 consumers e-mails providing a brief description of the survey and leading them to the website exhibiting the questionnaire. Nearly 2600 consumers returned completed surveys within the first 7 days after they were contacted by the NPD via e-mails.[1]

Contingent valuation of non-biotech foods: payment card format

Contingent valuation questions were included in the survey instrument to evaluate whether and how strongly consumers value the attribute

[1] A little fewer than 1000 consumers responded on the first day of opening the survey field on the Internet.

that foods are made of non-biotech ingredients. The contingent valuation part asked respondents to imagine the following situation:

> Suppose that you walk into a grocery story and want to buy a box of breakfast cereals. The grocery store carries breakfast cereals (e.g. corn flakes, frosted flakes, corn pops) of two types: (1) made from GMO crops, and (2) made from conventional nonGMO crops.

With this introduction to contingent valuation, the next step is to determine the type of elicitation format. The most widely used methods of eliciting willingness to pay (WTP) were the open-ended and closed-ended (dichotomous choice) questioning techniques. The open-ended format tends to produce an unacceptably large number of non-responses or protest zero responses because of the cognitive difficulties associated with choosing a dollar amount of the value for a public good. Besides, it was often associated with strategic bias. The closed-ended format avoids these problems by giving only two choices to respondents, although this format yields less information as compared to other formats.

The payment card questioning technique has been increasingly used in recent years to compromise the advantages and disadvantages associated with the open-ended and closed-ended formats. The payment card method was developed as an alternative to the bidding game, the oldest and the most widely used elicitation method until recently (Mitchell and Carson, 1989). The survey instrument for this study included contingent valuation questions in the payment card format:

> Consumers might have to pay a higher price for nonGM foods due to the costs of segregation in the production and marketing system plus the additional costs of testing, certification and labeling GM foods. Suppose the price of breakfast cereals made from GM crops is $4.00 per box. The price of conventional nonGM breakfast cereals will be higher than $4.00, but is not determined yet. What is the most above the current price of $4.00 you would be willing to pay to purchase a box of conventional nonGM breakfast cereals?

Contingent valuation questions in the form of a payment card contains an ordered set of threshold values (Cameron and Huppert, 1989). The payment card for this study includes various sizes of premium ranging from $0.00 to $3.00 for a box of breakfast cereals (with a base price of $4.00) made of non-biotech crops (see Table 8.2 for distribution of responses across the range of premiums). The payment card approach avoids the high rate of item non-response on open-ended valuation questions. In this approach, consumers are asked simply to go over the range of values and to circle the highest amount they would be willing to pay.

Empirical Model Specification

Estimation procedure

With the contingent valuation question generating value responses in the form of intervals rather than point valuations, mid-points as approximations to the true unobserved values may be used in fitting a univariate distribution of values. The mid-points can also be used as the dependent variable in ordinary least square (OLS) regression. In consideration of the fact that expected values within the intervals are not necessarily equal to the interval mid-points, Cameron and Huppert (1989)

Table 8.2. Distribution of responses for each size of premium in payment card format: USA and UK.

Interval ($)	US frequency	UK frequency
0.00–0.09	701 (22.9%)	433 (16.8 %)
0.10–0.19	143 (4.67%)	97 (3.77%)
0.20–0.29	150 (4.90%)	118 (4.59%)
0.30–0.39	120 (3.92%)	214 (8.33%)
0.40–0.49	60 (1.96%)	122 (4.75%)
0.50–0.74	429 (14.0%)	178 (6.93%)
0.75–0.99	83 (2.71%)	129 (8.91%)
1.00–1.24	274 (8.95%)	127 (4.94%)
1.25–1.49	21 (0.68%)	66 (2.57%)
1.50–1.74	58 (1.89%)	168 (6.54%)
1.75–1.99	11 (0.35%)	12 (0.46%)
2.00–2.49	48 (1.56%)	46 (1.79%)
2.50–2.99	29 (0.94%)	19 (0.73%)
3.00 or higher	50 (1.63%)	181 (7.04%)
'Don't know'	883 (28.8%	558 (21.7%)
Number of observations	3060 (100 %)	2568 (100%)

Source: NPD Genetically Modified Organisms (GMOs) Consumer Survey, 2000.

proposed the maximum likelihood (ML) procedure for estimating WTP valuation equations measured with interval data. The model underlying the ML estimator is given by the system (Stewart, 1983):

$$W_i^* = \beta \mathbf{X} + E_i$$
$$W_i = P_{j-1}, \text{ if } P_{j-1} \leq W_i^* < P_j, i = 1, 2, 3 \ldots n; j = 1, 2, 3 \ldots 14 \quad (1)$$

where W_i^* is the unobserved true WTP, $\mathbf{X}$ is a vector of explanatory variables that are hypothesized to influence consumer willingness to pay a premium for non-biotech foods, E_i is a random disturbance normally distributed; W_i is grouped observed WTP; and P_j represents observed threshold values for each WTP category. The likelihood function depicting the above model is given by

$$L = [\Phi(P_j - \beta \mathbf{X})/\sigma) - \Phi(P_{j-1} - \beta \mathbf{X})/\sigma)]^{D_{ij}} \quad (2)$$

where D_{ij} is 1 if W_i^* falls in the *j*th category and 0 otherwise. The likelihood function can be maximized with respect to the vector of parameters (β) using non-linear optimization algorithms.

Model specification

In equation (1), consumers' willingness to pay a premium (*W*) was hypothesized to be determined by the vector (**X**). In our study, the vector **X** is partitioned into four subgroups $\mathbf{X} = [X_1, X_2, X_3, X_4]$ including general attitudes toward food purchase, perceptions about attributes specific to agrobiotechnology, awareness of the agrobiotechnology issues, and labelling preference. Table 8.3 presents the definitions and descriptions of the variables included in the empirical model along with summary statistics

The first subgroup (X_1) includes attitudes toward the importance of food safety (*SAFE_1*) and price (*PRICE_2*) in food-purchasing decisions and consumption frequency of organic food products (*ORG_3*). We expect that the importance of food safety (*SAFE_1*) is positively associated with consumer WTP a premium for non-biotech foods. The importance of price (*PRICE_2*) in food shopping is anticipated to have a negative effect on WTP. The consumption frequency of organic food products (*ORG_3*) is hypothesized to be positively related to WTP.

The second subgroup (X_2) involves eight perceived attributes in relation to agrobiotechnology: (1) health risks; (2) environmental risks; (3) moral and ethical considerations; (4) the image of multinational corporations as the primary beneficiaries of biotechnology; (5) the growing control of multinational corporations over farming; (6) potential increase in yields; (7) reduced use of chemicals in crop production; and (8) potential improvement in nutritional composition. The first four attributes describe negative aspects of agrobiotechnology while the latter three capture positive sides of agrobiotechnology.

There were considerable correlations (ranging from 0.54 to 0.76) among consumer perceptions about the eight attributes of agrobiotechnology. To cope with potential multicollinearity problems in specifying the empirical model, the eight perceived attributes were reduced to four variables: (i) an index of risk perception (*RISK_1*) using stated perceptions about health and environmental risks[2]; (ii) consumer perception about moral and ethical aspects of agrobiotechnology (*MORAL_2*); (iii) an index of consumer perceptions about multinational corporations (*MULTI_3*) based on consumer perceptions about (4) and (5); and (iv) an index of benefit perception (*BENEFIT_4*) based on consumer perceptions about (5), (6) and (7).

The third subgroup (X_3) refers to consumer awareness (*AWARE*) as measured with a six-point scale ranging from 'Nothing' to 'A Great Deal' to the question 'how much have you read or heard about genetically modified organisms (GMOs)?' The fourth subgroup (X_4) represents consumer perceptions about labelling (*LABEL*) as reflected in response to the question 'A labelling system is necessary to differentiate conventional foods from genetically modified foods on the supermarket shelf'. Respondents were given a six-point Likert scale ranging from 'Disagree Completely' to 'Agree Completely'.

[2] The index is constructed by adding up the two variables. Hence, the index would range theoretically from 2 to 12 with 12 representing complete agreement with the health and environmental risks of agrobiotechnology. Other indexes are similarly constructed.

Table 8.3. Definitions and descriptive statistics of variables used in the empirical model.

Variable	Description	Mean	SD
X_1 Subgroup			
SAFE_1	Importance of food safety in food purchasing decisions	4.89 (4.37)	1.13 (1.43)
PRICE_2	Importance of food price in food purchasing decisions	4.95 (4.58)	1.09 (1.25)
ORG_3	Consumption frequency of organically grown food products (1 = Never; 6 = All the Time)	2.46 (2.86)	1.35 (1.48)
X_2 Subgroup			
RISK_1		6.89 (8.05)	2.64 (2.50)
Health risks	Biotech foods pose health hazards		
Environmental risks	Agrobiotechnology poses hazards on ecosystem		
MORAL_2	It is morally and ethically wrong to use biotechnology	3.11 (3.28)	1.65 (1.75)
MULTI_3		7.98 (9.12)	2.96 (2.59)
Beneficiary	Multinational corporations are primary beneficiaries of agrobiotechnology, while consumers assume most of the risks		
Control on farming	Multinational corporations are increasingly controlling farming		
BENEFIT_4		11.8 (10.6)	3.33 (3.50)
Increase in yields	Agrobiotechnology reduces world food shortages by increasing yields		
Chemical use	Agrobiotechnology reduces the use of chemicals in crop production		
Nutritional contents	Agrobiotechnology enhances nutritional composition		
X_3 Subgroup			
AWARE	How much have you heard about GMOs (1 = Nothing; 6 = A Great Deal)	3.21 (3.85)	2.45 (2.11)
X_4 Subgroup			
LABEL	Labelling is necessary to differentiate non-GMO from GMO foods in grocery stores	4.88 (5.24)	1.34 (1.28)

Unless otherwise indicated, questions are measured with a seven-point scale ranging from 1 = Disagree Completely; 6 = Agree Completely; 7 = Don't Know. Respondents who selected Don't Know are dropped from empirical estimations. Numbers in the parentheses represent means and standard deviations for UK responses.

Empirical Results

Three models were estimated based on the likelihood function for interval data as shown in equation (2). The first model is based on pooled data between the USA and UK, whereas the second and third models are estimated with regard to the US and UK data, respectively. Table 8.4 presents estimated parameters and asymptotic *t*-values for the three models along with other summary statistics. The hypothesis that all coefficients in the valuation models using payment card interval data are simultaneously equal to zero was tested using χ^2 statistics. The calculated χ^2 statistics are 772, 134.6 and 573 for the pooled, US and UK data, respectively. Given the critical value (25.19) at a 0.01 probability level with 10 degrees of freedom, the hypotheses were decisively rejected across the three models, suggesting that the specified models have the capabilities to explain the variation in the WTP a premium for non-biotech foods.

The results from pooled data clearly indicate that both general food shopping habits and and risk and benefit perceptions specific to agrobiotechnology play an important role in determining behavioural intentions as measured with WTP a premium to purchase non-

Table 8.4. Maximum likelihood estimates for willingness to pay a premium for non-biotech foods: payment card interval data approach.

	Pooled data		USA		UK	
Variable name	Parameter estimates	Asymptotic *t*-values	Parameter estimates	Asymptotic *t*-ratios	Parameter estimates	Asymptotic *t*-ratios
Constant	−0.0592	0.252	−0.4391	1.428	−0.2532	0.762
SAFE_1	0.0622**	2.936	0.0512*	1.630	0.0375	1.344
PRICE_2	−0.1671***	7.984	−0.0591*	1.860	−0.2174***	7.938
ORG_3	0.1155***	5.800	0.0665**	2.403	0.1292***	4.750
RISK_1	0.0707***	4.027	0.0560**	2.318	0.0843***	3.509
MORAL_2	0.0565**	2.500	0.0017	0.049	0.0815**	2.792
MULTI_3	0.0295**	2.016	0.0336*	1.674	0.0437**	2.142
BENEFIT_4	−0.0460***	4.996	−0.0191	1.540	−0.0558***	4.353
AWARE	−0.0310	1.578	−0.0604**	2.323	−0.0085	0.031
LABEL	0.1134***	4.430	0.0785**	2.229	0.1394***	3.911
COUNTRY[a]	−0.2648***	4.613	–	–	–	–
σ	0.9908***	45.41	0.882***	28.65	1.02***	35.196
Log *L*	−4269.9		−1672.5		−2560.3	
Log *L* ($\beta = 0$)	−4656.1		−1739.8		−2846.8	
χ^{2b}	772.4		134.6		573.0	
No. of observations	1887		785		1102	

*$P < 0.1$; ** $P < 0.05$; and ***$P < 0.01$.

[a]The variable, *COUNTRY*, is a binary variable equal to one if the respondent is from the USA.

[b]Critical value of chi-squared with 10 degrees of freedom is 25.19 at $\alpha = 0.01$.

biotech foods. *SAFE_1* had a positive effect on WTP a premium indicating that if consumers agree that food safety is an important consideration in making food purchasing decisions, they were likely to pay a higher premium than those who do not agree. The importance of price (*PRICE_2*) had a theoretically correct sign, confirming the negative linkage between price sensitivity and WTP. As hypothesized, consumption frequency of organically grown food products (*ORG_3*) was positively linked to the size of premium consumers would be willing to pay for non-biotech foods.

As expected, risk and benefit perceptions about agrobiotechnology proved to be major factors shaping consumer willingness to pay a premium for non-biotech foods. If consumers perceived risks on human health or environments from the use of biotechnology in crop and food production (*RISK_1*), then they were more likely to pay a premium to avoid consuming breakfast cereals made of biotech ingredients. In contrast, if consumers associate agrobiotechnology with various benefits (*BENEFIT_4*), they were not likely to pay a premium to purchase non-biotech foods. This result is significant in consideration of the fact that consumers are not direct beneficiaries of those benefits and some of them are yet to be realized. If the second-generation biotech food products with substantive private and social benefits are materialized and those benefits outweigh the risks posed by agrobiotechnology, the magnitude of the market for non-biotech foods is clearly influenced.

Along with risk and benefit perceptions, moral and ethical considerations (*MORAL_2*) and perceptions about multinational corporations had a measurable impact on shaping behavioural intentions with regard to purchasing non-biotech foods. If consumers perceive that it is ethically and morally wrong to use biotechnology to genetically modify food products, they tended to express more willingness to pay a premium for non-biotech breakfast cereals. Likewise, consumers were more willing to pay a premium for non-biotech foods if they believe that corporations are the main beneficiaries from agrobiotechnology, while consumers assume most of the risk or if they are concerned about the growing influence of multinational seed companies on farming. These results support the notion that consumer behaviours in regard to non-biotech food consumption are influenced not only by direct and indirect risk and benefit perceptions but also by other considerations including moral values and beliefs about the social value of farming.

Consumer awareness of GMO issues (*AWARE*) was negatively associated with WTP but the association was not statistically significant. Labelling preference (*LABEL*) was positively and statistically significantly linked to the amount of premium consumers would be willing to pay for non-biotech foods. If consumers agree that a labelling system is needed to segregate non-biotech from biotech foods in grocery stores, they were likely to pay a higher premium to purchase breakfast cereals made of non-biotech ingredients. Finally, consumers in the USA (*COUNTRY*) were willing to pay a significantly lower premium for non-biotech foods as compared to the UK consumers, holding other factors constant. This result may be reflective of structural differences in attitudes toward foods between US and UK consumers. Further, food scares such as mad cow disease (BSE) and foot and mouth disease that have swept European nations over the last few years may have made biotech foods a disproportionately sensitive issue.

If consumers in the USA and UK structurally differ in perceiving risks, benefits and other attributes of agrobiotechnology and in willingness to pay for non-biotech foods, pooling the data from both countries could generate misleading results. Therefore, this chapter splits the data into USA and UK subsets and estimates the valuation model for each country. Comparison of the estimated results from the two countries reveals important differences in terms of factors influencing behavioural intentions with regard to non-biotech foods. For example, moral and ethical consideration (*MORAL_2*) did not play an important role in shaping behavioural intentions with regard to non-biotech foods in the USA, whereas it did for UK consumers. The degree of awareness of the agrobiotechnology issues (*AWARE*) was negatively associated with WTP a premium in both countries but had a statistically significant effect on the USA consumers only. Benefit perception (*BENEFIT_4*) was positively linked to WTP a premium in the USA but not statistically significant. This result may be associated with the fact that the benefits are mostly yet to be materialized. In contrast, it was a major factor driving down the size of premiums that UK consumers would be willing to pay for non-biotech foods.

Predicted mean WTP for breakfast cereals of non-biotech ingredients

Using estimated parameters from the three models, fitted values ($\hat{W}$) were calculated conditional on ($\mathbf{X}\hat{\beta}$). First, based on the pooled data, the US mean WTP a premium for non-biotech breakfast cereals was \$0.3134, while consumers in the UK on average were willing to pay a substantially higher premium (\$0.5782). While the model using pooled data allows us to calculate mean WTPs for each country based on the binary indicator (*COUNTRY*), separate models would produce mean WTPs more consistent with the preferences of each country. Using separate models, the mean WTP was \$0.3838 and \$0.7316 for the US and UK consumers, respectively. These comparisons indicate that the demand for non-biotech foods would be greater in the UK than in the USA. The mean size of premium (\$0.3838) for the US consumers is translated into about 9.5% of the base price (\$4.00) of breakfast cereals, while \$0.7316 for the UK consumers is about 18% of the base price.

These fitted values provide information useful in assessing the magnitude of potential demand for non-biotech foods. The mean size of premium of 9.5% in the USA suggests that the strength of demand for non-biotech foods is neither immense nor negligible. As discussed earlier, estimates of segregation costs available to date range from 5 to 20% per bushel for maize or soybean. Assuming that the share of grain costs in manufacturing a box of breakfast cereals is 30%, 20% of additional segregation costs translate into a 6% increase in the price of the final products. The average size of premiums that certain segments of the US population would be willing to pay seems to be enough to cover additional costs associated with segregating non-biotech from biotech foods. The comparison indicates that a niche market for non-biotech foods could emerge if consumers are given the right to choose between biotech and non-biotech foods. For the UK consumers, the estimated mean size of premium of 18% indicates that the strength of demand for non-biotech foods is considerably greater than that in the USA.

Concluding Remarks

The WTP estimates are contingent upon several factors that were built into the specification of the empirical models. First, the estimates will differ depending on how the respondents who selected the 'Don't Know' option to various questions formulate their attitude toward agrobiotechnology in the future. Second, the explanatory variables included in the WTP valuation model are mostly subjective perceptions about private or social risks and benefits associated with agrobiotechnology. If consumer perceptions about various attributes of agrobiotechnology change over time, the estimates would change accordingly. More fundamentally, the estimated size of market is linked to uncertainties inherent with the new gene splicing techniques in crop production. Paraphrasing, reliable scientific evidence either confirming alleged risks or establishing the equivalence between biotech and non-biotech foods would dramatically change consumer perceptions and consequently the strength of the demand for non-biotech foods.

References

Babcock, B. and Beghin, J. (1999) Potential market for non-GMO corn and soybeans. *Economic Perspectives on GMO Market Segregation*. Staff Paper No. 298, Iowa State University, Ames, Iowa.

Bullock, D.S., Desquilbet, M. and Nitsi, E.I. (2000) *The Economics of Non-GMO Segregation and Identity Preservation*. Working Paper, November 2000, Department of Agricultural and Consumer Economics, University of Illinois, Urbana-Champaign.

Cameron, T.A. and Huppert, D.D. (1989) OLS versus ML estimation of non-market resource values with payment card interval data. *Journal of Environmental Economics and Management* 17, 230–246.

Caswell, J.A. (2000) An evaluation of risk analysis as applied to agricultural biotechnology (with a case study of GMO labeling). *Agribusiness: an International Journal* 16, 115–123.

Economic Research Service (ERS) (2000a) Biotechnology: U.S. grain handlers look ahead. *Agricultural Outlook*. US Department of Agriculture, Washington, DC.

Economic Research Service (ERS) (2000b) Biotech corn and soybeans: changing markets and the government's role. ERS Issues Center, US Department of Agriculture, Washington, DC

Good, D., Bender, K. and Hill, L. (2000) *Marketing of Speciality Corn and Soybean Crops*. Value Project Publication AE-4733, March 2000, University of Illinois, Urbana-Champaign.

Josling, T., Unnevehr, L., Hill, L. and Cunningham, C. (1999) Possible policy and market outcomes in WTO 2000. *The Economics and Politics of Genetically Modified Organisms in Agriculture: Implications for WTO 2000*. Bulletin 809, University of Illinois, Urbana-Champaign.

Mitchell, R. and Carson, R.T. (1989) *Using Surveys to Value Public Goods: the Contingent Valuation Method*. Resources for the Future, Washington, DC.

Stewart, M.B. (1983) On least squares estimation when the dependent variable is grouped. *Review of Economic Studies* 50, 737–753.

9 A Comparative Analysis of Consumer Acceptance of GM Foods in Norway and the USA[1]

Wen S. Chern[1] and Kyrre Rickertsen[2]
[1]*Department of Agricultural, Environmental, and Development Economics, The Ohio State University, Columbus, OH 43210-1067, USA;* [2]*Department of Economics and Social Sciences, Agricultural University of Norway, Ås, Norway*

Introduction

Genetically modified organisms (GMOs) have been developed from advanced biotechnology to achieve certain desirable traits in agricultural production such as weed and pest resistance. However, the adoption of genetically modified (GM) crops has been controversial because of the opposition by some consumer and environmental groups. If there are no direct tangible benefits to the consumer, the foods produced with GMO ingredients may be perceived as being inferior to their non-GM counterparts. This perception is critical for the acceptance of GM foods (Caulder, 1998; Hoban, 1998; Hallman and Metcalfe, 2001). There have been concerns about the consumer's acceptance of GM foods in many countries of the world such as those in the European Union (EU) and Japan, as no food manufacturers have dared to test the markets with specifically labelled GM foods under the mandatory labelling regulations.

Since the first commercialization of GM crops in 1996, the adoption of Roundup Ready soybeans and Bt maize has increased rapidly in the US (Darr, 2001; Darr and Chern, 2002). However, over this short time period, the use of these GM products has been regulated in the EU, Japan and other countries. In 1997, the EU imposed a mandatory labelling of GM foods with a 1% tolerance level, while Japan followed suit in 2001 with a 5% GM content limit. The debates on consumer acceptance and labelling regulations have attracted much interest concerning consumer attitudes toward GMOs and GM foods. There have been several consumer surveys conducted in the USA (Hoban, 1999; Hallman and Metcalfe, 2001; Moon and Balasubramanian, 2001; Mendenhall and Evenson, 2002), Europe (Boccaletti and Moro, 2000, for Italy; Burton *et al.*, 2001, for the UK; Spetsidis and Schamel, 2001, for Germany; and Verdurme *et al.*, 2001, for Belgium) and Japan (Macer and Ng, 2000; Ng *et al.*, 2000). Most of these studies are descriptive in nature and few deal with the estimation of the willingness to pay (WTP) for GM foods. Moon and Balasubramanian (2001) estimated the WTP for breakfast cereals made

[1] The authors would like to thank Arild Skogmo, Frode Alfnes and Dadi Kristofersson for help with the Norwegian survey, and Naoya Kaneko, Lewis Horner and the Center for Survey Research of The Ohio State University for research assistance and funding support for the US survey.

of non-GM ingredients in the USA and the UK. But their samples are not representative. Boccaletti and Moro (2000) also attempted to quantify the WTP for generic GM products with different hypothetical attributes in Italy and Burton *et al.* (2001) calculated the WTP for such products in the UK. Our study attempts to extend these previous works to design a survey instrument for eliciting the WTP for different GM foods based on national representative samples. Specifically, since 2000, a joint research project has been undertaken to conduct a multi-country analysis on consumer attitudes toward GM foods and on eliciting the consumer's WTP for GM versus non-GM foods in Japan, Norway, Taiwan and the USA. In this chapter, we only report the results from pilot national telephone surveys using a revised uniform questionnaire conducted in Norway and the USA in 2002.

For the remainder of this chapter, we will first compare the GM food regulations between the EU and the USA. We will then present the survey results and compare the estimated WTP for GM soybean oil, salmon fed with GM soybeans and GM salmon.

Consumer Concerns and Labelling

There is substantial resistance to GM crops in Europe and other parts of the world. Consumer organizations have expressed concerns regarding antibiotic-resistant marker genes, potential allergic reactions, ethical and religious concerns, and the lack of consumer choice due to inadequate labelling (Franks, 1999).

Most national labelling systems are still under development and different countries have taken different approaches. As noted earlier, the EU has imposed mandatory labelling systems. In the EU, a number of directives set the framework for the labelling systems in the member states. Directive 90/220 from 1990 establishes requirements for labelling GM crop varieties for seeds, the Novel Food Regulation 258/97 from 1997 sets a 1% tolerance level for whole and processed foods, and Regulation 1139/98 from 1998 covers GM varieties of maize and soybeans that were released before Regulation 258/97 was adopted. However, the EU directives and regulations do not come into effect until the member states enact them as national laws. Some member states also want to go beyond the base requirements. For example, Austria prefers a ban on GM foods (Phillips and McNeill, 2000).

Most recently, on 3 July 2002, the European Parliament backed the European Commission's two new proposals on GMOs which will establish 'a sound community system to trace and label GMOs and to regulate the placing on the market and labelling of food and feed products derived from GMOs' (European Commission, 2002). The proposals 'will require the traceability of GMOs throughout the chain from farm to table and provide consumers with information by labelling all food and feed consisting of, containing or produced from a GMO' (European Commission, 2002). Under these proposed regulations, the most notable changes from the existing provisions include that products such as soybean and rapeseed oils which were exempted from the existing labelling regulation due to their undetectibility of the presence of transgenic DNA or protein, are now subject to the newly proposed labelling regulation. Furthermore, the feeds containing GMOs such as GM soy meal are required to be labelled as such. However, the Parliament rejected the more extreme amendments that would have required labelling of products like meats or eggs produced with GM feed as GM products. The proposals were later adopted by the European Council in 2002.

Norway is a member of the European Economic Space and is in many cases bound by EU directives and regulations. Norway has adopted somewhat stricter requirements than those established in the EU's Novel Food Regulation. One major difference is that labelling is mandatory even if GM foods do not differ. Due to consumer opposition, none of the major Norwegian food retailers sells GM foods.

In the USA, the government made a decision on 3 May 2000 to reject biofood labels on the grounds that from a health and safety standpoint, these foods do not differ from their conventional counterparts. Since GM foods such as GM soybeans are nutritionally equivalent to the conventional ones, the Food and Drug Administration (FDA) does not require labelling of GM foods (Vogt and Parish, 1999). So far, there has not been notable consumer opposition to GM foods in grocery stores despite opposition from consumer groups who have accused the US government of catering to the biotechnology industry and ignoring the consumer's right to know.

The Survey

Two telephone surveys were conducted during March and April 2002 in Norway and the USA. We asked similar questions, however, the surveys were conducted in different languages creating some differences regarding the exact wording of questions. Some questions from the original English questionnaire were also omitted from the Norwegian survey. For example, adjectives like 'extremely' were toned down in the Norwegian translation and questions concerning 'race' or 'religion' (Norway is 95% white and Protestant) were omitted. We have used the US wordings of the alternatives and questions in the tables presented later. There are many advantages of doing a telephone survey. One is that the alternative choices of several questions can be randomly selected for each interview. The interviewers were trained to answer questions of the respondents and thus the quality of the responses should be higher than a typical mail survey. With the random digit dialling method used in the US survey, we can reach the entire country very easily.

The survey contained seven sections and included many questions on general food purchasing habits, attitude and perception towards GM foods, information sources, knowledge, willingness to buy GM foods with varying attributes, labelling regulation, contingent valuation (CV) for GM versus non-GM food products, and demographic information. The US survey contained 61 questions while the Norwegian survey had fewer questions because it included only soybean oil and salmon, but not cornflake cereal as in the US survey. The CV part of the survey consisted of sequential closed-ended binary choice questions (Carson and Mitchell, 1995). The specific design of these WTP questions will be described in more detail later. One of the most important features of the survey is that we did not assume *a priori* that GM foods are inferior to the conventional counterparts. Therefore, the respondents could state their preferences for GM foods even if they were priced higher than the conventional products. A copy of the questionnaire is available upon request.

The nationwide US survey consisted of 250 respondents aged 18 and over. The survey was conducted by telephone with the random digit dialling method. Even though the sample was relatively small, it covered 43 states in the continental USA. This pilot survey was funded and conducted by the Center for Survey Research (CSR) of The Ohio State University. We went through many rounds of revisions. Among them was a pre-test by a group of graduate students. The final version was given to the CSR for conversion to a telephone interview format. The CSR also conducted another pre-test and the feedback was used to change some of the questions and wordings. The US survey was conducted within a 3-week period in April 2002, with a mix of daytimes and evenings. The overall response rate was 28.7% while the cooperation rate (among households in which we could speak with the eligible respondent) was 80.6%. Average age of the US survey respondents is 47 while 77% are females. Note that in the US survey, we required the respondents to be a food shopper in the household. There are 4.3% of the respondents who are vegetarians.

The Norwegian survey was conducted by Skogmo (2002) and the Norwegian results from the public survey are based on his results. In the Norwegian survey, 100 respondents aged 18 and over living in Oslo (the capital) and 100 respondents living in Nordland (a county without any major cities in the northern part of Norway) were randomly selected from the telephone directory and interviewed. The directory covers about 97% of Norwegian households. The sample consists of 46% male and 54% female respondents. The average age of the respondents was 49 years or about 4 years above the national average for the age group 20–80 years. The high mean age was partly a result of 40% of the interviews being conducted during daytime when many retired people answered the telephone. Furthermore, four out of five calls were rejected, pointing to a potential self-selection problem with less participation among people with valuable time.

Comparison of Survey Results

The results in Table 9.1 show that about 45% of the respondents considered themselves 'not informed' and about 45% considered themselves 'somewhat informed ' about GM foods or GMOs. A somewhat larger percentage of Americans (14.1%) than Norwegians (8.0%) claimed to be 'very well informed'.

The high proportions of not informed respondents correspond well with the proportion of correct answers to our two knowledge statements. Only 37.5% of the Norwegian and 43.8% of the American respondents thought it was false that 'Non-genetically modified soybeans do not contain genes while genetically modified soybeans do' while 36.0% of Norwegians and 61.3% of Americans believed it was false that 'By eating GM foods, a person's genes could be altered'. The responses to the knowledge questions are used to identify the knowledgeable consumers as those answering both questions correctly. These questions provide important explanatory variables used in the econometric models discussed later.

The results in Table 9.2 show that a majority of Norwegians (59.5%) and close to half of Americans (48.9%) believed that GM foods were risky to human health while 23.5% of Norwegians and 20.7% of Americans thought they were safe. A third of the Norwegians considered them extremely risky.

Less than a third of Norwegian (30.5%) and 43.0% of American respondents claimed that they were willing to consume foods produced with GM ingredients. The American resistance is unexpected given that about 70% of the foods on the retail food store shelves are said to contain some form of GMO ingredient in the USA (Kinsey, 2001). A larger proportion of the Norwegian than the US respondents were either 'extremely unwilling' (45.5%) or more surprisingly 'extremely willing' (13.0%) to consume GM foods.

The opposition against GM foods was reduced when some benefits associated with them were explicitly mentioned in the questions suggesting that GM foods can grow in popularity when consumers become aware of the potential benefits. Benefits offered in our surveys were reduced use of pesticides, improved nutritional qualities or lower price. Close to 40% of Norwegians and around 70% of Americans were willing to consume GM foods conditional on those benefits. When we asked which of these potential benefits was the most important about 65% of the Norwegian and 55% of the American respondents answered reduced use of pesticides and below 10% answered reduced price. More than half of Norwegians found reduced price to be 'extremely unimportant' for their decision to buy or not to buy GM foods. The insensitivity to price may be caused by the hypothetical nature of the choice (i.e. no real goods or payments) as discussed in much of the experimental economics literature (e.g. List and Shogren, 1998).

Table 9.1. Consumer information and knowledge, percentage distribution for each question.

Question	Alternative	Norway	USA
Before this survey, how well were you informed about GM foods or organisms?	Very well	8.0	14.1
	Somewhat	45.0	41.0
	Not informed	47.0	44.9
Non-genetically modified soybeans do not contain genes while genetically modified soybeans do.	True	16.0	23.4
	False	37.5	43.8
	Don't know	46.5	32.8
By eating GM foods, a person's genes could be altered.	True	28.0	22.3
	False	36.0	61.3
	Don't know	36.0	16.4

Table 9.2. Consumer attitudes toward GM foods, percentage distribution for each question.

		Alternatives					
Question	Country	Extremely (1)	Somewhat (2)	Neither (3)	Somewhat (4)	Extremely (5)	Don't know
How risky would you say GM foods are in terms of risk to human health?	Norway	33.5	26.0	8.0	13.0	10.5	9.0
1, 2 = Risky and 4, 5 = Safe	USA	9.4	39.5	16.0	15.2	5.5	14.5
How willing are you to consume foods produced with GM ingredients?	Norway	13.0	17.5	4.0	18.0	45.5	2.0
1, 2 = Willing and 4, 5 = Unwilling	USA	4.7	38.3	13.7	23.8	16.4	3.1
How willing would you be to consume GM foods if they reduced the amount of pesticides applied to crops?	Norway	17.0	21.5	9.5	11.5	35.5	5.0
1, 2 = Willing and 4, 5 = Unwilling	USA	13.7	54.7	9.4	11.3	9.0	2.0
How willing would you be to purchase GM foods if they were more nutritious than similar foods that are not GM?	Norway	17.5	19.5	7.5	10.0	39.0	6.5
1, 2 = Willing and 4, 5 = Unwilling	USA	18.0	53.9	5.1	9.4	10.9	2.7
How important is the price factor when you decide whether or not to buy GM foods?	Norway	16.0	20.0	6.0	7.0	50.5	0.5
1, 2 = Important and 4, 5 = Unimportant	USA	29.7	37.5	7.0	12.1	12.5	1.2
How willing would you be to purchase GM foods if it posed a risk of causing allergic reactions for some people?	Norway	1.5	8.5	2.0	4.0	83.5	0.5
1, 2 = Willing and 4, 5 = Unwilling	USA	3.5	21.5	5.9	26.2	41.4	1.6
How important are ethical or religious concerns when you decide whether or not to consume GM foods?	Norway	21.5	8.0	3.5	2.5	62.5	2.0
1, 2 = Important and 4, 5 = Unimportant	USA	12.5	23.8	15.2	18.0	28.9	1.6
How important is it to you that food products are specifically labelled as GM or non-GM?	Norway	94.0	4.5	0.5	0.0	1.0	0.0
1, 2 = Important and 4, 5 = Unimportant	USA	58.6	28.5	4.3	5.9	1.6	1.2

We also asked about some potential sources of concern. More than 80% of Norwegians and 40% of Americans were 'extremely unwilling' to purchase GM foods if it posed a risk of causing allergic reaction for some people. Only 10.0% of Norwegians and 25.0% of Americans were willing to take such a risk. Ethical and religious concerns were important for 29.5% of Norwegians and 36.3% of Americans while such concerns were 'extremely unimportant' for as much as 62.5% of Norwegians and 28.9% of Americans.

The majority of Norwegian (98.5%) and American (87.1%) consumers demand labelling. These results are in line with the results in the Eurobarometer (European Commission, 2001) where 94.6% of the 16,029 respondents in the 15 member states of the EU wanted to have the right to choose between GM and non-GM foods. Support for labelling was reduced when the respondents are reminded that labelling may increase food prices, however, 55% of Norwegians support labelling even if prices were increased by 5% or more. The insensitivity to price may again be partly explained by the hypothetical nature of the question.

The results indicate more favourable attitudes to GM foods in the USA than in Norway; however, the opinions in the USA are also quite mixed. This general conclusion is consistent with Priest (2000) who found that the USA increasingly resembles Europe in having significant amounts of reservation towards biotechnology.

Methodology

The main objective of this study is to estimate the willingness to pay for different qualities of soybean oil (non-GM and GM) and salmon (non-GM, GM-fed and GM). GM foods are not sold in Norway but GM soybean oil is commonly sold in the USA. Salmon can potentially be fed with GM soybeans (GM-fed salmon) and a GM salmon has been developed by the Canadian company Genesis. The GM salmon grows faster than wild salmon (but not necessarily faster than farmed salmon) and the feeding costs are lower (*Aftenposten*, 9 September 2001). None of these qualities of salmon is yet for sale. Nevertheless, there is a considerable interest for consumers' acceptance and WTP for these qualities in the aquaculture sector.

To calculate WTP we use a stated choice method (SCM), which is based upon buyers' hypothetical choice for GM food purchases. We used a simple design developed for the telephone survey and the only attributes included were prices of GM versus non-GM implying that attributes like reduced use of pesticides or improved nutritional values were not considered. Note that in the survey we defined non-GM products as those with GM content of less than 3% since, as stated, 'it is nearly impossible to ensure 100% purity'.

A disadvantage of the SCM (and other stated preference methods) is that peoples' behaviour in a hypothetical setting may not fully reflect actual behaviour, i.e. the respondents may not act on their stated choices. However, given that none of the GM qualities of salmon are available, we could not use experimental auctions or other incentive compatible techniques.

In the Norwegian survey, we had two alternatives of soybean oil and three alternatives of salmon. The choice experiment consisted of two steps and each step consisted of one binary choice for soybean oil and two binary choices for salmon. In step one, we asked the respondents if they would choose: (i) non-GM or GM-fed salmon, (ii) non-GM or GM salmon, and (iii) non-GM or GM soybean oil given identical prices for each of the two choices. In the US survey, we also included cornflake cereal. Therefore, we had four versions of the questionnaire with each including only two of the three products at a time. In this chapter, we analyse only the results pertaining to soybean oil and salmon.

The base prices we used reflected prices found for the non-GM products in stores (i.e. located in Oslo, Norway and Columbus, USA). The percentage distributions of the respondents' choices are shown in Table 9.3. More than 80% of Norwegians chose the non-GM alternative for each of the three choices. For the American respondents, 45.1% chose non-GM soybean oil, 59.2% chose non-GM salmon (over GM-fed) and 68.9% chose non-GM salmon (over GM salmon). Not for any of the choices did more

Table 9.3. Stated choices at identical prices, percentage distribution for each choice.

		First choice					
Choices	Country	Non-GM	GM	GM-fed	Indifferent	None	Don't know
Salmon	Norway	81.8		1.0	8.6	8.1	0.5
Non-GM/GM-fed	USA	59.2		6.5	24.9	8.3	1.2
Salmon	Norway	86.4	1.0		4.0	7.6	1.0
Non-GM/GM	USA	68.9	3.6		21.0	5.4	1.2
Soybean oil	Norway	85.4	2.5		7.0	4.5	0.5
Non-GM/GM	USA	45.1	8.7		24.9	19.1	2.3

than 10% of the respondents prefer a GM product but in the USA close to a quarter of the respondents are indifferent between the GM and non-GM alternatives. For estimation, the choices of indifferent respondents are weighted with a half on each of the two indifferent alternatives.

In step two, each respondent was given the same choices as in step one but offered price reductions for the commodity he did not choose. The price reductions were in the interval 5–50% for GM soybean oil and GM-fed salmon and 10–60% for GM salmon. Respondents that were indifferent between some alternatives in step one were randomly offered a reduced price for one of the alternatives. Note that the choices made in both steps are taken as two separate observations in the econometric analysis.

Following Ben-Akiva and Lerman (1985) and Haab and McConnell (2002), we specify a random utility model that is linear in parameters:

$$V_{in}=\beta_{i0}+\beta_1 p_{in}+\beta_{i2}x_{n2}+\ldots+\beta_{ik}x_{nk}+\varepsilon_{in}, \tag{1}$$

where V_{in} is respondent n's utility of choosing alternative i, p_{in} is the price offered to respondent n for alternative i, $x_{n2} \ldots x_{nk}$ are the individual specific characteristics (e.g. gender or education) of respondent n, and the error terms ε_{in} are assumed to be independently, identically and extreme value (Gumble) distributed. The estimated parameters, except the utility of money (β_1), are allowed to vary across the alternatives allowing the personal characteristics to have non-constant effects for the alternatives and thereby an impact on the choices made. For identification, the parameters of the first equation (except (β_1) are normalized to zero. Letting the scale parameter μ = 1, the probability of choosing alternative i for respondent n is estimated by the logit model:

$$\pi_n(i)=\frac{e^{V_{in}}}{\sum_j e^{V_{jn}}} \tag{2}$$

For soybean oil we use a binary (i = 1 is non-GM and i = 2 is GM) and for salmon a multinomial (i = 1 is non-GM, i = 2 is GM-fed and i = 3 is GM) logit model.

The estimated parameters can be combined to identify monetary values associated with changes in each attribute and characteristic level. Since the utility of the non-GM alternative (i = 1) is $V_{1n} = \beta_1 p_{1n} + \varepsilon_{1n}$, the WTP_{in} for the GM alternatives (i = 2,3) can be calculated from the expression:

$$\beta_1 p_{1n}+\varepsilon_{1n}=\beta_{i0}+\beta_1(p_{in}+WTP_{in})+\beta_{i2}x_{n2}+\ldots+\beta_{ik}x_{nk}+\varepsilon_{in}. \tag{3}$$

Assuming that $E(\varepsilon_{1n}) = E(\varepsilon_{2n}) = E(\varepsilon_{3n}) = 0$, the average consumer's willingness to pay for each alternative is

$$\overline{WTP_i}=\frac{1}{\beta_1}\left(\beta_{i0}+\beta_{i2}\overline{x_2}+\ldots+\beta_{ik}\overline{x_k}\right), \tag{4}$$

where $\overline{x_k}$ denotes the mean value of the individual specific characteristic k. The marginal change in WTP for alternative i associated with a change in characteristic k is

$$\frac{\partial WTP_i}{\partial x_k}=-\frac{\beta_{ik}}{\beta_1}. \tag{5}$$

Regression Results

The models are first estimated using the Norwegian data and the LIMDEP program version 7. Specifically, the binary logit model is used for soybean oil while the multinomial logit model is employed for salmon. We estimate models using different sets of characteristics; however, the average consumer's WTP for each alternative is reasonably robust for choice of variables. Later we use the same specification and estimate the corresponding models with the US data. The definitions and sample means of the variables used are presented in Table 9.4. Since each stated choice in the survey is taken as one observation in the regression analysis, the number of observations (*n*) is larger than the sample size from the survey. The number of observations varies slightly among the products depending upon the stated choices. For example, respondents with 'indifferent' between the alternatives would result in two observations while a decisive choice is counted as only one observation. The list of variables includes many typical demographic variables such as age, gender, education, income and region. In this specification, knowledge about GMOs is among the key determinants of the WTP for GM vs. non-GM products. There are many other variables which can be constructed from the survey data and we continue to search for a more refined specification with the data we have.

Table 9.5 compares the regression results for soybean oil. The coefficient estimates from Norway are much more satisfactory than those obtained from the US survey. Consider first the results from Norway. The price coefficient has a negative sign, as expected, and it is significant. It basically means that the prices had impact on which alternative the respondent chose (i.e. price significantly influenced the utility gained from each alternative). A significant price coefficient is, of course, a necessity if WTP values are to be accurately calculated. If it is not significant, it can be interpreted as the respondents being price insensitive.

The constant term is the 'starting point' of utility gained from the alternative, and in this case it is negative. The GM alternative has *on average* a 'negative' utility, other variables not taken into account, indicating that it is of lesser value to most consumers as compared to the non-GM oil, whose constant term is normalized to zero. Both *ZONE* and *ORDER* are significant at the 10% level. The finding is that people in Nordland have a higher probability of *not* choosing the GM soybean oil, compared to people in Oslo. Since there are limitations in a sample this small, no general conclusions can be drawn, but there is a difference between respondents in this sample. The *ORDER* variable tells us that more people chose to go for the GM alternative if this product (soybean oil) was the *second* product asked about after salmon in the survey. This result suggests that the order of products may affect the respondent's choice of GM vs. non-GM products in the survey. Only one of two knowledge variables (*KNOW2*) is significant. This result is interesting, since the variable is constructed from an objective test of knowledge. Respondents who answered question A4 in the survey correctly are more positive towards genetically modified soybean oil. It is very logical that you are more negative if you believe consuming GM oil will affect your genes, but it can also be interpreted in a more general fashion: the more you know about GMO, the more likely you are to be positive towards it. This 'fear of the unknown' is a quite common phenomenon with humans, and clearly also valid in the context of GM foods. The variable *OIL* (coded 1 if the respondent believes GM oil to be on the market already) is not significant at any respectable probability level. Furthermore, *AGE* negatively affects the utility gained from the GM alternative. In other words, young people are more likely to buy GM soybean oil. The education variable (*EDU*) is also significant at the 5% level. A higher level of education will positively affect utility gained from GM oil. The education level is expectedly highly correlated with age, as obtaining higher education has been an increasing trend in Norway. It is also common to suspect that age and education are correlated with knowledge level, but this need not necessarily hold true.

The fit of the model for soybean oil with the US data is considerably poorer as the McFadden R^2 is only 0.07 compared with 0.303 for the Norwegian model. As shown in Table 9.5, the price coefficient in the US model is not significant, making it difficult to produce reliable estimates of the WTP for a premium to avoid GM soybean oil. The only significant variables in the US model for

Table 9.4. Definition and sample mean of variables.

		Norway sample mean		US sample mean	
Variable	Definition	Oil (n = 384)	Salmon (n = 374)	Oil (n = 245)	Salmon (n = 168)
PRICE	Price of the product [a]	35.80	71.74	1.79	5.56
MIDWEST	1 if Midwest; 0 otherwise	NR	NR	0.22	0.22
SOUTH	1 if South; 0 otherwise	NR	NR	0.32	0.31
WEST	1 if West; 0 otherwise	NR	NR	0.21	0.26
NORTHEAST	1 if Northeast; 0 otherwise (dropped)	NR	NR	0.25	0.21
ZONE	1 if Nordland; 0 if Oslo	0.52	0.51	NR	NR
*VER*1[b]	1 if survey version 1 (oil and cereal); 0 otherwise. Irrelevant for salmon	NR	NR	0.30	NR
*VER*2[b]	1 if survey version 2 (cereal and oil); 0 otherwise. Irrelevant for salmon	NR	NR	0.26	NR
*VER*3[b]	1 if survey version 3 (oil and salmon); 0 otherwise. Dropped for soybean oil. Irrelevant for cornflake cereal	NR	NR	0.44	0.40
VER4[b]	1 if survey version 4 (cornflakes cereal and salmon); 0 otherwise. Dropped fpr salmon. Irrelevant for soybean oil	NR	NR	NR	0.60
ORDER[b]	1 for survey version 1 (salmon and oil); 0 for version 2 (oil and salmon) for Norway	0.49	0.52	NR	NR
*KNOW*1	−1 if wrong answer to A3; 0 if don't know; 1 if correct answer to A3[c]	0.20	0.21	0.30	0.43
*KNOW*2	Defined similarly to KNOW1 but related to A4[d]	0.11	0.08	0.47	0.54
OIL	1 if one believes that GM ingredients are used in the production of soybean oil; 0 otherwise.	0.43	NR	0.58	NR
AGE	0.1 × (age – mean age)	−0.007	−0.01	−0.27	−0.02
GENDER	1 if male; 0 if female (USA); −1 if female; 1 if male (Norway)	−0.047	−0.07	0.31	0.24
EDU	Education level coded from 1 (elementary school) to 8 (doctoral/advanced degree) in the USA and from 1 to 7 in Norway	3.297	3.33	3.94	4.03
KIDS	1 if living with children (17 years old or younger)	0.45	0.45	0.39	0.36
ln (*INCOME*)	Logarithm of income categories from 1 to 10 measured in increments of US$10,000 in the USA; from 1 to 11 in increments of NOK 100,000 in Norway	1.40	1.39	1.54	1.61
FREQ	Frequency of salmon consumption coded 1–4 (1 if once a week; 2 if once a month; 3 if once every 3 months; 4 if once a year)	NR	2.53	NR	2.46

n, number of observations; NR, not relevant.

[a]Prices are in US$ per 32 fl. oz for oil, and per pound for salmon in the US survey and NOK per litre for oil and per kilo for salmon in the Norwegian survey.

[b]We have four versions of questionnaires in the US survey. The Norwegian survey has only two versions for oil and salmon. All order effects are taken care of by the version variables (*VER*1, *VER*2, *VER*3 and *ORDER*).

[c]A3 states 'Non-GM soybeans do not contain genes while genetically modified soybeans do'.

[d]A4 states 'By eating GM foods, a person's genes could be altered'.

Table 9.5. Regression results for soybean oil.

	Norway		US	
Variable	Coefficient estimate	*t*-Ratio	Coefficient estimate	*t*-Ratio
PRICE	−0.187***	−5.61	−0.543	−1.05
CONSTANT	−3.038***	−4.88	−1.900***	−2.85
MIDWEST	NR	NR	0.955*	1.89
SOUTH	NR	NR	0.694	1.51
WEST	NR	NR	−0.070	−0.13
ZONE	−0.810*	−1.86	NR	NR
*VER*1	NR	NR	0.763*	1.91
*VER*2	NR	NR	0.208	0.46
ORDER	−0.831*	1.96	NR	NR
*KNOW*1	−0.172	−0.58	−0.163	−0.71
*KNOW*2	0.880***	2.92	0.329	1.49
OIL	0.614	1.57	0.389	1.06
AGE	−0.269*	−1.85	0.085	0.66
GENDER	0.851***	3.61	0.463	1.18
EDU	0.317**	2.04	0.048	0.45
KIDS	0.198	0.43	0.400	1.03
ln (*INCOME*)	−0.848**	−2.18	−0.373	−1.47
McFadden R^2	0.303		0.07	
n	384		245	

n, number of observations; NR, not relevant.
*Significant at 10%; **significant at 5%; ***significant at 1%.

soybean oil are *CONSTANT* and *VER*1. These results show that American consumers have a negative utility associated with GM soybean oil without taking into account any other factors. Furthermore, the order of the products asked in the survey affects the valuation or the WTP for GM vs. non-GM attributes. However, the US model shows the opposite results from the Norwegian model as placing soybean oil first in the questionnaire appears to increase rather than decrease the probability of choosing GM soybean oil.

Table 9.6 presents and compares the regression results for salmon from the USA and Norwegian survey data. The multinomial model results for the US are much more satisfactory than those for soybean oil. The price coefficient is statistically significant at the 5% level and the McFadden R^2 is considerably higher (0.16). The estimated coefficients for non-GM vs. GM salmon are much stronger than those for non-GM vs. GM-fed salmon, indicating that American consumers are more concerned or fearful about GM salmon than GM-fed salmon. The results for GM salmon show that the constant term, regional dummies, two knowledge variables and gender are all significant determinants for choosing GM vs. non-GM salmon. The Norway model for salmon also shows very similar results as the knowledge variables and gender are statistically significant. The price variable is highly significant. In addition, income is found to be significant for GM-fed salmon. The salmon models with significant price coefficients provide a much stronger bases for computing the WTP estimates than the soybean oil model for the USA.

Estimated Willingness to Pay

The WTP estimates are obtained from the estimated regression models according to equation (4). The sample means of the explanatory variables are used in these computations. We may also compute the WTP by respondents and investigate the distribution of estimated WTP among respondents. The computed WTP may be interpreted as the WTP a premium for a non-GM product or as the WTP to avoid a GM product. In other

Table 9.6. Regression results for salmon.

Variable	Norway Coefficient estimate	Norway *t*-Ratio	US Coefficient estimate	US *t*-Ratio
PRICE	−0.068***	−7.42	−0.421**	−2.71
GM-fed salmon				
CONSTANT	−2.130**	−2.46	−1.496	−1.45
MIDWEST	NR	NR	0.149	0.23
SOUTH	NR	NR	0.357	0.62
WEST	NR	NR	0.184	0.28
ZONE	−0.630	−1.62	NR	NR
VER3	NR	NR	0.493	1.15
ORDER	−0.102	−0.27	NR	NR
KNOW1	0.118	0.44	−0.371	−1.25
KNOW2	0.858***	3.32	0.737**	2.20
FREQ	0.115	0.51	0.195	1.00
AGE	−0.107	−0.87	−0.020	−0.12
GENDER	0.536***	2.67	0.062	0.11
KIDS	0.163	0.39	0.340	0.65
EDU	0.144	1.00	−0.154	−1.10
ln (*INCOME*)	−0.926***	−2.64	−0.167	−0.39
GM salmon				
CONSTANT	−3.565***	−3.59	−5.778***	−3.79
MIDWEST	NR	NR	2.057**	2.35
SOUTH	NR	NR	2.015**	2.48
WEST	NR	NR	0.222	0.25
ZONE	−0.366	−0.86	NR	NR
VER3	NR	NR	−0.308	−0.57
ORDER	0.381	0.93	NR	NR
KNOW1	0.059	0.20	−0.821**	−2.32
KNOW2	1.207***	4.10	1.367***	2.94
FREQ	−0.021	−0.09	0.694***	3.03
AGE	−0.105	−0.76	0.309	1.60
GENDER	0.583***	2.63	1.389**	2.02
KIDS	−0.173	−.0.38	0.461	0.75
EDU	0.190	1.18	−0.255	−1.59
ln (*INCOME*)	−0.505	−1.33	0.829	1.53
McFadden R^2	0.261		0.159	
n	374 + 364 = 738		168 + 165 = 333	

n, number of observations; NR, not relevant.
*Significant at 10%; **significant at 5%; ***significant at 1%.

words, the WTP may be interpreted as the amounts that we would have to reduce the price of the non-GM alternative to let the average consumer be equally well off.

The results are shown in Table 9.7. Consider first the results for Norway. The price of non-GM soybean oil was NOK 40 and the price of GM soybean oil has to be reduced by NOK 22.13 to NOK 17.87 per liter to make the average Norwegian consumer equally well off. In a similar way, the price of GM-fed salmon has to be reduced by NOK 43.42 and GM salmon by NOK 53.96 from the base price of NOK 80. This corresponds to price reductions of 55%, 54% and 67% for GM soybean oil, GM-fed salmon and GM salmon, respectively. As expected, the required reduction in price is larger for GM salmon than for the other GM alternatives. There is a distinction between direct and indirect GM consumption and there is also a difference between plant and animal genes.

Table 9.7. Estimated WTP values to avoid GM alternatives.

		Alternative		
Country	Item	GM soybean oil	GM-fed salmon	GM salmon
Norway	Mean, NOK	22.13	43.42	53.96
	Mean, US$[a]	2.77	5.43	6.75
	% reduction	55	54	67
USA	Mean, US$	1.82	2.75	4.49
	% reduction	84	46	71

[a]The exchange rate is set to NOK 8.00 per US$.

The WTP values to avoid GM products for American consumers are estimated to be $1.82 per 32 fl. oz for soybean oil, $2.75 per pound for GM-fed salmon and $4.49 for GM salmon. The corresponding percentages of price reduction are 84%, 46% and 71% for soybean oil, GM-fed salmon and GM salmon, respectively. These figures are much higher than expected. With respect to the WTP for soybean oil, the estimate is based on the price coefficient which is not statistically significant as noted earlier. Thus we cannot place any confidence on this estimate. The unexpectedly high value of the WTP for soybean oil in the USA may be caused by the high standard errors associated with the price coefficient and other parameters in the model. The high errors may cause the estimated WTP to be either extremely large or extremely small. The price coefficient for salmon is statistically significant and thus the resulting WTP estimates should be more credible. The results show that American consumers are willing to pay substantial amounts of premium to avoid GM-fed or GM salmon.

The marginal WTP values reported in Table 9.8 show how much a change in one of the individual specific characteristics will affect the WTP to avoid the different GM alternatives. For most parts, the effects of the characteristics are consistent across the various GM alternatives. Since many coefficient estimates are not statistically significant in the US models, the signs of the effects are not entirely consistent. In Norway, the age effects are always positive and significant for GM soybean oil and GM-fed salmon. If the age of the respondent increases by 10 years, then the respondent demands an extra price reduction of NOK 1.88, 3.52 and 3.54 for GM soybean oil, GM-fed salmon and GM salmon, respectively. The gender effects are always negative and significant. Females are coded as −1 and males as 1 implying that female consumers demand price reductions of NOK 4.48, 9.32 and 11.72, as compared with the average Norwegian consumer, for GM soybean oil, GM-fed salmon, and GM

Table 9.8. Marginal WTP values to avoid GM alternatives[a].

		Alternative		
Country	Variable	GM soybean oil	GM-fed salmon	GM salmon
Norway	Age	1.88	3.52	3.54
	Gender	−4.48	−9.32	−11.72
	Education	−2.87	−5.29	−5.85
	Income	1.05	3.71	3.03
USA	Age	−0.16	0.05	−0.73
	Gender	−0.85	−0.15	−3.30
	Education	−0.09	0.37	0.61
	Income	0.19	0.10	−0.49

[a]NOK for Norway and US$ for the USA.

salmon respectively. The effect of education is always negative and significant. The more education the less price reductions are needed. If the educational level (from one to six) increases by one, then the respondent requires NOK 2.87, 5.29 and 5.85 less compensation for consuming GM soybean oil, GM-fed salmon and GM salmon, respectively. The effect of income is always positive and significant implying that respondents with higher incomes demand larger price reductions. Since the log of income is used as a variable, there is always a positive but decreasing effect of income, and the estimates reported are for changes from mean income. If the mean household income increases by one class (or NOK 100,000), then the respondent demands an additional price reduction of NOK 1.05, 3.71 and 3.03 for GM soybean oil, GM-fed salmon and GM salmon, respectively.

In the USA, the only significant demographic variable is gender (coded 1 for male and 0 for female). The results are very similar to those found in Norway, indicating that females are willing to pay more than males in order to avoid GM products. Specifically, females are willing to pay \$0.85, \$0.15 and \$3.30 more than males to avoid GM soybean oil, GM-fed salmon and GM salmon, respectively. Even though the estimated coefficients of education are not statistically significant, the results show the opposite effects as compared with those from Norway. It appears that more-educated American consumers are willing to pay more premiums for non-GM salmon than less-educated consumers. In other words, more-educated people would demand higher price reductions for consuming GM products than less-educated ones.

The reported WTP figures are quite substantial, indicating a strong opposition to GM foods in Norway and the USA. Given the potential hypothetical bias mentioned above they must be interpreted as upper bounds. However, we may note that the reported WTP values are identically and inversely related to the estimated price parameter, β_1, implying that any hypothetical bias affects the levels of the WTP and not the relative price effects between the GM and GM-fed salmon.

Conclusions

This chapter presents survey results and analyses from a joint research project to conduct a multi-country study on consumer acceptance of GM foods. The results indicate more favourable attitudes to GM foods among US than Norwegian consumers. However, the opinions in the USA are also quite mixed and only 43% of the American respondents claimed that they are willing to consume foods produced with GM ingredients. The opposition against GM foods is reduced when some benefits associated with them are introduced into the questions, suggesting that GM foods have a potential to become more popular. Reduced use of pesticides and improved nutritional qualities are perceived as more important potential benefits than reduced price. Health concerns are apparently more important than ethical or religious concerns in explaining the negative attitudes towards GM foods. The support for mandatory labelling is overwhelming in the surveys even when labelling may increase food prices.

There is a substantial WTP to avoid GM alternatives. In the surveys, 80% of the Norwegian respondents chose the non-GM alternatives in each case and for the American respondents 45% chose non-GM soybean oil, 59% non-GM salmon over GM-fed salmon, and 69% chose non-GM salmon over GM salmon. These figures indicate that there are differences between direct and indirect GM consumption and between animal and plant genes.

The WTP for avoiding the GM alternatives indicates that the average Norwegian consumer demands price reductions of 55%, 54% and 67% for GM soybean oil, GM-fed salmon and GM salmon, respectively, as compared with the conventional alternatives. For American consumers, the estimated price reductions demanded for GM alternatives are 84%, 46% and 67% for soybean oil, GM-fed salmon and GM salmon, respectively. It is surprising to have such a high WTP for vegetable oil in the USA. As noted earlier, since the US estimate is based on an insignificant price coefficient for soybean oil, the WTP estimate is not credible for vegetable oil. Overall, the high percentage premiums may, at least to some extent, be due to the hypothetical nature of the choices without any real payments. Note also that the estimated WTP values obtained in this study are substantially higher than those estimated by Chen and Chern (2002) using a mail survey conducted in Columbus, Ohio, USA. It is important to validate these estimates with larger samples.

Future research will focus on searching for a better modelling specification so that we can use more information from the surveys to construct our econometric models and to test econometrically the structural differences between Norway and the USA. The surveys reported in this chapter were pilot surveys. These pilot surveys were also used as a basis for re-designing our survey questionnaire for a large-scale public survey with 1000 telephone interviews conducted in 2003. We also plan to do similar surveys in Japan, Spain and Taiwan.

References

Ben-Akiva, M. and Lerman, S.R. (1985) *Discrete Choice Analysis: Theory and Application to Travel Demand*. The MIT Press, Cambridge, Massachusetts.

Boccaletti, S. and Moro, D. (2000) Consumer willingness to pay for GM food products in Italy. *AgBioForum*. 3, 259–267.

Burton, M.L., Rigby, D., Young, T. and James, S. (2001) Consumer attitudes to genetically modified organisms in food in the UK. *European Review of Agricultural Economics* 28, 479–498.

Carson, R.T. and Mitchell, R.C. (1995) Sequencing and nesting in contingent valuation surveys. *Journal of Environmental Economics and Management* 28, 155–173.

Caulder, J. (1998) Agricultural biotechnology and public perceptions. *AgbioForum* 1, 38–39.

Chen, H.-Y. and Chern, W.S. (2002) Willingness to pay for GM foods: results from a public survey in the U.S. Paper presented at the 6th International Conference on 'Agricultural Biotechnology: New Avenues for Production, Consumption, and Technology Transfer', Ravello, Italy, 11–14 July 2002.

Darr, D. (2001) Econometric analysis of genetically modified organism adoption: a study of Ohio grain farmers. Unpublished Masters' thesis, Department of Agricultural, Environmental, and Development Economics, The Ohio State University, Columbus, Ohio, USA.

Darr, D. and Chern, W.S. (2002) Estimating adoption of GMO soybeans and corn: a case study of Ohio, U.S.A. In: Santaniello, V., Evenson, R.E. and Zilberman, D. (eds) *Market Development for Genetically Modified Foods*. CAB International, Wallingford, UK, pp. 141–158.

European Commission (2001) *Europeans, Science and Technology. Eurobarometer* 55(2). European Commission, Brussels.

European Commission (2002) Press release, 3 July. Available at www.europa.eu.int under the category of 'environment'.

Franks, J.R. (1999) The status and prospects for genetically modified crops in Europe. *Food Policy* 24, 565–584.

Haab, T.C. and McConnell, K.E. (2002) *Valuing Environmental and Natural Resources: the Econometrics of Non-market Valuation*. Edward Elgar Publishers, Cheltenham, UK.

Hallman, W.K. and Metcalfe, J. (2001) Public perceptions of agricultural biotechnology: a survey of New Jersey residents. US Department of Agriculture. Available at http://www.nalusda.gov/bic/pubpercept.

Hoban, T., J. (1998) Trends in consumer attitudes about agricultural biotechnology. *AgBioForum* 1, 3–7.

Hoban, T.J. (1999) Public perceptions and understanding of agricultural biotechnology. US Information Agency. Available at www.usia.gov/journals/ites/1099/ijee/bio-toc.htm.

Kinsey, J.D. (2001) The new food economy: consumers, farms, pharms, and science. *American Journal of Agricultural Economics* 83, 1113–1130.

List, J.A. and Shogren, J.F. (1998) Calibration of the difference between actual and hypothetical valuations in a field experiment. *Journal of Economic Behavior and Organization* 37, 193–205.

Macer, D. and Ng, M.A.C. (2000) Changing attitudes to biotechnology in Japan. *Nature Biotechnology*. 18, 945–947.

Mendenhall, C.A. and Evenson, R.E. (2002) Estimates of willingness to pay a premium for non-GM foods: a survey.' In: Santaniello, V., Evenson, R.E. and Zilberman, D. (eds) *Market Development for Genetically Modified Foods*. CAB International, Wallingford, UK, pp. 55–61.

Moon, W. and Balasubramanian, S.K. (2001) Estimating willingness to pay for nonbiotech foods: a comparison across US and UK consumers.' Paper presented at the Annual Meeting of the American Agricultural Economics Association, Chicago, Illinois, 5–8 August.
Ng, M.A.C., Takeda, C., Watanabe, T. and Macer, D. (2000) Attitudes of the public and scientists to biotechnology in Japan at the start of 2000. *Eubios Journal of Asian and International Bioethics* 10, 106–113.
Phillips, P.W.B. and McNeill, H. (2000) A survey of national labeling policies for GM foods. *AgBioForum* 3, 219–224.
Priest, S.H. (2000) U.S. public opinion divided over biotechnology? *Nature Biotechnology* 18, 939–942.
Skogmo, A. (2002) Consumer attitudes toward genetically modified foods: a study of consumer acceptance in two regions of Norway. MSc thesis, Department of Economics and Social Sciences, Agricultural University of Norway, Aas, Norway.
Spetsidis, N.M. and Schamel, G. (2001) A survey over consumer cognitions with regard to product scenarios of GM foods in Germany. Paper presented at the 71st EAAE Seminar on 'The Food Consumer in the Early 21st Century', Zaragoza, Spain, 19–20, April.
Verdurme, A., Gellynck, X. and Viaene, J. (2001) Consumer's acceptance of GM food. Paper presented at the 71st EAAE Seminar on 'The Food Consumer in the Early 21st Century, Zaragoza, Spain, 19–20 April.
Vogt, D.U. and Parish, M. (1999) Food biotechnology in the USA: science, regulation, and issues. CRS Report to Congress, 2 June, Available at http://www.usinfo.state.gov/topical/global/biotech/crsfood.htm.

10 Comparing Consumer Responses towards GM Foods in Japan and Norway[1]

Jill J. McCluskey,[1] Kristine M. Grimsrud[2] and Thomas I. Wahl[3]
[1]*Department of Agricultural Economics, Washington State University, 211J Hubert Hall, Pullman, WA 99163, USA;* [2]*Department of Economics, University of New Mexico, Albuquerque, NM 87131-0001, USA;* [3]*International Marketing Program for Agricultural Commodities and Trade (IMPACT) Center, Washington State University, Hulbert Hall, Rm 123, PO Box 646214, Pullman, WA 99164-6210, USA*

Introduction

The introduction of genetically modified (GM) crops to world markets has created a new division between the crop trading countries. The USA and Canada have great economic interests in exporting their transgenic crops, however, lack of public acceptance of GM food products in the European Union (EU) and Japan has already resulted in reduced or curbed demand for GM food products. Many European and Japanese consumers believe GM foods pose a threat to human health. They fear short- and long-term consequences for their own health and their offspring.

A suggested remedy has been consumer education about GM food safety. However, the results of the 1996 Eurobarometer survey suggested that more knowledgeable people do not necessarily have a more positive opinion; they just have a more definite opinion about biotechnology (Biotechnology and the European Public Concerted Action Group, 1997). A later survey, the 2001 Eurobarometer survey, showed that 70.9% of Europeans simply do not want genetically modified organisms (GMOs).

Consumer attitudes may be just as important as consumer knowledge. The Eurobarometer survey showed that moral doubts were more important than health risks in shaping public acceptance of gene technology (Biotechnology and the European Public Concerted Action Group, 1997). Environmental concerns are also important-in the 2000 Eurobarometer survey, 59.4% of EU citizens interviewed said that they thought GMOs could have a negative impact on the environment.

Macer and Ng (2000) report that support for biotechnology and genetic engineering in Japan is decreasing, especially for agricultural applications. They used a mail survey spanning the years 1991, 1993, 1997 and 2000 and found that Japanese interest in science and biotechnology increased from 30% in 1991 to 47% in 2000. From the 2000 survey, 97% of respondents are familiar with the term 'biotechnology', which implies that awareness of biotechnology has increased significantly among the Japanese public. Also

[1] The authors wish to thank Hiromi Ouchi for excellent research assistance. Further, the authors thank without implicating Phil Wandschneider for helpful discussion and advice. The authors gratefully acknowledge financial support from the IMPACT Center at Washington State University.

from the 2000 survey, only 31% are likely to support GM foods, and only 20% say they are willing to buy GM fruits. They conclude that although the majority of Japanese consumers are optimistic about biotechnology, there are increasingly negative views towards its application for agriculture.

Consumer attitudes and behaviour towards GM food products are complex and differ across cultures. A better understanding is essential for designing market strategies. We investigate factors that affect consumer acceptance of GM food in Japan and Norway and estimate the discounts necessary for consumers to be willing to purchase GM food or the premium consumers are willing to pay for GM food. We compare consumer preferences across countries.

Mandatory labelling of GM foods has obvious implications for trade. The EU has imposed mandatory labelling for some foods that contain GM ingredients. In October 1999, the EU gave preliminary approval to a law that requires labels on all foods containing more than 1% GM ingredients. In Japan, authorities have ordered mandatory labelling for 29 categories of food if they contain any GM ingredients. The USA has argued that there is no health-related or scientific reason to reject GM commodities and food products and has challenged the EU's mandatory GM labelling as a non-tariff trade barrier.

The Codex committees of the World Trade Organization (WTO) are working on harmonizing international standards and resolving trade disputes associated with food labelling to promote fair trade of foods while protecting consumer health. Since different countries have different attitudes towards GM food products, the Codex frameworks allow each country to develop its own standards. The challenge of Codex is to set international standards for GM food labelling that both promotes fair trade and allows for consumer choice. An important issue in GM labelling policy is scientific versus consumer sovereignty. Although the scientific consensus may be that GM foods are completely safe for consumption aside from potential allergens, it may be the case that a majority of the population in a given country prefers to avoid GM foods.

Mandatory labelling forces US producers to segregate crops to claim food products are 'GM-free'. This would be difficult and costly. For example, many grain elevators are not physically equipped to segregate crops. US producers may lose market share because consumers can reject their GM crops.

Data

In August 2001, we conducted 400 in-person interviews in Japanese at the Seikatsu Club Consumer Cooperative (Seikyou), a grocery store-like setting, in Matsumoto City, Japan. Matsumoto is a relatively agricultural area with about 13% of the population coming from farm households compared with 2% in all of Japan. Consumer cooperatives usually focus on offering 'safe foods', as a marketing strategy, and target members who are more willing to purchase safe foods. The Seikyou has significant market power in the Japanese market place.

In January 2002, we conducted 400 in-person interviews in Norwegian at the RIMI Liertoppen grocery store in the Oslo region of Norway. This region is the most populated part of Norway and one of the Norwegian centres of economic activity. The RIMI chain of grocery stores has chosen a low-price/limited selection niche in the market and has in this way gained significant power in the Norwegian market place.

The surveys solicited demographic information, respondents' attitudes about the environment and food safety, and their self-reported knowledge and perceptions about biotechnology. Our surveys included contingent valuation questions regarding willingness to accept a discount to purchase food products made from GM wheat. The hypothetical market for the good in question must be as close as possible to a real market in order to reveal people's true preferences if an actual market existed (Pearce and Turner, 1990). In Japan, we asked about noodles made with GM wheat, and in Norway we asked about bread made with GM wheat.

The food products (bread and noodles) used in this study are appropriate for examination since they are frequently consumed food products.

First, consumers were asked if they were willing to purchase the GM food product if it was offered at the same price as the corresponding, non-GM product. If the respondent's answer to this question was 'no', a follow-up question was asked if the respondent was willing to purchase the GM food product if offered a percentage discount compared with the corresponding non-GM product. The discount was set at one of the following levels: 5%, 10%, 25%, 40% and 50%. Each level of discount was used for one-fifth of the surveys. That is, 80 of the 400 surveys had a 5% discount for GM bread, another 80 surveys had a 10% discount for GM bread, and so on. The assignment of survey version (and thus, discount) was random to the respondent. The maximum discount was set at 50%, assuming that people who would not prefer the GM food product at such at large discount would not choose the product at any discount. No follow-up question was asked if the customer's answer was 'yes' to the initial question, and he/she was willing to purchase the GM food product at no discount. The rationale for no follow-up to a 'yes' response is that the type of genetic modification associated with these GM food products is a *process attribute*, which reduces production costs – as opposed to a *product-enhancing* attribute. An example of a GM product with a product-enhancing attribute is the Flavr Savr tomato. Proponents claim that the GM products with process attributes are identical to non-GM products. Opponents view genetic modification as a negative attribute. Therefore, it would not make economic sense after an initial 'yes' to respond with a follow-up question that involves paying a premium for these GM products that only have cost-reducing attributes.

Although Norwegian consumers were generally sceptical of GM foods, more of the Norwegian consumers chose the GM products compared with the Japanese consumers. Of the 400 Norwegian respondents, more than one-quarter of the sample would buy bread with GM wheat flour when offered no discount compared to conventional bread. Further, 39% of consumers in the sample replied that they would be willing to purchase the GM product if it was cheaper than the conventional product. Of the 400 Japanese respondents, only 3% said that they would be willing to purchase the GM noodles at the same price as the corresponding non-GM product. Further, only 17% of consumers in the sample stated that they would be willing to purchase the GM product if it was less expensive than the corresponding non-GM product. The rest of the Japanese respondents, that is 80%, were not willing to purchase the GM noodles even with the discount.

Empirical Results

The model that is applicable for examining the outcomes of our survey can be considered a special case of the double-bounded logit model (Hanemann *et al.*, 1991). In this model, the initial bid (B_0) equals zero and implies no price difference between the GM food product and the non-GM food product. The second bid (B_D), is the GM food product offered at a random percentage discount relative to the non-GM food product. This bid is only given to individuals who answer that they would not buy a GM food product at the same price as a non-GM food product.

Let *WTA* denote an individual's willingness to accept (WTA) compensation (or bid function) for GM food products, relative to the non-GM food products, and let B_D # 0 denote the discount bid on GM food products relative to non-GM food products. The following discrete outcomes of the bidding process are observable:

$$D = \begin{cases} 1 \ WTA \le B_0 \\ 2 \ B_0 < WTA \le B_D. \\ 3 \ \ B_D < WTA \end{cases} \quad (1)$$

The WTA function for each food product for individual *i* is

$$WTA_i = \alpha + \rho B_i + \lambda' \mathbf{z}_i + \varepsilon_i \ i = 1, \ldots, n \quad (2)$$

where B_i is the ultimate discount bid individual i faces, $\mathbf{z}_i$ is a column vector of observable characteristics of the individual, ε_i is a random variable accounting for random noise and possibly unobservable characteristics. Unknown parameters to be estimated are α, ρ and λ. Linearity in $\mathbf{z}$ and ε is assumed for all individuals.

The bid information and other demographic information were used to estimate the magnitude of factors that affect consumers' WTA for GM food products and how much of a relative discount consumers will require to purchase GM food products. Our estimation results for Japan show that variables representing food safety and environmental attitudes, self-reported knowledge about biotechnology, self-reported risk perceptions towards GM foods, income, and education all significantly increase the necessary discount required for Japanese consumers to choose GM foods. With the Norwegian data, increasing self-reported risk perceptions toward GM foods and preferences for domestically produced food both significantly increase the discount required for Norwegian consumers to choose GM foods.

The mean willingness to accept for both GM food products, *WTA*, was estimated by restricting $\lambda = 0$ (Hanemann *et al.*, 1991). The empirical mean *WTA* can then be calculated as $-\tilde{\alpha}/\tilde{\rho}$. Our results indicate that Japanese Seikyou members, on average, want a 60% discount on GM noodles compared to non-GM noodles. Our results indicate that, on average, the Norwegian consumers in our sample want a 49.5% discount on GM bread compared to conventional bread. (Further details of each country analysis are presented in Grimsrud *et al.*, 2003, and McCluskey *et al.*, 2003.)

Conclusions

The Japanese and Norwegian cultures both place a great deal of value on tradition. This worldview extends to the food they eat and feed to their children. Therefore, it is not surprising that most Japanese and Norwegian consumers want to avoid GM foods. A better understanding of Japanese and Norwegian consumers' attitudes and behaviour towards GM food products and how these attitudes affect the purchasing choices for such food products is essential for marketing GM food products in those countries. We found that although the majority of Norwegian consumers want to avoid GM food, they are not as negative as the Japanese.

The Norwegian policy restricting GM food products and the results of surveys show that there is considerable scepticism in the Norwegian population towards GM foods. The results of this study suggest that our sample of Norwegian consumers is, on average, willing to purchase GM bread with a 49.5% discount compared to the corresponding non-GM product. Consumers' perceptions and attitudes towards GM food and respondents' age increase *WTA* (that is, a greater discount would be required) for GM food.

The strong scepticism in the Norwegian population may be fading, leaving a potential future market for GM foods. People younger than the average age of the sample were willing to purchase the GM food products with an average discount that is half (or less) of what the customers above the average age of the sample needed. The gap in between generations for *WTA* indicate that younger people may be more open to GM foods and that it may a question of time before the Norwegian market may be more open to GM food products.

More Japanese respondents rejected the GM food products compared with the Norwegians. An implication of these studies is that there is an opportunity to market food segregated from GM products in both countries, but especially in Japan. Japanese consumers need to be convinced of the safety of GM foods if they are to be marketed successfully there.

References

Biotechnology and the European Public Concerted Action Group (1997) Europe ambivalent on biotechnology. *Nature* 387, 845–847.

Grimsrud, K.M., McCluskey, J.J., Loureiro, M.L. and Wahl, T.I. (2003) Consumer attitudes toward genetically modified food in Norway. IMPACT Center Technical Working Paper.

Hanemann, W.M., Loomis, J. and Kanninen, B.J. (1991) Statistical efficiency of double-bounded dichotomous choice contingent valuation. *American Journal of Agricultural Economics* 73, 1255–1263.

Macer, D. and Ng, M.A.C. (2000) Changing attitudes to biotechnology in Japan.' *Nature Biotechnology* 18, 945–947.

McCluskey, J.J., Grimsrud, K.M., Ouchi, H. and Wahl, T.I. (2003) Consumer response to genetically modified food products in Japan. *Agricultural and Resource Economics Review* 32(2), 222–231.

Pearce, D.W. and Turner, R.K. (1990) *Economics of Natural Resources and the Environment*. Harvester Wheatsheaf, London.

11 Willingness to Pay for GM Foods: Results from a Public Survey in the USA

Hsin-Yi Chen and Wen S. Chern
Department of Agricultural, Environmental, and Development Economics, The Ohio State University, Agricultural Admin Building, 2120 Fyffe Road, Columbus, OH 43210-1067, USA

Introduction

Introducing biotechnology into agricultural production is one of the most prominent benchmarks in the history of agricultural development. The application of genetic modification technology on agricultural crops and the resulting genetically modified organisms (GMOs) are considered one of the most important yet controversial advancements in science and technology. Despite all the promises and benefits proclaimed by many biotech companies and the governments such as reduced pesticide usage, higher crop yields, enhanced nutritional values and many more, the controversy surrounding its application to food production persists in many countries.

Studies have shown that many consumers in countries like those in the European Union (EU) and Japan have difficulties accepting genetically modified (GM) products (Macer and Ng, 2000). Consumers are hesitant to buy GM foods largely because of concerns about the uncertain effects of GM foods on human health. There are also religious and ethical concerns about the possible intake of genes from animals contained in GM foods, and consumers have no means to identify these products. These concerns in turn have generated a strong demand for the labelling of GM products in the EU and countries like Japan and Taiwan. In addition, the potential environmental endanger of GM products appears to be one of the major hindrances for consumers to accept GM foods.

Consumer acceptance of GM products, therefore, has become a vital factor on how prosperous the market for GM foods will be in the future. It will affect the future course of private and public investment in the development and use of GM technology. Thus, consumer perception of and acceptance of GM technology and GM foods are crucial for the global market of GM products, agricultural trade and the future development of agricultural biotechnology.

Despite the seemingly ample information available, research on consumer perception and attitudes towards biotechnology in general or GM foods in particular, has been limited. In most cases, the methodology adopted relied heavily on qualitative questions and descriptive comparisons. Few attempted to quantify the consumer's purchase behaviour and to investigate the willingness to buy GM foods. Also, the constantly changing attitudes expressed by consumers signify the need to investigate and obtain more up-to-date data on consumer attitudes toward GM foods. Furthermore, few studies have attempted to estimate the consumer's willingness to pay

(WTP) for GM or non-GM foods. This study, therefore, is aimed at providing the WTP estimates based on consumer attitude towards GM foods and other characteristics. Actually this study is part of a research project attempting to provide a more global perspective by conducting a multi-country survey in four countries, Japan, Norway, Taiwan and the USA. This chapter, however, presents only the analysis of a survey conducted in the Columbus Metropolitan Area, Ohio, under this project.

Specifically, the objectives of this study are to analyse an Ohio survey on the consumer acceptance of GM foods and to conduct a contingent valuation of the willingness to pay for three selected products with and without GMO ingredients, i.e. vegetable oil, salmon and cornflake breakfast cereal.

Background and Literature

Consumer acceptance towards GM foods varies greatly among countries. Studies in the USA mostly show that its consumers have a higher acceptance rate towards biotechnology and GM foods than those in other countries. One of Hoban's studies (1998) indicates that two-thirds of American consumers are positive about plant biotechnology, and this support is the highest among men and people with more formal education. National surveys conducted in recent years also show that roughly between 35 and 45% of American consumers consider they have heard or read a lot or some about biotechnology (International Food Information Council Foundation, 2001). In Europe, on the other hand, the support for GM technology in general and GM foods in particular are relatively low as compared to the USA. Based on a European consumer survey conducted in 1999, Gaskell (2000) shows that Europeans are mostly neutral about agricultural biotechnology but opposed to both GM foods and cloning of animals, especially in countries such as Greece, Austria and Luxemburg. Specifically, the survey indicates that only 22% of the European respondents are supporters of GM foods, 25% are risk-tolerant supporters and up to 53% are opponents. A study in the UK shows that while 11% of the respondents would not try GM foods, 42% say they might still try, indicating that UK consumers are not highly opposed to GM foods (Loader and Henson, 1998). Another study conducted by the Nordic Industrial Fund (2000) for Denmark, Finland, Norway and Sweden also suggests that Nordic consumers are united in their negative attitudes toward GM foods. Another study by Burton *et al.* (2001), based on a survey of residents in Manchester, UK, shows that the average willingness to pay to achieve a reduction in risk associated with GM products is 9.8% of the respondent's food expenditure, and the average willingness to pay to achieve a GM-free diet is 13% of their food expenditure. Their study also finds that gender is a significant determinant of attitudes towards GM technology and female respondents are willing to pay more to reduce risk.

In Asian countries, a study by Hoban (1996) concludes that Japanese consumers have optimistic attitudes towards GM biotechnology similar to American consumers, as 87% of the Japanese respondents in his study were positive about the use of biotechnology. However, a recent study by Macer and Ng (2000) indicates that a small majority of Japanese respondents in the past 3 years have been favourably inclined to GM technology and consider it a means of improving the quality of life (54% in 1997 and 59% in 2000), implying the changing attitudes among the Japanese consumers. In Taiwan, even though knowledge of GM foods is not pervasive, a Gallup Poll (2000) shows that Taiwanese consumers are not against the use of GM technology in food production and are willing to purchase GM foods.

Studies on other agricultural technologies have been done. Wang *et al.* (1997) attempted to measure the WTP for bovine somatotropin (rBST)-free milk in Vermont and their results show that for rBST-free milk, 37.4% of the respondents do not want to pay any premium, 50.6% are willing to pay a premium up to 40 cents per gallon and 12.0% would pay a premium of 41 cents or more. Also, the authors find that the consumer WTP is affected by several demographic variables, such as income, education and gender. Halbrendt *et al.* (1995) conducted a nationwide contingent valuation (CV) survey to measure the consumer willingness to purchase pork with lower saturated fats, and their results show that the survey respondents are willing to pay an average 16–23 cents more per pound for fresh pork with reduced levels of saturated fats. Buzby *et al.* (1995) investigated the consumer willingness to pay for grapefruit with reduced chemical residue. Their results indicate that respondents are willing to pay an average of 31% more for grapefruit under the 50% risk reduction scenario, and 38% more if there is a 99% reduction in the risk associated with chemical residue.

Methodology

CV has been used to elicit consumer willingness to pay for non-market goods, such as water quality improvement (Carson and Mitchell, 1981) or air pollution control (Loehman and De, 1982). CV is also widely used in evaluating consumer willingness to pay for food safety, such as reduced food-borne risks (Hammitt, 1986). Although there are several economic tools to value non-market goods, CV is generally considered to be the most appropriate choice for measuring food safety (Buzby *et al.*, 1995).

Among the most important tasks for a CV analysis are questionnaire design and survey procedure (Haab and McConnell, 2001). The CV method uses surveys in which people are asked how much they are willing to pay for a change in the condition of some environmental resources or a service that is meaningful to the respondent in a hypothetical situation (Diamond *et al.*, 1993; Haab and McConnell, 2002). Early CV designs involve open-ended questions such as 'What is the maximum amount you would pay for...?' More recently, the commonly used methods include iterative bidding, payment cards and dichotomous choice questions. Several studies reveal that different techniques of asking CV questions provide significantly different estimates of Hicksian surplus (Boyle and Bishop, 1988).

In this study, the CV is employed to estimate the WTP for non-GM food products. The CV scenario in the survey conducted in this study contains both dichotomous choice and polychotomous choice questions. Food products used in the survey include vegetable oil, salmon and cornflake breakfast cereal. Vegetable oil and cornflake breakfast cereal used a dichotomous response question, as the respondents were asked to choose between GM and non-GM alternatives given the price scenario. For salmon, a polychotomous response question was designed, as surveyed respondents were asked to rank within three different types of GM, non-GM but fed with feed containing GM maize or soybeans, and non-GM fed with non-GM feed alternatives. In this study, we only analyse a model for the dichotomous choice between GM and non-GM products. Thus, the GM salmon and GM-fed salmon are combined as the same category in this study.

The random utility model

We adopt a random utility model for analysing dichotomous CV responses. Following Haab and McConnell (2002), the indirect utility function for respondent j can be written as:

$$U_{ij} = u\,(y_j, Z_j, \varepsilon_{ij}) \tag{1}$$

where i is the dichotomous choice (1 as the preferred state and 0 the status quo) and j refers to the respondent. The determinants of utility are y_j, the jth respondent's income, Z_j, a vector of respondent characteristics and attributes of the choice, and ε_{ij}, a component of preferences known to the individual respondent but not observed by the researcher.

Based on this model, respondent j chooses the non-GM food if the utility of non-GM food exceeds the utility of the status quo (GM food), given prices:

$$U_{1j}(Z_j,\ y_j - Pngm_j, 1,\ \varepsilon_{1j}) > U_{0j}\,(Z_j,\ y_j - Pgm_j,\ 0,\ \varepsilon_{0j}) \tag{2}$$

where 1 denotes the respondent j choosing the non-GM food, 0 denotes the respondent j choosing the status quo (GM food), $Pngm_j$ is the price of non-GM food, Pgm_j is the price of GM food. All other variables were defined previously.

Therefore, the probability that the respondent thinks he/she is better off by choosing the non-GM food, given its price can be expressed as:

$$\text{Prob (Respondent } j \text{ chooses non-GM food)} = \text{Prob}\,[U_{1j}(Z_j, y_j - Pngm_j, 1, \varepsilon_{1j}) > U_{0j}(Z_j, y_j - Pgm_j, 0, \varepsilon_{0j})]. \quad (3)$$

The logistic model

With a further assumption of the linear form for the utility function, the utility of respondent j choosing the non-GM food can be specified as:

$$U_{1j} = \alpha_1 Z_j + \beta_1 (Y_j - Pngm_j) + \varepsilon_{1j} \quad (4)$$

And the utility of respondent j choosing the GM food is:

$$U_{0j} = \alpha_0 Z_j + \beta_0 (Y_j - Pgm_j) + \varepsilon_{0j} \quad (5)$$

If respondent j chooses the non-GM food, it implies that the utility of choosing the non-GM food is greater than that of choosing GM food:

$$U_{1j} > U_{0j} \quad (6)$$

By assuming the marginal utilities of money (income) for non-GM food and GM food are identical, i.e. $\beta_1 = \beta_0 = \beta$, the probability of choosing non-GM food is:

$$\begin{aligned}\text{Prob (non-GM)} &= \text{Prob}\,[\alpha_1 Z_j + \beta_1 (Y_j - Pngm_j) - \alpha_0 Z_j - \beta_0 (Y_j - Pgm_j) > 0] \\ &= \text{Prob}\,[(\alpha_1 - \alpha_0) Z_j - \beta(Pngm_j - Pgm_j) + (\varepsilon_{1j} - \varepsilon_{0j}) > 0]\end{aligned}$$

This can be written more compactly as:

$$\text{Prob (Non-GM)} = \text{Prob}\,[\alpha Z_j - \beta(\Delta P) + \varepsilon_j > 0] \quad (7)$$

where,

$$\alpha = (\alpha_1 - \alpha_0)$$
$$\Delta P = (Pngm - Pgm)$$
$$\varepsilon_j = (\varepsilon_{1j} - \varepsilon_{0j}).$$

Assume further that the error term has a logistic distribution and it is symmetrical. Therefore, we can derive the probability of choosing non-GM food as:

$$\begin{aligned}\text{Prob (non-GM)} &= \text{Prob}\,[\alpha Z_j - \beta\Delta P + \varepsilon > 0] \\ &= \text{Prob}\,[-(\alpha Z_j - \beta\Delta P) < \varepsilon] \\ &= 1 - \text{Prob}\,[-(\alpha Z_j - \beta\Delta P) > \varepsilon] \\ &= \text{Prob}\,[\varepsilon < (\alpha Z_j - \beta\Delta P)] \quad (8)\end{aligned}$$

Furthermore, with a logistic distribution, ε has a mean of zero and variance $\pi^2\sigma^2/3$. Normalizing by σ creates a logistic variable with mean zero and variance $\pi^2/3$. Equation (8) becomes:

$$\begin{aligned}\text{Prob (non-GM)} &= \text{Prob}\,[\theta < (\alpha Z_j - \beta\Delta P)/\sigma] \\ &= \Psi\,[\alpha Z_j/\sigma - (\beta\Delta P/\sigma)]\end{aligned} \quad (9)$$

where $\theta = \varepsilon/\sigma$, σ is the standard error and Ψ is the cumulative distribution function.

Therefore, by using a logistic distribution, the probability of choosing the non-GM product is:

$$\text{Prob (non-GM)} = [1 + \exp(-(\alpha Z_j/\sigma - \beta\Delta P/\sigma))]^{-1}. \quad (10)$$

Calculating willingness to pay

For calculating the WTP, we need to estimate the parameters α and β for the vector of explanatory variables (Haab and McConnell, 2002). A CV question induces the respondent to choose between the proposed condition at the required payment, and the current state. The required payment therefore states the respondent's willingness to pay in order to achieve the proposed scenario. In our case, the WTP is the proposed price of a non-GM product that would make the respondent indifferent between consuming GM (paid with the current price of the GM product) and the non-GM product. Based on this principle, the WTP for the non-GM food product can be defined as:

$$\alpha_1 Z_j + \beta(y_j - WTPngm_j) + \varepsilon_{1j} = \alpha_0 Z_j + \beta(y_j - WTPgm_j) + \varepsilon_{0j}. \quad (11)$$

Solving equation (11) for WTP yields:

$$WTPngm_j - WTPgm_j = \alpha Z_j/\beta + \varepsilon_j/\beta \quad (12)$$

where:

$$\alpha = \alpha_1 - \alpha_0$$
$$\varepsilon_j = \varepsilon_{1j} - \varepsilon_{0j}.$$

However, the parameters are unknown and therefore must be estimated. In the expression for mean, only the ratio of parameter estimates is required. Relying on Slutsky's theorem on consistency, the logit maximum likelihood estimates for $\theta = \{\alpha/\sigma, \beta/\sigma\}$ are consistent (Haab and McConnell, 2002). Therefore, a consistent estimate of expected willingness to pay for a non-GM food product derived from equation (12) is:

$$E\,(WTPngm_j - WTPgm_j \mid \alpha, \beta, Z_j) = \alpha Z_j/\beta \quad (13)$$

where α is the vector of the estimated coefficients of the explanatory variables and β is the estimated coefficient of the price difference between a non-GM and GM food product. Note that the estimated price coefficient obtained from equation (9) is $(-\beta)$, and therefore the calculation of WTP needs to reverse the sign for β.

By adopting the logistic model to estimate the probability of choosing the non-GM food, the econometric model can be specified as the logit model:

$$y = \alpha k + \beta p + \varepsilon \quad (14)$$

$$\text{where } y = \begin{cases} 1 & \text{if the respondent chooses the non-GM food product} \\ 0 & \text{otherwise.} \end{cases}$$

Also, k is a vector of explanatory variables and p is price factor. In our empirical model, the price factor is defined as the 'price difference' between non-GM and GM food in order to capture the price effect and WTP can be estimated as the expected premium for non-GM food.

The Survey and Data

A mail survey was conducted in the Columbus Metropolitan Area, Ohio, in March 2001. A three-wave procedure combined with mailing and telephone was used in order to maximize the response rate. The sampling frame was obtained from the Center for Survey Research at The Ohio State University. The Center randomly selected 650 telephone subscribers in the Columbus Metropolitan Area based on the zip code. The questionnaires and postage-paid return envelopes were mailed to these randomly selected households. In total, 141 completed survey questionnaires were returned, along with 120 undeliverable returning questionnaires, yielding an overall response rate of 26.6%. Four versions of the questionnaire with different prices in the CV section were equally distributed among the 650 mailings. Among the 141 returned respondents, we collected 39 copies of version 1, 28 of version 2, 26 of version 3 and 48 of version 4. (These four versions of prices will be discussed later.)

Survey questionnaire

In the first part of the questionnaire, consumer knowledge and awareness of biotechnology and GM foods are being elicited. Next, respondents were asked about their attitudes and acceptance toward GM foods, as well as other GM food-related issues such as environmental concern and pesticide usage. For most of these questions, five options for the response are typically given along with an option of 'Do not know'. For example, in the question, 'To what extent do you feel that GM foods are risky, or safe, for human health?', the respondent was given the choices of 'Extremely risky', 'Very risky', 'Somewhat risky/Somewhat safe', 'Very safe', 'Extremely safe' and 'Do not know'. Thirdly, respondents were asked about their support for GM food labelling and type of labelling.

Afterwards, a CV scenario was presented along with the food products and price combinations. Surveyed respondents were first asked about their consumption habit and frequency of the food product. Then, they were asked to select or rank the food products according to different GM contents, given the prices. In designing the price matrix, we assumed that GM food products are cheaper than their non-GM counterparts. Therefore, we specified the prices of GM food products by taking a discount of those of non-GM food products, which were based on the market prices. The discount ranged from 10% to 25%. Table 11.1 presents the four price scenarios used in the survey. Note that the market prices observed in Columbus, Ohio, at the time of survey were used as the base prices. In two versions, these base prices were changed slightly to provide more variations in prices. Even though this range of differences between the GM and non-GM products seems reasonable, the specific price variations chosen are somewhat subjective.

The last part of the questionnaire contained the demographic information such as age, sex, race, income, education, religion, occupation, etc. (A copy of the questionnaire is available upon request.)

Table 11.1. Price matrix for CV design.

	Vegetable oil price ($ per 32 fl oz)		Salmon price ($ per pound)			Cornflake cereal price ($ per 18 oz)	
Version	Non-GM	GM	Non-GM	GM-fed	GM	Non-GM	GM
1 (10% difference)	$2.49[a]	$2.24	$6.99[a]	$6.29	$5.66	$4.39[a]	$3.95
2 (20% difference)	2.49[a]	1.99	6.99[a]	5.59	4.47	4.39[a]	3.51
3 (15% difference)	2.19	1.86	5.99	5.09	4.33	3.79	3.22
4 (25% difference)	2.19	1.64	5.99	4.49	3.37	3.79	2.84

[a]These are the observed reference market prices of the products during the survey period.

Variables

In the logistic model, the dependent variable is a binary variable being one if the respondent chose the non-GM food and zero otherwise. Note that in the case of salmon, our survey collected data on the ranking of the three types of salmon, i.e. GM salmon, non-GM but fed with GM feed, and non-GM fed with non-GM feed. However, in the regression model, we simply estimated the probability of choosing non-GM versus GM salmon. We grouped GM salmon and non-GM salmon but fed with GM feed together as GM salmon in the analysis. An extension of the model dealing with three separate choices in a multinomial logit model would be desirable for future research.

From the survey data, various explanatory variables can be grouped into six categories: 'Knowledge and Awareness', 'Attitude', 'Perception', 'Labelling', 'Demographic' and 'Price'. Variable definitions and the sample means used in the econometric model are presented in Table 11.2. Since the models for vegetable oil and cornflakes are based on a slightly different sample size than the model for salmon, different descriptive statistics are shown. It is interesting to note that only 58% of the respondents considered themselves as either 'very well' or 'somewhat' informed about GMOs or GM foods. Furthermore, a majority of the respondents thought GM foods are 'somewhat risky' to human health (53%), while only 19% replied 'extremely or very safe'.

Empirical Results

Table 11.3 presents the regression results for the three food products: vegetable oil, salmon and cornflake breakfast cereal. The results show that the variables related to Attitude, Perception, Labelling, Demographic, and Price have significant effects on consumer choices between GM and non-GM food products. The knowledge and awareness variables, however, appear to be not statistically significant. Let us discuss these findings in more detail.

Attitude

Results indicate that the risk perception of GM foods places an important impact on GM food consumption, as higher risk perception generates lower GM food consumption. The percentage of organic food purchase, used as an indicator of attitude towards risk, is insignificant in the cornflake cereal model but significant in the vegetable oil and salmon models. Environmental concern of GM foods is also a significant factor determining GM food consumption, so is religious or ethical concern. Further, the perceived difference between GM and non-GM food affects the willingness to consume GM foods, implying that if the perceived difference is not huge, consumers are more willing to consume GM foods.

Perception

Price attribute is a significant determinant for willingness to consume GM foods, as the respondent's concern on price tends to induce them to consume more GM food, which is assumed to be cheaper than their traditional counterpart. Interestingly, those who think it is most important to reduce saturated fats in GM vegetable oil still tend to consume more non-GM vegetable oil. On the other hand, those who believe the most important

Table 11.2. Variable definition and sample means.

Variable/category	Definition and coding	Sample mean (vegetable oil and cornflake cereal)	Sample mean (salmon)
Knowledge and awareness			
KNOW	1 if very well/somewhat informed of GMOs or GM foods; 0 otherwise	0.58	0.59
OIL	1 if aware of GM vegetable oil; 0 otherwise	0.35	
CF	1 if aware of GM vegetable cornflake cereal; 0 otherwise	0.36	
GMS	Percentage of GM foods sold in the market place (guessed)	38%	38%
Attitude			
RP1	1 if GM foods are extremely/very risky; 0 otherwise	0.09	0.08
RP2	1 if GM foods are somewhat risky; 0 otherwise	0.53	0.57
RP3	1 if GM foods are extremely/very safe; 0 otherwise	0.19	0.20
(Focus group: do not know)			
O1	1 if 1–20% organic food purchase; 0 otherwise	0.57	0.58
O2[a]	1 if more than 20% (or 21–40%) organic food purchase; 0 otherwise purchase; 0 otherwise	0.10	0.06
O3[b]	1 if more than 40% organic food purchase; 0 otherwise		0.06
(Focus group: 0% and do not know)			
EN1	1 if GM technology is extremely/very beneficial to the environment; 0 otherwise	0.21	0.22
EN2	1 if GM technology is extremely/very risky to the environment; 0 otherwise	0.11	0.12
(Focus group: Somewhat risky)			
REL	1 if religious concerns are extremely/very important; 0 otherwise	0.18	0.21
PESTICID	1 if large/some pesticide decrease after applying GM technology; 0 otherwise	0.43	0.48
DIF1	1 if GM and non-GM foods are extremely/very different; 0 otherwise	0.39	0.39
DIF2	1 if GM and non-GM foods are not very/not at all different; 0 otherwise	0.26	0.28
(Focus group: I have no idea)			
CON	1 if excellent/good government performance in food safety; 0 otherwise	0.38	0.38
Perception			
SATFB	1 if the respondent believes the potential to reduce saturated fats in foods is the most important benefit of GM foods; 0 otherwise	0.17	
PESTB	1 if the respondent believes the potential to reduce pesticides in foods is the most important benefit of GM foods; 0 otherwise	0.67	0.74
PRICE	1 if the respondent ranks 'price' as the first and second important food attribute; 0 otherwise	0.25	
SAFETY	1 if the respondent ranks 'safety' as the first and second important food attribute; 0 otherwise	0.45	
TASTE	1 if the respondent ranks 'taste' as the first and second important food attribute; 0 otherwise	0.56	
Labelling			
LABEL	1 if labelling of GM foods is extremely/very important; 0 otherwise	0.67	0.65

Continued

Table 11.2. *Continued.*

Variable/category	Definition and coding	Sample mean (vegetable oil and corn-flake cereal)	Sample mean (salmon)
Demographic			
AGE1	If <34 years old; 0 otherwise	0.11	0.11
AGE2	If 35–60 years old; 0 otherwise	0.57	0.65
(Focus group: >60 years old)			
GENDER	1 if males; 0 otherwise	0.44	0.44
MARITAL	1 if married; 0 otherwise	0.53	0.56
EDU1	1 if some college and associate degree; 0 otherwise	0.27	0.25
EDU2	1 if bachelor degree, some graduate school and graduate degree; 0 otherwise	0.44	0.46
(Focus group: No high school, some high school and high school diploma)			
IN1	1 if income $30,001–$50,000; 0 otherwise	0.27	0.27
IN2	1 if income $50,001–$70,000; 0 otherwise	0.24	0.25
IN3	1 if income more than $70,001; 0 otherwise	0.21	0.23
(Focus group: Less than $30,000)			
RACE	1 if Caucasian; 0 otherwise	0.86	0.86
RELIGION	1 if Protestant; 0 otherwise	0.45	0.42
CHD	1 if there is one or more children under 18 years old in the household; 0 otherwise	0.31	0.33
Price			
POIL	Price difference between non-GM and GM oil	0.42	
PSAL	Price difference between non-GM and GM salmon		1.62
PCF	Price difference between non-GM and GM cornflake cereal	0.72	
Sample size		105[c]	92[c]

[a]O2 = 1 if 21–40% organic food purchase in the salmon model.
[b]O3 = 1 if more than 40% organic food purchase in the salmon model.
[c] The usable sample sizes are smaller than 141 because of missing data in the CV section of the survey.

benefit of GMOs is to reduce pesticide usage tend to consume more GM salmon and GM cornflake breakfast cereal.

Labelling

The opinion on labelling is a significant factor in the salmon and maize flake breakfast cereal models, showing that the more important the respondents think that GM food labelling is, the more non-GM salmon and maize flake breakfast cereal they are going to consume.

Demographic

Demographic characteristics turned out to be insignificant with respect to age, gender, marital status and education, except age and education in the salmon model. Income dummies are highly significant in the cornflake breakfast cereal model and have negative effects, implying that the people with higher income tend to consume more GM cornflake breakfast cereal. This result is somewhat surprising. Note that the number of children within the household has a significant negative effect on respondents' willingness to consume GM foods, as the concern for younger children in the household would certainly decrease the consumption of GM foods.

Price

Price is highly significant in the three models, suggesting that lower prices of GM foods encourage consumers to consume more GM products.

Table 11.3. Regression results.

	Vegetable oil		Salmon		Cornflake cereal	
Variable/category	Coeff.	*t*-Ratio	Coeff.	*t*-Ratio	Coeff.	*t*-Ratio
Constant	0.2420	0.069	10.6209	1.806	8.6643	1.744
Knowledge and awareness						
KNOW	−0.8311	−0.758	1.6416	1.008	0.9009	0.662
OIL	−0.3423	−0.331				
CF					−0.0907	−0.082
GMS			0.0322	1.059		
Attitude						
RP1	6.0563	2.073***	2.7625	0.756	9.3633	1.879**
RP2	3.9027	1.985**	−1.3237	−0.870	1.7853	1.173
RP3	−0.6010	−0.260	−6.0145	−1.518*	−2.4177	−0.916
O1	−0.2036	−0.172	0.4022	0.250	1.3369	1.074
O2	−2.9776	−1.370*	−10.9137	−2.631***	−1.8885	−0.894
O3			0.6134	0.205		
EN1	−8.5130	−2.051***	−4.3476	−1.512*	−4.1532	−1.697*
EN2	3.7732	1.444*	−0.0407	−0.014	−0.1628	−0.079
REL	7.5893	2.148***	5.2668	1.972**	2.6617	1.165
PESTICIDE	1.2412	1.021	1.9062	1.312*	−0.2328	−0.249
DIF1	−0.3020	−0.277	−0.5385	−0.387	−0.5233	−0.465
DIF2	−2.4975	−2.011**	−2.7618	−1.515*	−3.8577	−1.767**
CON	−0.5099	−0.523	1.4127	1.003	−0.1416	−0.140
Perception						
PRICE	−5.3223	−2.591***			−6.2551	−2.317***
SAFETY	2.5639	1.525*			−0.2956	−0.208
TASTE	−0.6826	−0.624			0.3861	0.268
SATFB	2.8779	2.016**				
PESTB			−2.8351	−1.551*	−1.7446	−1.487*
Labelling						
LABEL	0.6757	0.641	3.8607	1.871**	2.6327	2.085***
Demographic						
AGE1	−1.3395	−0.592	−4.4977	−1.033	−0.2374	−0.070
AGE2	0.3099	0.173	−2.6670	−1.509*	−1.7804	−0.955
GENDER	−1.3487	−1.157	0.5547	0.313	0.3989	0.341
MARITAL	0.5832	0.592	1.3714	0.816	1.2198	1.071
EDU1	−0.8810	−0.522	−1.8517	−0.826	−0.2328	−0.151
EDU2	1.1396	0.823	−2.2280	−1.324*	1.2022	0.866
IN1	−1.7423	−1.127	−0.2659	−0.147	−6.8875	−2.247***
IN2	−2.8542	−1.646*	0.4146	0.166	−5.7335	−2.021**
IN3	−1.1893	−0.753	−1.8905	−0.861	−5.6907	−1.940**
RACE	0.7852	0.499	−0.7364	−0.446	−1.6685	−1.004
RELIGION	−1.4551	−1.254	−5.1121	−2.044**	−1.4260	−1.056
CHD	2.6626	1.969**			3.9305	2.491***
Price						
POIL	−7.7657	−1.775**				
PSAL			−5.8891	−2.215***		
PCF					−5.8173	−1.702**
Number of observations	105		92		105	
McFadden R^2	0.5944		0.6537		0.6177	

Coeff., coefficient.

***2.5% significance level; **5% significance level; *10% significance level.

McFadden R^2 values in these models range from 0.5944 to 0.6537, which are actually quite high for this type of cross-sectional data. In general, the results indicate that the willingness to purchase GM foods is heavily influenced by the risk perception of GM foods to human health, environmental concern and religious concern when consuming GM foods, as well as the perceived difference between GM and non-GM foods. Also, the importance of food characteristics such as 'price' will affect consumers on their GM food consumption. The sensitivity to price is also reflected by the significance of the price factor, showing that more GM food products will be chosen if the price difference between non-GM and GM foods increases.

Demographic variables are not very significant. Only income and the number of children in the household affect the consumer's purchase decision. Surprisingly, the respondents with higher income tend to consume more GM cornflake breakfast cereal, implying that wealthy people are more confident on this GM product, and would not view it as particularly risky. We are not sure whether or not this result is caused by the fact that higher-income households in the USA tend to consume more breakfast cereals. Furthermore, breakfast cereal is considered to be a relatively expensive food item.

Willingness to Pay for Non-GM Foods

Based on the methodology described above, the WTP for the three non-GM foods can be computed for the entire sample and the results are presented in Table 11.4. Note that we compute first the WTP household by household. The figures presented in the table are simply means or averages of all households in the sample. The WTP for a non-GM product reflects the premium for the non-GM food that the consumer is willing to pay. We also compute the percentage of premium using the price of the GM food as the base. Since there are different prices for GM foods used in the four versions of price scenarios, the percentage figures vary depending on the base price. The results show that the survey respondents are willing to pay a premium of 5–8% for non-GM vegetable oil, 15–28% for non-GM salmon and 12–17% for non-GM cornflake breakfast cereal.

Table 11.5 shows the computed WTP premiums for various demographic groups by sex, age and race. It is interesting to observe that the WTP premiums for non-GM foods vary by demographic group. Note that even though some demographic variables may not be significant in the logit model, the computed WTP premiums can still be different among demographic groups. This is because the WTP is based on the entire model and the entire set of estimated parameters, not just the coefficient related to a particular demographic variable. The results are very telling that female respondents are always willing to pay a higher premium for non-GM food products than male respondents, especially in the case of vegetable oil and cornflake breakfast cereal. This finding is in accordance with previous studies regarding consumer WTP on organic food produce (Huang, 1993).

Survey respondents between 35 and 60 years old tend to pay higher premiums for non-GM salmon and non-GM vegetable oil than those who are younger than 35 or older than 60. Furthermore, the respondents younger than 35 are willing to pay more for non-GM cornflake breakfast cereal than the

Table 11.4. Estimated WTP premiums for non-GM food products.

Item	Vegetable oil ($ per 32 fl oz)	Salmon ($ per pound)	Cornflake breakfast cereal ($ per 18 oz)
WTP premium	$0.13	$0.96	$0.49
Percentage of premium[a]	5~8%	15~28%	12~17%

[a]Percentage of premium = (WTP premium/price of GM food product) ×100.

Table 11.5. Estimated WTP premiums for non-GM food products by demographic groups.

	Sex		Age			Race	
Product/item	Female	Male	<35 years old	35–60 years old	60 years >old	Non-white	White
Vegetable oil (32 fl oz)							
WTP premium	$0.34	−$0.16	−$0.27	$0.39	−$0.31	$0.54	$0.08
Percentage of premium	15~21%	(−7) ~ (−10)%	(−12)~(−16)%	17~24%	(−14)~(−19)%	24~33%	4~5%
Salmon (per pound)							
WTP premium	$1.13	$0.74	$0.51	$1.05	$0.90	$1.53	$0.86
Percentage of premium	18~34%	12~22%	8~15%	17~31%	14~27%	24~45%	14~26%
Cornflake breakfast cereal (18 oz)							
WTP premium	$0.70	$0.16	$0.66	$0.59	$0.15	$1.28	$0.38
Percentage of premium	18~25%	4~6%	17~23%	15~21%	4~5%	32~45%	10~13%

other age groups. The senior respondents, however, are the least willing to pay higher prices for non-GM vegetable oil and cornflake breakfast cereal, but willing to pay more premiums for non-GM salmon than those who are younger than 35. This finding suggests that middle-aged consumers tend to put more concern on food safety issues than those who are younger or older and therefore are willing to pay more for non-GM foods. Besides, their income sources are more stable as compared to younger and older generations, and therefore middle-aged consumers are more willing to pay a premium for non-GM food products. On the other hand, senior citizens are less willing to pay more for non-GM vegetable oil and cornflake breakfast cereal than that for non-GM salmon which indicates that senior citizens are less sensitive to food safety, especially those food products that are less relevant to their consumption compared to younger generations, such as cornflake breakfast cereal.

Results also show that non-White respondents are more likely to pay a premium for non-GM food products than White respondents. Note that the difference in WTP between the two racial groups is dramatic. Non-White respondents are willing to pay a premium of at least 24% for the three food products, while White respondents are only willing to pay a maximum premium of 26%. This finding is somewhat surprising and it may suggest that the non-White respondents may lack confidence on food safety and therefore are willing to pay a higher premium for non-GM foods than the White respondents.

Implications and Discussion

The purposes of this study are to conduct an analysis on GM food consumption and to measure consumer willingness to pay for non-GM vs. GM foods. The empirical results show that the willingness to consume GM foods depends on risk perception, environmental concern, religious and ethical concern of GM foods, opinion on labelling of GM foods, and perceived difference between GM and non-GM foods. Also, the number of children within the household is a key determinant on GM food consumption. In addition, the price factor of GM foods is fairly significant to the respondents in the survey, suggesting that by advertising GM food products with a lower price, the consumption of GM foods might be substantially increased.

These results imply that in order to gain consumer acceptance of GM foods, it is important to change their risk perception of GM foods and to deviate their other concerns. The survey results show that only 59% of the respondents indicated that they are either very well or somewhat informed on GMOs or GM foods. In fact, the majority of consumers are still not very well informed about GM foods. Therefore, how can we change the consumer's perception? The government, food industry and consumer groups have to provide unbiased information to the consumer. If the information can change consumers' perceptions, then the willingness to buy GM foods would increase. Therefore, the effectiveness of the information is very crucial to the success of GM foods in the future.

The econometric results also show that the respondents are willing to pay a premium, ranging from 5% for non-GM vegetable oil to 28% for non-GM salmon. Clearly, these results imply that the consumer must see a tangible benefit in order for them to buy GM foods. Therefore, the future of GM foods is critically dependent upon the ability to reduce the price for GM foods as compared to their traditional non-GM counterparts. Therefore, the stress on the indifference between GM and non-GM foods is unlikely to induce the consumer's willingness to buy GM foods.

Our results also show that the consumer WTP for non-GM foods varies among demographic characteristics. Specifically, female respondents, those aged between 35 and 60, and non-White respondents are willing to pay a higher premium for non-GM foods than other groups. This finding is useful to the government, food industry and consumer groups for designing appropriate programmes to educate the consumer about GMOs and GM foods targeted to different demographic groups.

Conclusions

In this study, we attempt to investigate consumer attitudes towards GM foods and to elicit WTP for non-GM foods. The empirical results indicate that the consumer acceptance towards GM foods is affected by attitudinal factors, such as risk perception, environmental impacts, opinion on GM food labelling, perceived difference between GM and non-GM foods, and the potential benefits of GM foods. Overall, the high risk associated with GM foods as perceived by the respondents is found to be the main hindrance to the consumer's acceptance of such foods, which reinforces the necessity to educate the general public to be more aware of GM foods with more unbiased scientific information. Also, the result points to the importance of GM food labelling, implying the need to provide the consumer with more information on GM foods so that consumer confidence can be established. Moreover, the price factor is significant in determining the purchase of GM foods, suggesting that lower price can be a useful tool to stimulate GM food consumption.

The results of WTP indicate that the survey respondents are willing to pay a premium in order to differentiate between GM and non-GM foods. This implies that producers of non-GM foods might benefit from the labelling policy. If consumers are willing to bear the premium for non-GM foods, producers do not need to fully absorb the cost of segmenting the market. From the government standpoint, labelling of GM foods might cause a warfare loss to the society in the long run if the market is not competitive for both GM and non-GM food products. That is if non-GM food products were produced by a few producers. Consumers would pay a higher price in order to avoid GM foods, but the prices of non-GM foods were higher than those in a competitive market. The welfare loss in the long term might discourage the government from enforcing a mandatory labelling policy regarding GM foods. Therefore, it is crucial to educate the general public about the characteristics of GM foods so that the risk perception associated from consuming such foods can be mitigated.

References

Boyle, K.J. and Bishop, R.C. (1988) Welfare measurements using contingent valuation: a comparison of techniques. *American Journal of Agricultural Economics* 70, 20–28.

Burton, M., James, S., Ridby, D. and Young, T. (2001) Consumer attitudes to genetically modified organisms in food in the UK. Paper presented at the 71st EAAE Seminar 'The Food Consumer in the Early 21st Century', Zaragoza, Spain, 19–20 April.

Buzby, J.C., Skees, J.R. and Ready, R.C. (1995) Using contingent valuation to value food safety: a case study of grapefruit and pesticide residues. In: Caswell, J.A. (ed.) *Valuing Food Safety and Nutrition.* Westview Press, Boulder, Colorado, pp. 219–256.

Carson, R.T. and Mitchell, R.C. (1981) An experiment in determining willingness to pay for national water quality improvements. Office of Policy Analysis Draft report, US Environmental Protection Agency, Washington, DC.

Diamond, P.A., Hausman, J., Leonard, G.K. and Denning, M.A. (1993) Does contingent valuation measure preferences? Experimental evidence. In: Hausman, J.A. (ed.) *Contingent Valuation: A Critical Assessment*. North Holland, New York, pp. 41–89.

Gallop Market Survey Corp., Taiwan (2002) *Public Opinion Survey of Genetically Modified Foods*. Report publishe by Department of Health, The Executive Yuan, Republic of China (Taiwan), August 22.

Gaskell, G. (2000) Agricultural biotechnology and public attitudes in the European Union. *AgBioForum* 3, 87–96.

Haab, T. and McConnell, K.E. (2002) *Valuating Environmental and Natural Resources: the Econometrics of Non-Market Valuation*. Edward Elgar, Cheltenham, UK.

Halbrendt, C., Sterling, L., Snider, S. and Santoro, G. (1995) Contingent valuation of consumer willingness to purchase pork with lower saturated fat. In: Caswell, J.A. (ed.) *Valuing Food Safety and Nutrition*. Westview Press, Boulder, Colorado, pp. 313–339.

Hammitt, J.K. (1986) *Estimating Consumer Wilingness-to-Pay to Reduce Foodborne Risk*. Prepared by Rand Corporation for US Environmental Protection Agency, R-3447-EPA, Washington, DC.

Hoban, T.J. (1996) How Japanese consumers view biotechnology. *Food Technology* July, 85–88.

Hoban, T.J. (1998) Trends in consumer attitudes about agricultural biotechnology. *AgBioForum* 1, 3–7.

Huang, C. (1993) A simultaneous system approach for estimation of consumer risk perceptions, attitudes, and willingness to pay for residue-free produce. Paper presented at the American Agricultural Economics Association Meeting, Orlando, Florida.

International Food Information Council Foundation (IFIC) (2001) More U.S. consumers see potential benefits to food biotechnology.' Wirthlin Group Quorum Surveys, January 2001. Available at http://www.ific.org/proactive/newsroom/release.vtml?id=19241 (accessed March 2001).

Loader, R. and Henson, S. (1998) A view of GMOs from the UK. *AgBioForum* 1, 31–34.

Loehman, E. and VoHu, D. (1982) Application of stochastic choice modeling to policy analysis of public goods: a case study of air quality improvements. *Review of Economics and Statistics* 54, 474–480.

Macer, D. and Ng, M.A.C. (2000) Changing attitudes to biotechnology in Japan'. *Nature Biotechnology* 18, 945–947.

Nordic Industrial Fund (2000) *Negative attitude to gene-modified food*. Available at http://www.nordicinnovation.net.

Wang, Q., Halbrendt, C., Kolodinsky, J. and Schmidt, F. (1997) Willingness to pay for rBST-free milk: a two-limit tobit model analysis'. *Applied Economics Letters* 4, 619–621.

12 A Comparison of Consumer Attitudes towards GM Food in Italy and the USA

Marianne McGarry Wolf,[1] Paola Bertolini[2] and Jacob Parker-Garcia[1]
[1] *Agribusiness Department, California Polytechnic State University, San Luis Obispo, CA 93407, USA;* [2] *Facoltà di Economia, Modena, Italy*

Introduction

In 1973, science made a very significant advancement when a gene was transferred between plants for the first time. Just 14 years later, the first outdoor genetically engineered plants were grown. By 1995, crops of these plants were cultivated on commercial acreage (Lamb, 2000). Between 1995 and 1998, biotech companies introduced 16 new genetically modified (GM) crops into the USA for use by farmers. In the USA, GM foods are very much a part of American nutrition. According to the US Department of Agriculture (USDA), one-third of the maize and more than half the soybeans and cotton grown in the USA are in some way products of biotechnological applications. Total hectares of biotech crops grew worldwide from 39.9 million in 1999 to 44.2 million in 2000 (Stickman, 2001). The US share of biotech crops was 69% in 2000.

To date, the USA has managed to avoid food scares such as mad cow disease and the other food safety scares that have plagued Europe. Americans are much more confident about the safety of their food supply and trust government regulation more. This confidence has led American consumers to be more accepting of GM foods. A recent study by *The Packer* (2001a) found that American consumers felt it was appropriate to modify food items genetically to: be more resistant to plant disease and less reliant on pesticides, 70%; help prevent disease, 64%; improve nutritional value, 58%; improve flavour, 49%; and extend shelf life, 48%.

By contrast, in the European Union (EU) the consumer generally views GM foods as unhealthy. The politically active 'Green Movement' has done much to publicize and put the issue of genetically altered food on the European continent in a negative light. For example, it has derisively nicknamed genetically altered food 'Frankenfoods'. A survey cited by the EU found that most Europeans see GM food as health hazards, despite assurances from producers (Wielaard, 2001). In November 1999, the European Commission passed a law requiring all European retailers to label food containing more than 1% genetically modified ingredients. The Commission also required restaurants to inform consumers if meals contained GM ingredients. Wolf *et al.* (2001) found that the Irish consumer was more likely than the US consumer to indicate that mandatory labelling of GM food was very important in their first phase of research, October 1999 and January 2000. However, after the media coverage of the recall of 2.5 million boxes of Taco Bell brand taco shells produced by Kraft Foods that contained StarLink™ maize in the USA in September 2000 (Copple,

2000), Wolf *et al.* (2001) found that the proportion of consumers in the USA that indicated that mandatory labelling of GM food was very important grew to a level similar to that in Ireland.

Although both the US and Irish consumer indicated that mandatory labelling of GM food was very important, Wolf *et al.* (2001) also found that the Irish consumer is less likely to purchase food that has been genetically modified than the US consumer. Almost one-half of the Irish respondents indicated they were likely to purchase food that has been genetically modified, more than two-thirds of the US consumers indicated they were likely to purchase food that has been genetically modified. These findings are similar to the results reported by *The Times* in April 2001. A survey in the UK indicated that 48% of respondents would eat GM food (*The Times*, 2001). The results for the USA are similar to the findings of *The Packer's* survey of 1000 US consumers (*The Packer*, 2001a). The Packer's survey found that 60% of consumers were extremely, very, or somewhat likely to purchase a fresh produce item that has been genetically modified.

Attitudes in Italy appear to be much the same as those of other EU countries. In some cases, Italian law is stricter towards GM foods. For example, Italian law forbids the use of GM seeds in open fields. They are considered possible health hazards. In April 2001, a fire destroyed a Monsanto factory in northern Italy. Authorities know it was an arson attack but are not sure who set the fire. There has been speculation that places responsibility on the anti-GM, Green Party activists. Before the fire, protesters had gathered at the Monsanto factory protesting about the production of genetically modified seeds. This incident caused Italian Farm Minister Alfonso Pecoraro Scanio, a member of the Green Party, to seize some 400 tons of suspect Monsanto soybean and maize seeds. Tests revealed the seeds were genetically modified and that they had been imported from the USA and were not Italian (*Deutsche Presse-Agentur*, 2001). The objective of this research is to compare consumer attitudes towards GM food in the USA and Italy.

Methodology

This research uses a survey instrument that was administered through the use of a personal interview during the autumn of 2001. The random sample of 232 food shoppers for the USA was collected in San Luis Obispo County, California. San Luis Obispo County was designated the best test market in the USA by *Demographics Daily* (Thomas, 2001). San Luis Obispo was found to be the best of 3141 counties to represent a microcosm of the USA based on 33 statistical indicators. The random sample of 252 food shoppers for Italy was collected in Modena, Italy, during the autumn of 2001. Modena, in Emilia Romagna, is a rich industrial area that represents one of the most important areas of food production in Italy for both industrial food and for typical traditional quality food such as parmesan cheese, Modena ham, Parma ham and Modena vinegar. In addition, Modena is important to the food distribution system of Italy since the largest distribution group, Coop, resides in Modena. These characteristics make Modena an important area to represent consumers' attitudes towards food in Italy, especially in northern Italy.

This research examines differences in the following between the US and Italian respondents: attitudes towards science, food purchasing behaviour, organic food purchasing behaviour, knowledge of organic foods, familiarity with GM food, and consumer attitudes towards GM food and labelling. Consumer attitudes toward purchasing GM food are examined based on the purpose for the use of biotechnology: to help plants withstand weedkillers, to improve nutrition, to kill pests and allow farmers to use less pesticide, and to improve taste. Attitudes towards government agencies and food safety and attitudes towards food producers and food safety and the environment are examined in relation to consumers' attitudes toward GM food.

Attitudes Toward Science and Technology Use

Both the US and Italian consumers agree that scientific research is an important factor in the quality of life (Table 12.1). However, most US

Table 12.1. Scientific research is an important factor in the improvement of the quality of life.

	USA (n = 232)	Italy (n = 252)	Chi squared
Strongly agree	45.7%	50.0%	
Agree	46.6%	44.4%	
Disagree	4.7%	4.4%	
Strongly disagree	3.0%	1.2%	2.58

consumers are using the Internet and e-mail at home and less than half of the Italian consumers are using these technologies at home (Table 12.2).

Food Purchasing Behaviour

Approximately two-thirds of consumers in Italy and the USA have purchased organic food within the past year (Table 12.3). However, a higher proportion of consumers in the USA have purchased organic milk, fresh fruits, fresh vegetables and wine.

In research concerning organic lettuce, Wolf *et al.* (2002) have shown that there appears to be confusion in consumers' understanding of the properties of organic food in the USA. For example, in the examination of organic lettuce, it was found that consumers value the organic characteristics of lettuce such as environmentally friendly as somewhat to very desirable, while they rate organically grown and certified as only slightly to somewhat desirable. Thus, Wolf hypothesized that consumers do not understand the properties of organic foods (Wolf *et al.*, 2002). This research has attempted to address the possible

Table 12.2. Technology usage at home.

	USA (n = 232)	Italy (n = 252)	Chi squared
Internet	81.6%	40.2%	85.72**
E-mail	76.8%	31.9%	97.11**
None	22.2%	57.5%	69.77**

**Significant difference at 0.05 level.

Table 12.3. Organic food consumption within past year.

Phase 1	USA (n = 196)	Italy (n = 252)	Chi squared
Any organic food	67.6%	64.6%	0.475
Meats	20.7%	15.4%	2.35
Milk	31.2%	16.1%	15.29**
Other dairy products (excluding milk)	24.2%	21.3%	0.614
Fresh fruits	57.8%	44.9%	8.10**
Fresh vegetables	57.8%	44.9%	8.10**
Wine	9.1%	4.7%	3.64*
Bakery items including bread	20.3%	20.5%	0.003
Other	12.6%	13.4%	0.064

*Significant difference at 0.10 for all tests; **significant difference at 0.05 level.

misconceptions of consumers by examining their responses to the question: 'How strongly do you agree or disagree that all produce sold at a farmers' market is organic?' The farmers' markets in the research region in the USA were observed to sell primarily conventionally grown produce. Therefore, respondents that either agree or strongly agree are consumers that are likely confused about the attributes of organic produce. Over a quarter of consumers in the USA agree that all produce sold at farmers' markets is organic (Table 12.4). Only 12.4% of consumers in Italy agree that all produce sold at farmers' markets is organic. Therefore, it appears that the Italian consumer has a better understanding of organic food than the consumer in the USA.

A higher proportion of consumers in Italy plan to increase their purchase of organic food in the next year. The majority of consumers in the USA, 54.7%, expect the quantity of organic food purchased to stay the same (Table 12.5).

In the USA, over three-quarters of food purchasers read the label for nutritional information very or somewhat often when making a purchase decision (Table 12.6). However, slightly more than half of Italian purchasers read the label for nutritional information very or somewhat often when making a purchase decision. Approximately two-thirds of respondents in both countries read the label for ingredient information very or somewhat often when making a purchase decision (Table 12.7).

Table 12.4. All produce sold at a farmers' market is organic.

	USA (n = 232)	Italy (n = 252)	Chi squared
Strongly agree	2.6%	2.0%	
Agree	24.5%	10.4%	
Disagree	52.0%	80.0%	
Strongly disagree	21.0%	7.6%	43.35**

**Significant difference at 0.05 level.

Table 12.5. In the next year, the quantity of purchases of organic food products.

	USA (n = 232)	Italy (n = 252)	Chi squared
Stay the same	54.7%	33.5%	
Increase	26.7%	44.5%	
Decrease	4.3%	2.8%	
Will not purchase organic food products	14.2%	19.3%	25.89**

**Significant difference at 0.05 level.

Table 12.6. Nutritional label readership and purchase decision.

	USA (n = 232)	Italy (n = 252)	Chi squared
Very often	45.3%	25.6%	
Somewhat often	33.2%	29.5%	
Not very often	15.9%	28.3%	
Not at all	5.6%	16.5%	35.04**

**Significant difference at 0.05 level.

Table 12.7. Ingredient label readership and purchase decision.

	USA (n = 232)	Italy (n = 252)	Chi squared
Very often	34.5%	32.3%	
Somewhat often	29.7%	36.6%	
Not very often	27.5%	21.3%	
Not at all	8.3%	9.8%	4.166

Familiarity with GM Foods

Approximately one-half of consumers in the USA are familiar with GM food (Table 12.8). However, less than a third of Italian consumers indicated that they are familiar with GM food. These awareness levels for Italy are significantly lower than those found by Wolf *et al.* (2001) in their examination of familiarity with GM food in the USA and Ireland. Approximately half of the respondents in the USA and 40% of respondents in Ireland were at least somewhat familiar with GM food.

Television is the number one source of information about GM food (Table 12.9). More than half of Americans and Italians get information about GM food from newspapers and television. Consumers in the USA are more likely to get information about GM food from radio news; news magazines; discussions with family, friends or colleagues; the Internet; and by working in farming or food processing.

Television news reporters are considered the most appropriate source of information about GM food in the USA and Italy (Table 12.10). However, there are very different

Table 12.8. Familiarity with genetically modified food.

	USA (n = 232)	Italy (n = 252)	Chi squared
Very familiar	8.3%	2.4%	
Somewhat familiar	41.3%	25.2%	
Not very familiar	34.8%	55.5%	
Not at all familiar	15.7%	16.9%	29.14**

**Significant difference at 0.05 level

Table 12.9. Sources of genetically modified food awareness.

	USA (n = 232)	Italy (n = 252)	Chi squared
Television news	62.8%	78.3%	14.23**
Newspaper	54.3%	51.6%	0.373
Radio news	21.6%	9.4%	13.92**
News magazines	31.9%	13.0%	25.24**
Consumer Reports magazine	12.2%	17.0%	2.23
Discussion with family, friends or colleagues	38.4%	16.9%	28.15**
Internet	14.7%	3.9%	16.92**
Employment, work in farming or food processing	7.8%	2.8%	6.22**
Other	3.0%	2.0%	0.565

**Significant difference at 0.05 level.

Table 12.10. Appropriate sources of information concerning genetically modified food.

	USA (n = 232)	Italy (n = 252)	Chi squared
Television news reporters	62.80%	55.10%	2.92*
Newspaper reporters	60.20%	29.90%	44.86**
Local government agencies	44.20%	22.00%	26.92**
Farmers	41.40%	6.30%	84.14**
Discussions with family, friends, or colleagues	35.90%	5.90%	67.64**
Education seminar	34.50%	0%	104.85**
University professors	29.40%	40.60%	6.55**
Radio news reporters	29.30%	5.90%	46.90**
Internet websites	25.00%	4.30%	42.52**
Science teachers	24.10%	0%	69.30**
Representatives from food processors	22.40%	4.70%	33.18**
Visits to food production facilities	17.70%	0%	49.25**
Local politicians	15.50%	0%	42.57**
Reports from a seed producer	12.50%	0%	33.77**
E-mail	8.70%	0%	23.14**
Other	4.7%	0.4%	9.52**

*Significant difference at 0.10 level; ** significant difference at 0.05 level.

opinions between the consumers in the USA and Italy concerning the appropriateness of other sources of information. The second most important source of information for the consumers in Italy is the university professor, the seventh most important source of information for those in the USA. Less than a third of the consumers in Italy indicated that newspaper reporters are appropriate sources of information, while 60% of consumers in the USA indicated that newspaper reporters are appropriate sources of information. Therefore, an educational campaign concerning GM food will need to include television news in both countries. However, other sources of information will differ between the two countries.

Attitudes Towards Government Agencies and Food Producers

Respondents were asked how strongly they agree or disagree with the following statements: 'government agencies in my country have done a very good job at ensuring food safety in the past'; 'I trust government agencies in my country to ensure food safety in the future'; 'global food producers have done a very good job at ensuring food safety in the past'; 'global food producers are producing food using environmentally safe methods'; and 'computer technology is an important factor in the improvement of the quality of life'. The following rating scale was used to evaluate these statements: strongly agree = 4; agree = 3; disagree = 2; strongly disagree = 1 (Table 12.11).

Respondents in the USA are more likely than the Italian respondents to agree that: 'government agencies in my country have done a very good job at ensuring food safety in the past'; 'I trust government agencies in my country to ensure food safety in the future'; 'global food producers have done a very good job at ensuring food safety in the past' and 'global food producers are producing food using environmentally safe methods'.

Attitudes Towards Genetically Modified Food

Most consumers in both countries indicated that government imposition of mandatory labelling is important, 98% in Italy and 89.6% in the USA (Table 12.12). This shows that most Italians agree with the current law of mandatory labelling GM foods and Americans want the government to label if the foods are

Table 12.11. Mean ratings.

Statement	USA (n = 232)	Italy (n = 252)	t-Test
Government agencies in my country have done a very good job at ensuring food safety in the past	2.99	2.12	14.41**
I trust government agencies in my country to ensure food safety in the future	2.94	2.57	6.02**
Global food producers have done a very good job at ensuring food safety in the past	2.44	2.04	6.5**
Global food producers are producing food using environmentally safe methods	2.29	2.10	3.06**

**Significant difference at 0.05 level.

Table 12.12. Government imposition of mandatory labelling for genetically modified food.

	USA (n = 232)	Italy (n = 252)	Chi squared
Very important	61.3%	84.35%	
Somewhat important	28.3%	13.4%	
Not very important	10.0%	2.0%	
Not at all important	0.4%	0.4%	35.19**

**Significant difference at 0.05 level.

genetically modified. More Americans felt mandatory labelling was not very important, 10%, compared to Italians, 2%. A comparison of the results observed by Wolf *et al.* (2001) in Ireland and the USA indicates that the Italian consumer rates the importance of labelling higher than both the Irish and US consumers.

Most Americans, 71.6%, indicate that they would at least *maybe* buy GM foods (Table 12.13). Only 43.1% of Italians will possibly buy GM foods. More than half of Italians would not purchase GM foods. The results observed by Wolf *et al.* (2001) indicate that the consumers in Ireland are slightly more likely to purchase GM food than the Italian consumers, half indicate that they would at least *maybe* buy GM foods. The consumers in the USA observed in the research generated by Wolf *et al.* (2001) and in this research are more likely to consume GM food than their European counterparts.

Attitudes Towards Different Attributes of Genetically Modified Food

Respondents were asked how likely they were to purchase GM food to improve nutrition, kill pests allowing farmers to use less pesticides, improve taste and help plants withstand weedkillers. The rating scale used

Table 12.13. Likelihood to buy genetically modified food.

	USA (n = 232)	Italy (n = 252)	Chi squared
Definitely, probably, maybe	71.6%	43.1%	
Probably not, definitely not	28.4%	56.9%	39.95**

**Significant difference at 0.05 level.

to evaluate purchase interest is: 5 = definitely; 4 = probably; 3 = maybe; 2 = probably not; and 1 = definitely not. The US respondents were more likely to purchase GM food products overall and for the purposes of improving nutrition, helping plants withstand weedkillers and improving taste (Table 12.14). The US and Italian consumers evaluated the importance of using biotechnology to kill pests and allow farmers to use less pesticides similarly. A similar proportion of consumers in both countries purchased organic food in the past year.

A comparison of the importance of specific uses of biotechnology is given in Table 12.15 for each country. The attributes are listed from high to low based on their means and paired to examine differences between the means for each country. The only attribute that achieves a *maybe will purchase* among the Italian consumers is that to kill pests and allow farmers to use less pesticides. Over 40% of the Italian consumers expect to increase their purchases of organic foods in the next year. The consumers in Italy are less willing to purchase a GM food product that helps plants withstand weedkiller, improves nutrition, or improves taste. However, the consumer in the USA, that is more likely to read a nutrition label, rates improved nutrition the highest attribute for a GM food product. The second most important attribute is to kill pests and allow farmers to use less pesticides for the consumer in the USA.

Attitudes Towards Government Agencies and Producers, and Willingness to Purchase GM Foods

The responses to the questions concerning trust in government and global food producers to ensure food safety were compared with the

Table 12.14. Likelihood to buy – attribute mean rating.

Statement	USA (*n* = 232)	Italy (*n* = 254)	*t*-Test
How likely are you to purchase a food product that has been genetically modified?	2.98	2.26	7.50**
To kill pests and allow farmers to use less pesticides?	3.26	3.20	0.53
To help plants withstand weedkillers?	3.00	2.71	2.88**
To improve nutrition?	3.43	2.70	6.96**
To improve taste?	3.16	1.83	13.23**

**Significant difference at 0.05 level.

Table 12.15. Likelihood to buy – attribute mean rating.

	Mean rating	Paired *t*
Italy (*n* = 254)		
To kill pests and allow farmers to use less pesticides?	3.20	
To help plants withstand weedkillers?	2.71	7.38**
To improve nutrition?	2.70	0.012
To improve taste?	1.83	12.72**
USA (*n* = 232)		
To improve nutrition?	3.43	
To kill pests and allow farmers to use less pesticides?	3.26	3.63**
To improve taste?	3.16	1.66*
To help plants withstand weedkillers?	3.00	2.74**

*Significant difference at 0.10 level; **significant difference at 0.05 level.

willingness of a respondent to purchase GM food. The results show that for the US respondents, those who believe the government and global food producers ensure food safety are more willing to purchase GM food (Tables 12.16 and 12.17). The Italian respondents who believe the global food producers ensure food safety are more willing to purchase GM food. However, for the Italian respondents there is no relationship between the belief that government agencies ensured food safety in the past and willingness to purchase a GM food product. Most Italian respondents evaluated the government and global food producers lower than the US respondents. Both the Italian and US consumers are more likely to purchase a GM food product when they agree that global food producers are using environmentally safe methods (Table 12.18).

Table 12.16. Mean willingness to purchase a genetically modified food product: government agencies in my country have done a very good job at ensuring food safety in the past.

	Mean	*t* statistic
Italy ($n = 252$)		
Strongly agree or agree	2.46	1.62
Strongly disagree or disagree	2.19	
USA ($n = 232$)		
Strongly agree or agree	3.08	2.83**
Strongly disagree or disagree	2.57	

**Significant difference at 0.05 level.

Table 12.17. Mean willingness to purchase a genetically modified food product: Global food producers ensure food safety.

	Mean	*t* statistic
Italy ($n = 252$)		
Strongly agree or agree	2.70	2.89**
Strongly disagree or disagree	2.16	
USA ($n = 232$)		
Strongly agree or agree	3.31	4.90**
Strongly disagree or disagree	2.69	

**Significant difference at 0.05 level.

Table 12.18. Mean willingness to purchase a genetically modified food product: Global food producers are producing food using environmentally safe methods.

	Mean	*t* statistic
Italy ($n = 252$)		
Strongly agree or agree	2.59	2.75**
Strongly disagree or disagree	2.14	
USA ($n = 232$)		
Strongly agree or agree	3.34	4.60**
Strongly disagree or disagree	2.77	

**Significant difference at 0.05 level.

Conclusions

The objective of this research is to use a case study to compare consumer attitudes toward GM food in the USA and Europe using two communities. The results of this research indicate that half of the consumers in the USA are familiar with GM food, while only 28% of the consumers in Italy are familiar with GM food. This level of familiarity in Italy is lower than that observed in Ireland by Wolf *et al.* in 2001. Both the consumers in the USA and Italy agree that the most appropriate source for information about GM food is television news reporters. The consumers in the USA indicate that there are many appropriate sources of information for GM food. However, the Italian consumers indicate that there are few appropriate sources and rate university professors high relative to the US consumers.

The Italian consumer is less likely to read nutrition labels, equally likely to read ingredient labels and believes it is more important than the US consumer to label GM food. Most Americans, 71.6%, indicate that they would at least *maybe* buy GM foods. Only 43.1% of Italians will possibly buy GM food. The results observed by Wolf *et al.* (2001) indicate that the consumers in Ireland are slightly more likely to purchase GM food than the Italian consumers; half indicate that they would at least *maybe* buy GM foods. The consumers in the USA observed in the research generated by Wolf *et al.* (2001) and in this research are more likely to consume GM food than their European counterparts.

Respondents are asked how likely they are to purchase GM food to improve nutrition, kill pests allowing farmers to use less pesticides, improve taste and help plants withstand weedkillers. The US respondents are more likely to purchase GM food products overall and for the purposes of improving nutrition, helping plants withstand weedkillers and improving taste. The US and Italian consumers evaluate the importance of using biotechnology to kill pests and allow farmers to use less pesticides similarly. There are also similar proportions of consumers that purchased organic food in the past year in both countries.

The only attribute that achieves a *maybe will purchase* among the Italian consumers is the attribute, to kill pests and allow farmers to use less pesticides. Approximately 40% of the consumers in Italy expect to increase their organic purchases in the next year. The consumers in Italy are less willing to purchase a GM food product that helps plants withstand weedkillers, improves nutrition or improves taste. However, the consumer in the USA, who is more likely to read a nutrition label, rates improved nutrition the highest attribute for a GM food product. The second most important attribute for the consumer in the USA is to kill pests and allow farmers to use less pesticides.

The consumers responded to questions concerning trust in government and global food producers to ensure food safety. These questions are compared with the willingness of a respondent to purchase GM food. The results show that for the US respondents, those who believe the government and global food producers ensure food safety are more willing to purchase GM food. The Italian respondents who believe the global food producers ensure food safety are more willing to purchase GM food. However, for the Italian respondents there is no relationship between the belief that government agencies ensured food safety in the past and willingness to purchase a GM food product. Both the Italian and US consumers are more likely to purchase a GM food product when they agree that global food producers are using environmentally safe methods.

This research shows that the consumers in the USA and Italy have different attitudes toward GM food. An educational programme concerning GM foods must be different for the consumer in the USA and Italy. The consumers in both countries have different opinions concerning the appropriate sources of information for GM food. Further, they have different opinions concerning the attributes of GM food. The consumers in the USA indicate that improved nutrition is the most important attribute. The consumers in Italy indicate that reduced pesticides is the most important attribute and are more likely to increase their purchases of organic foods than the consumer in the USA.

References

Bates, B. (2001) Will Frankenfood feed the world? *Time Magazine* 19 June.
Berger, M. (2000) Public health and agricultural biotechnology: a review of the legal, ethical, and scientific controversies presented by genetically altered foods. Published dissertation, Emory University.
Bocker, A. and Hanf, C.-H. (2002) Is European consumers' refusal of GM food a serious obstacle or a transient fashion? In: Santaniello, V., Evenson, R.E. and Zilberman, D. (eds) *Market Development for Genetically Modified Food.* CAB International, Wallingford, UK.
Copple, B. (2000) Scientist, activist, yogi? *Forbes* 30 October, pp. 54–56.
Cray, C. (2000) Biosafety truce reached. *Multinational Monitor* 21, pp. 6–8.
Deutsche Presse-Agentur (2001) Italian Monsanto seed test positive to GM. *Deutsche Presse-Agentur*, 5 April.
Foltz, T. (2002) Organic enter mainstream in San Francisco region. *The Packer*, 28 January, p. B8.
Hayden, T. (2002) Bad seeds in court. *U.S. News and World Report*, 4 February.
Lamb, M. (2000) Brave new food. *Mother Earth News*, pp. 54–62 179.
Moore, S.K. and Scott, A. (1999) Waging a war for public approval. *Chemical Week*, 15 December, pp. 23–39.
Spanier, O. and Tamura (1993) *Food Flavor and Safety*. American Chemical Society, San Francisco.
Stickman, A. (2001) New markets for biotech. *Technology Review*, July/August.
Thayer, A.M. (2002) Ag-biotech industry is gambling on an information campaign, continued farmer acceptance, and promises for the future. *Chemical and Engineering News*, 2 October, pp. 21–29.
The Packer (2001a) Fresh Trends 2001 profile of the fresh produce consumer.
The Packer (2001b) EU likely to retain biotech food ban. 31 December, p. B3.
The Times (2001) GM tide turns. (London), 3 April.
Thomas, G.S. (2001) Playing in San Luis Obispo. *Demographics Daily*, 6 February.
Wielaard, R. (2001) Wary of industry law suits, EU to move this month on biotech. *BC News*, 8 March.
Wolf, M.M., McDonnell, L. and Domegan, C. (2001) A comparison of consumer attitudes toward genetically modified food in Ireland and the USA. Paper presented at the International Conference on Agricultural Biotechnology Research, Ravello, Italy.
Wolf, M.M., Johnston, B., Cochran, K. and Hamilton, L. (2002) Consumer attitudes toward organically grown lettuce. *Journal of Food Distribution Research* Volume 32 (1).

13 Consumer Attitudes Towards GM Food in Ireland and the USA

Marianne McGarry Wolf,[1] Juliana McDonnell,[2] Christine Domegan[2] and Heidi Yount[1]

[1]*Agribusiness Department, California Polytechnic State University, San Luis Obispo, CA 93407, USA;* [2]*National University of Ireland, Galway, Ireland*

Introduction

Biotechnology is being used to produce genetically modified organisms (GMOs) that are used in food production. Food producers have used the new biotechnologies. Negative consumer response to these products in Europe, Japan and Australia has caused farmers to question whether or not to adopt the new technologies. There is concern that American consumer attitudes may follow those of the Europeans (*The Packer*, 2001). The European Union, Japan and Korea have labelling laws (Olson, 2000). Senator Barbara Boxer has introduced a GMO food-labelling law for the USA (Gregerson, 2000).

The objective of this research is to use a case study to compare consumer attitudes towards genetically modified food in the USA and Europe using two communities and two time periods. The results of the first phase were presented in Ravello, Italy, during August 2000 at the 4th International Conference on the 'Economics of Agricultural Biotechnology'.

Research Methodology

The research uses a survey instrument that was administered through the use of a personal interview. A simulated before and after experimental design was used to conduct this research to eliminate the impact of pre-measurement error. The first phase of this research examines 882 randomly selected food purchasers interviewed in October 1999 and January 2000. The second phase of research commenced in October 2000 with a random sample of 324 respondents. All phases of research were conducted in San Luis Obispo County, California, and Galway, Ireland. San Luis Obispo County was designated the best test market in the USA by *Demographics Daily* (Thomas, 2001). San Luis Obispo was found to be the best of 3141 counties to represent a microcosm of the USA based on 33 statistical indicators.

Attitudes towards Science and Food Purchasing Behavior

During both phases of the research US respondents appear to have more positive attitudes toward science (Table 13.1). During the first phase US consumers were more likely to have purchased organic food in the past year than Irish respondents (Tables 13.2). However, during the second phase organic purchasing increased and is similar in both countries.

 Consumer Acceptance of Genetically Modified Food (eds R.E. Evenson and V. Santaniello)

Table 13.1. Scientific research is an important factor in the improvement of the quality of life.

	Ireland	USA	Chi squared
Phase 1	$n = 197$	$n = 682$	
Strongly agree	39.6%	48.5%	
Agree	54.8%	43.3%	
Disagree	4.6%	6.0%	
Strongly disagree	1.0%	2.2%	8.754**
Phase 2	$n = 100$	$n = 224$	
Strongly agree	32.00%	45.50%	
Agree	60.00%	45.50%	
Disagree	4.00%	7.10%	
Strongly disagree	4.00%	1.80%	8.435**

**Significant difference at 0.05 level.

Table 13.2. Organic food consumption within past year.

	Ireland	USA	Chi squared
Phase 1	$n = 196$	$n = 678$	
Yes	52.6%	62.9%	
No	47.4%	30.8%	18.579**
Phase 2	$n = 100$	$n = 224$	
Yes	70.10%	72.30%	
No	29.90%	27.70%	6.945

Food labelling appears to be more important to US respondents than Irish respondents when purchasing food since nutritional labels and ingredient information are read more often by US respondents during both phases of the research (Tables 13.3 and 13.4).

Familiarity with GM Foods

The results of the first phase indicated that there was a similar level of familiarity with GM food in Ireland and the USA (Table 13.5). Approximately 43% of respondents in both countries indicated that they were familiar with

Table 13.3. Nutritional label readership and purchase decision.

	Ireland	USA	Chi squared
Phase 1	$n = 197$	$n = 683$	
Very often	27.9%	47.7%	
Somewhat often	28.4%	30.9%	
Not very often	28.9%	14.8%	
Not at all	14.7%	6.9%	41.755**
Phase 2	$n = 97$	$n = 224$	
Very often	18.00%	46.90%	
Somewhat often	27.00%	33.90%	
Not very often	37.00%	14.30%	
Not al all	18.00%	4.90%	46.211**

**Significant difference at 0.05 level.

Table 13.4. Ingredient label readership and purchase decision.

	Ireland	USA	Chi squared
Phase 1	*n* = 195	*n* = 681	
Very often	22.6%	38.3%	
Somewhat often	31.8%	30.8%	
Not very often	30.3%	21.3%	
Not at all	15.4%	9.5%	20.860**
Phase 2	*n* = 99	*n* = 224	
Very often	16.00%	38.80%	
Somewhat often	27.00%	33.00%	
Not very often	42.00%	21.00%	
Not at all	14.00%	7.10%	29.021**

**Significant difference at 0.05 level.

Table 13.5. Familiarity with genetically modified food.

	Ireland	USA	Chi squared
Phase 1	*n* = 196	*n* = 681	
Very familiar	5.1%	7.0%	
Somewhat familiar	38.3%	36.6%	
Not very familiar	39.3%	39.4%	
Not at all familiar	16.8%	17.0%	1.076
Phase 2	*n* = 98	*n* = 224	
Very familiar	6.10%	12.90%	
Somewhat familiar	34.30%	37.50%	
Not very familiar	45.50%	33.00%	
Not at all familiar	13.10%	16.50%	8.837*

*Significant at 0.10 level

GM food. This level of familiarity is similar to that observed by the Wirthlin Group Quorum Surveys conducted for the International Food Council in the USA in May 2000. The Wirthlin Group found that 43% of adults were aware that there were foods produced through biotechnology in the supermarket now. An increase in familiarity was observed in the USA during the second phase of this research, with half of the respondents indicating familiarity with GM food. However, the level of familiarity in Ireland remained at a level similar to that observed during the previous phase.

It is possible that the consumers in the USA became more familiar during the second phase due to the media coverage of the recall of 2.5 million boxes of Taco Bell brand taco shells produced by Kraft Foods that contained StarLink™ maize in September 2000 (Copple, 2000). The recall of the taco shells was followed by a recall of approximately 300 varieties of tacos, tortillas, tostadas and chips made by Mission Foods on 13 October 2000 (The Associated Press and Reuters, 2000).

An examination of the sources of information about GM food among respondents who indicated they were at least somewhat familiar shows that the Irish respondents have learned about GM food from a wide variety of sources (Table 13.6). Almost all of the familiar Irish respondents have heard about GM food from the newspaper or television news in both phases; while only two-thirds of familiar US respondents had heard about GM food from the newspaper or television news. Slightly over one-quarter of familiar US respondents had heard about GM food from the radio in the first phase. Radio increased as a source of information in the USA to over a third of respondents. Over four-fifths of the familiar Irish respondents had heard about GM food from radio news during the first phase and two-thirds during the

Table 13.6. Sources of genetically modified food awareness among very or somewhat familiar.

	Phase 1			Phase 2		
	Ireland (n = 86)	USA (n = 297)	Chi squared	Ireland (n = 40)	USA (n = 113)	Chi squared
Newspaper	97.6%	68.3%	28.93**	90.0%	70.8%	5.94**
Television news	97.6%	64.2%	34.88**	95.0%	61.9%	15.55**
Radio news	86.3%	27.9%	90.82**	62.5%	36.3%	8.28**
Discussion with family, friends, etc.	71.8%	36.7%	30.43**	37.5%	38.9%	0.026
News magazines	52.2%	40.4%	3.63**	35.0%	45.1%	1.24
Employment, work in farming or food processing	27.6%	16%	4.35**	15.0%	15.9%	0.02
Consumer Reports magazine	25.4%	20.5%	10.40**	15.0%	20.4%	0.03
Internet	22.8%	15.1%	2.04	20.0%	26.5%	0.68
Other	0%	11.3%	6.01**	5.0%	15.0%	2.74*

*Significant difference at 0.10 level; **significant difference at 0.05 level.

second phase. It is clear that GM food was an important issue to the familiar Irish respondent during the first phase because almost three-quarters indicated that they engaged in discussions with family, friends, and colleagues. However, only one-third of familiar US respondents indicated they have engaged in discussions with family, friends and colleagues about GM food. During the second phase fewer Irish respondents indicated discussions. A similar proportion of familiar US and Irish respondents indicated that they became aware of GM food from discussions with family, friends and colleagues. News magazines became a more important source of information for US respondents and a less important source of information for Irish respondents. Leading magazines in the US such as *Time* discussed the use of biotechnology in food production. *Time*'s article, 'Grains of hope' was a high-profile front-page article that discussed the biotechnology debate.

The second phase asked respondents to indicate appropriate sources of information concerning GM food. The Irish indicated more sources were appropriate to provide information concerning GM food (Table 13.7). The top two sources that respondents in both countries thought appropriate were TV news and newspaper reporters. Less than half of the respondents from both countries indicated that experts such as science teachers, representatives from food processors or university professors were appropriate sources of information concerning GM food.

Attitudes Towards Government Agencies, Producers and Technology

Respondents were asked how strongly they agree or disagree with the following statements: 'government agencies in my country have done a very good job at ensuring food safety in the past'; 'I trust government agencies in my country to ensure food safety in the future'; 'global food producers have done a very good job at ensuring food safety in the past'; 'global food producers are producing food using environmentally safe methods'; and 'computer technology is an important factor in the improvement of the quality of life'. The following rating scale was used to evaluate these statements: strongly agree = 4; agree = 3; disagree = 2; strongly disagree = 1 (Table 13.8).

Respondents in the USA are more likely than the Irish respondents to agree that: 'government agencies in my country have done a very good job at ensuring food safety in the past'; they trust 'government agencies in my country to ensure food safety in the future'; and 'global food producers have done a very good job at ensuring food safety in the past'. However, respondents in both countries are more likely to disagree that 'global food producers are producing food using environmentally safe methods'. Computer technology is perceived to improve the quality of life in both the USA and Ireland.

Table 13.7. Appropriate sources of information concerning genetically modified food.

	Ireland (n = 93)	USA (n =223)	Chi squared
TV news reporters	88.20%	55.60%	30.673**
Newspaper reporters	76.30%	53.80%	13.936**
Radio news reporters	69.90%	36.30%	25.040**
Local government agencies	55.60%	36.30%	8.322**
Educational seminars	55.20%	34.40%	8.404**
Internet websites	48.30%	30.50%	6.680**
Science teachers	42.60%	28.60%	3.978**
Representatives from food processors	39.70%	25.00%	4.907**
University professors	38.00%	34.50%	0.216
Discussion with friends, family, etc.	37.10%	30.80%	0.881
Farmers	33.30%	41.10%	1.04
Local politicians	24.60%	12.50%	5.2
Visits to food production facilities	24.40%	17.40%	1.225
E-mail	20.80%	13.80%	1.51
Reports from a seed producer	15.90%	16.10%	0.001
Other	7.50%	9.90%	0.221

**Significant difference at 0.05 level.

Table 13.8. Mean ratings.

Statement	Ireland (n = 100)	USA (n = 223)	t-Test
Government agencies in my country have done a very good job at ensuring food safety in the past	2.37	2.77	4.13**
I trust government agencies in my country to ensure food safety in the future	2.42	2.68	2.69**
Global food producers have done a very good job at ensuring food safety in the past	2.15	2.38	2.60**
Global food producers are producing food using environmentally safe methods	2.13	2.27	1.62
Computer technology is an important factor in the improvement of the quality of life	3.02	3.04	0.17

**Significant difference at 0.05 level.

Attitudes Towards GM Food

Most consumers in both countries indicated that government imposition of mandatory labelling is important, 95% in Ireland and 81% in the USA during the first phase (Table 13.9). Although the Irish respondents were less likely than the US respondents to read nutritional labels or ingredient labels when making a food purchase decision, they were more likely than the US respondents to indicate that government imposition of mandatory labelling of GM food is important during the first phase. Almost three-quarters of the Irish respondents indicated that mandatory labelling of GM food was very important, while approximately one-half of US respondents indicated that mandatory labelling of GM food was very important. During the second phase US consumers felt mandatory labelling was as important as the Irish consumers.

The Irish consumer is less likely to purchase food that has been genetically modified than the US consumer (Table 13.10). Almost one-half of the Irish respondents indicated they were likely to purchase food that has been genetically modified, more than two-thirds of the US consumers

Table 13.9. Government imposition of mandatory labelling for genetically modified food.

	Ireland	USA	Chi squared
Phase 1	$n = 197$	$n = 681$	
Very important	70.6%	52.3%	
Somewhat important	25.4%	28.9%	
Not very important	3.6%	12.2%	
Not at all important	0.5%	6.6%	31.709**
Phase 2	$n = 100$	$n = 223$	
Very important	68.0%	65.0%	
Somewhat important	27.0%	23.3%	
Not very important	4.0%	6.7%	
Not at all important	1.0%	4.9%	4.222

**Significant difference at 0.05 level.

Table 13.10. Likelihood to buy genetically modified food.

	Ireland	USA	Chi squared
Phase 1	$n = 197$	$n = 679$	
Definitely, probably, maybe	47.7%	72.8%	
Probably not, definitely not	27.0%	23.3%	43.38**
Phase 2	$n = 100$	$n = 223$	
Definitely, probably, maybe	49.5%	62.5%	
Probably not, definitely not	50.5%	37.5%	4.78**

**Significant difference at 0.05 level.

indicated they were likely to purchase food that has been genetically modified. The proportion of Irish consumers that indicated a positive likelihood to purchase remained stable during the two periods, however, the proportion of US respondents that were positive declined between the two phases. The findings of this research are similar to the results reported by *The Times* in April 2001. A survey in the UK indicated that 48% of respondents would eat GM food (*The Times*, 2001). The results for the US are similar to the findings of *The Packer*'s survey of 1000 US consumers (*The Packer*, 2001). *The Packer*'s survey found that 60% of consumers were extremely, very or somewhat likely to purchase a fresh produce item that has been genetically modified.

Attitudes Towards Different Attributes of GM Food

Respondents during the second phase were asked how likely they were to purchase GM food to improve nutrition, kill pests allowing farmers to use less pesticides, improve taste, and help plants withstand weedkillers. The US respondents were more likely to purchase GM food products overall and for the purposes of improving nutrition and improving taste (Table 13.11), while the US and Irish consumers evaluated the importance of using biotechnology to kill pests and allow farmers to use less pesticides and help plants withstand weedkillers similarly. The rating scale used to evaluate purchase interest is: 5 = definitely; 4 = probably; 3 = maybe; 2 = probably not; and 1 = definitely not.

A comparison of the importance of specific uses of biotechnology is given in Table 13.12 for each country. The attributes are listed from high to low based on their means and paired to examine differences between the means for each country. For the Irish respondents using biotechnology to kill pests and allow farmers to use less pesticides and improve nutrition are the top reasons. Nutrition is the most important reason to use biotechnology followed by

Table 13.11. Likelihood to buy – attribute mean rating.

	Ireland ($n = 100$)	USA ($n = 223$)	*t*-Test
How likely are you to purchase a food product that has been genetically modified?	2.53	2.80	2.10**
To improve nutrition?	3.03	3.30	1.82*
To kill pests and allow farmers to use less pesticides?	3.05	3.10	0.33
To improve taste?	2.56	2.91	2.50**
To help plants withstand weedkillers?	2.64	2.76	0.94

*Significant difference at 0.10 level; **significant difference at 0.05 level.

Table 13.12. Likelihood to buy – attribute mean rating.

	Mean rating	Paired *t*
Ireland ($n = 98$)		
To kill pests and allow farmers to use less pesticides?	3.05	
To improve nutrition?	3.03	0.29
To help plants withstand weedkillers?	2.64	3.58**
To improve taste?	2.83	0.46
USA ($n = 223$)		
To improve nutrition?	3.36	
To kill pests and allow farmers to use less pesticides?	3.17	3.57**
To improve taste?	3.02	2.89**
To help plants withstand weedkillers?	2.83	2.47**

**Significant difference at 0.05 level.

the use of less pesticides for the US consumers. Improving taste and helping plants withstand weedkillers were benefits that were rated lower. *The Packer* (2001) also found that consumers evaluated improved nutritional content as a more appropriate reason to genetically modify food items than improving flavour.

Attitudes Towards Government Agencies and Producers, and Willingness to Purchase GM Foods

The responses to the questions concerning trust in government and global food producers to ensure food safety were then compared with the willingness of a respondent to purchase GM food. The results show that for the US respondents, those who trust the government and global food producers to ensure food safety are more willing to purchase GM food. However, for the Irish respondents there is no relationship between trust in government agencies and global food producers ensuring food safety and willingness to purchase a GM food product. Most Irish respondents evaluated the government and global food producers lower than the US respondents. Both the Irish and US consumers are more likely to purchase a GM food product when they agree that global food producers are using environmentally safe methods (Tables 13.13–13.21).

Conclusions

The objective of this research is to use a case study to compare consumer attitudes toward GM food in the USA and Europe using two communities and two time periods. The research used a survey instrument that was administered through the use of a personal interview. The first phase of this research examines 882 randomly selected food purchasers interviewed in October 1999 and January 2000. The second phase of research commenced in October 2000 with a random

Table 13.13. Mean willingness to purchase a genetically modified food product: 'Government agencies in my country have done a very good job at ensuring food safety in the past'.

	Mean	*F* statistic
USA (*n* = 222)		
Strongly disagree	2.06	16.09**
Disagree	2.18	
Agree	2.97	
Strongly agree	3.53	
Ireland (*n* = 99)		
Strongly disagree	2.42	2.15
Disagree	2.45	
Agree	2.67	
Strongly agree	2.00	

**Significant difference at 0.05 level.

Table 13.14. USA (*n* = 222): 'Government agencies in my country have done a very good job at ensuring food safety in the past'.

		Mean difference
Strongly disagree	Disagree	−0.1244
	Agree	−0.9111**
	Strongly agree	−1.4739**
Disagree	Strongly disagree	0.1244
	Agree	−0.7867**
	Strongly agree	−1.3494**
Agree	Strongly disagree	0.9111**
	Disagree	0.7867**
	Strongly agree	−0.5627**
Strongly agree	Strongly disagree	1.4739**
	Disagree	1.3494**
	Agree	0.5627**

**Significant difference at the 0.05 level.

Table 13.15. Mean willingness to purchase a genetically modified food product: 'I trust government agencies in my country to ensure food safety in the future'.

	Mean	*F* statistic
USA (*n* = 222)		
Strongly disagree	2.21	37.11**
Disagree	2.29	
Agree	2.98	
Strongly agree	3.48	
Ireland (*n* = 98)		
Strongly disagree	2.08	5.77
Disagree	2.78	
Agree	2.50	
Strongly agree	2.00	

**Significant difference at the 0.05 level.

Table 13.16. USA ($n = 222$): 'I trust government agencies in my country to ensure food safety in the future'.

	Willingness to purchase a GM food product	Mean difference
Strongly disagree	Disagree	−0.007
	Agree	−0.7578**
	Strongly agree	−1.2641**
Disagree	Strongly disagree	0.007
	Agree	−0.6811**
	Strongly agree	−1.874**
Agree	Strongly disagree	0.7578**
	Disagree	0.6811**
	Strongly agree	−0.5063
Strongly agree	Strongly disagree	1.2641**
	Disagree	1.874**
	Agree	0. 5063

**Significant difference at the 0.05 level.

Table 13.17. Mean willingness to purchase a genetically modified food product: 'Global food producers have done a very good job at ensuring food safety in the past'.

	Mean	*F* statistic
USA ($n = 222$)		
Strongly disagree	2.25	31.31**
Disagree	2.67	
Agree	3.14	
Strongly agree	4.14	
Ireland ($n = 99$)		
Strongly disagree	2.11	5.88
Disagree	2.64	
Agree	2.55	
Strongly agree	3.50	

**Significant difference at the 0.05 level.

Table 13.18. USA ($n = 222$): 'Global food producers have done a very good job at ensuring food safety in the past'.

		Mean difference
Strongly disagree	Disagree	−0.4167
	Agree	−0.8851**
	Strongly agree	−1.8929**
Disagree	Strongly disagree	0.4167
	Agree	−0.4685**
	Strongly agree	−1.4762**
Agree	Strongly disagree	0.8851**
	Disagree	0.4685**
	Strongly agree	−1.0078
Strongly agree	Strongly disagree	1.8929**
	Disagree	1.4762**
	Agree	1.0078

**Significant difference at the 0.05 level.

Table 13.19. Mean willingness to purchase a genetically modified food product: 'Global food producers are producing food using environmentally safe methods'.

	Mean	*F* statistic
USA (*n* = 222)		
Strongly disagree	2.59	30.04**
Disagree	2.72	
Agree	3.10	
Strongly agree	3.75	
Ireland (*n* = 99)		
Strongly disagree	2.00	7.68*
Disagree	2.54	
Agree	2.81	
Strongly agree	3.00	

*Significant difference at the 0.10 level; **significant difference at the 0.05 level.

Table 13.20. USA (*n* = 222): 'Global food producers are producing food using environmentally safe methods'.

		Mean difference
Strongly disagree	Disagree	−0.1300
	Agree	−0.5060**
	Strongly agree	−1.160**
Disagree	Strongly disagree	0.1300
	Agree	−0.3760**
	Strongly agree	−1.031**
Agree	Strongly disagree	0.5060**
	Disagree	0.3760**
	Strongly agree	−0.6538
Strongly agree	Strongly disagree	1.156**
	Disagree	1.031**
	Agree	0.6538

**Significant difference at the 0.05 level.

Table 13.21. Ireland (*n* = 99): 'Global food producers are producing food using environmentally safe methods'.

		Mean difference
Strongly disagree	Disagree	−0.5385
	Agree	−0.8148**
	Strongly agree	−1.000
Disagree	Strongly disagree	0.5385
	Agree	−0.2764
	Strongly agree	−0.4615
Agree	Strongly disagree	0.8148**
	Disagree	0.2764
	Strongly agree	4615
Strongly agree	Strongly disagree	1.0000
	Disagree	0.4615
	Agree	0.1852

**Significant difference at the 0.05 level.

sample of 324 respondents. All phases of research were conducted in San Luis Obispo, California, and Galway, Ireland. San Luis Obispo has a population of approximately 42,000 and Galway has a population of approximately 57,000.

The results of the first phase indicated that there was a similar level of familiarity with GM food in Ireland and the USA. Approximately 43% of respondents in both countries indicated that they were familiar with GM food. The second phase of research was conducted shortly after the recall of many food products in the USA containing maize that was grown using the StarLink™ seed. An increase in familiarity was observed in the USA, with half of the respondents indicating familiarity with GM food. However, the level of familiarity in Ireland remained at a level similar to that observed during the previous phase.

Results of the first phase indicated a difference in attitudes between the Irish consumer and consumers in the USA toward GM food. The familiar US respondents perceived GM food to have neutral or positive attributes. The Irish consumer attributed more negative attributes to GM food. Further, the Irish consumers were more likely to indicate that mandatory labelling is important and less likely to purchase a GM food product. During the second phase of research, the Irish consumer continued to be less likely to purchase a GM food product. However, the Irish and consumers in the USA indicated similar attitudes toward the labelling of GM food during the second phase of research since more respondents in the USA indicate that mandatory labelling is important.

Additional questions were added during the second phase of research to examine consumer attitudes toward GM food based on the purpose for the use of biotechnology: to help plants withstand weedkillers, to improve nutrition, to kill pests and allow farmers to use less pesticide, and to improve taste. Respondents in the USA indicated that the most important reason to purchase a GM food product is to improve nutrition followed by modifying it to kill pests and allow farmers to use less pesticides. Genetically modifying food products to improve taste and help plants withstand weedkillers were less important criteria when purchasing a GM food product. The Irish respondents indicated that using biotechnology to kill pests and allow farmers to use less pesticides and to improve nutrition are equally important. Using biotechnology to help plants withstand weedkillers and improve taste were less important to the Irish respondent.

Questions were added during the second phase of research to examine attitudes toward government agencies and food safety and attitudes toward food producers and food safety and the environment. Perceptions of government agencies and food safety and perceptions of food producers and food safety and the environment are more positive in the USA than in Ireland. Perceptions of government agencies and food safety and perceptions of food producers and food safety are related to a consumer's attitudes toward GM food in the USA. However, such attitudes are not related to a consumer's attitudes toward GM food in Ireland. Consumers in the USA that have more positive attitudes toward government agencies and food safety and of food producers and food safety are more likely to purchase GM food. Such a relationship does not exist in Ireland. However, there are less positive attitudes toward government regulators and global food producers in Ireland. It was found that a positive relationship exists between perceptions of global food producers using environmentally safe methods and the purchase probability for GM food for consumers in both countries.

References

Alexander, N. and Toner, C. (2001) More consumers see potential benefits to food biotechnology. International Food Information Council. Available at www.ific.org/procative/newsroom/release.vtml?id=19241.

Central Statistics Office Home Page, Principal Statistics (1999) Available at www.cso.ie/principalstat/pristat2.html.

Copple, B. (2000) Scientist, activist, yogi? *Forbes*, 30 October, pp. 54–56.

Gregerson, J. (2000) What do we know about GMO's? *Food Engineering*, March.
Hoban, T. (2000) U.S. food consumption is largely unaffected by StarLink corn recall. *News Services*, 27 November.
Nash, J.M. (2000) Grains of hope. *Time* 156 (5).
Olson, J. (2000) GMO free zone. *Farm Industry News*, February.
The Associated Press and Reuters (2000) Scope of biotech corn product recall revealed. *CNN.com*, 2 November.
The Packer (2001) Fresh Trends 2001 profile of the fresh produce consumer.
The Times (London) (2001) GM tide turns. 3 April.
Thomas, G.S. (2001) Playing in San Luis Obispo. *Demographics Daily*, 6 February. Available at wysiwyg://44/http://bizjournals.bcentral.com/journals/demographics/.
Turcsik, R. (2001) Still life in biotech. *Progressive Grocer*, April, pp. 16–22.
US Bureau of the Census (1991) *State and Metropolitan Data*.

14 Attitudes Towards GM Food in Colombia[1]

Douglas Pachico[1] and Marianne McGarry Wolf[2]
[1]*International Center for Tropical Agriculture (CIAT), AA 6713, Cali, Colombia;*
[2]*Agribusiness Department, California Polytechnic State University, San Luis Obispo, CA 93407, USA*

Transgenic or genetically modified crops are widely grown, covering over 50 million hectares in 2001 (James, 2002), and transgenic food is widely consumed, entering an estimated 60% of processed foods in the USA (Hopkins, 2001). Some see genetically modified (GM) crops as critical to improving agricultural productivity and ensuring food supplies especially for the poor and malnourished in developing countries (Evans, 1998; Oxfam Policy Department, 1999). Others argue that this is a myth and that there are significant health and environmental risks from GM crops and food (Altieri, 2001). In this context, consumer attitudes towards GM foods have become a factor both in the market demand for GM foods, and in their regulation. In some markets there is no doubt that consumer attitudes have slowed the utilization of GM crops (Charles, 2001).

Because consumer attitudes have become such a key factor in the acceptance of GM food, and because these attitudes seem to vary so substantially between countries, increasing attention has been paid to understanding consumer attitudes towards GM food (Bredahl *et al.*, 1998; Sheehy *et al.*, 1998). Particular attention has been paid to understanding differences in consumer attitudes between the USA and Europe (Nelson, 2001; Wolf *et al.*, 2001; Wolf and Domegan, 2002). In general, European consumers have a stronger sense of the potential risks of GM foods than do US consumers. Little if any similar research has examined consumer attitudes towards GM food in low-income countries where hunger and malnutrition are most common and where, therefore, GM crops might have their greatest contribution to the welfare of consumers.

This chapter makes an initial examination of consumer attitudes in Colombia toward GM food. It follows on from research previously conducted in the USA and Ireland. First, the methods of the study are briefly described and some characteristics of the sample population noted. Second, some general background attitudes of the Colombian consumer with respect to food safety, science and government regulation are reviewed. Third, levels and sources of consumer knowledge about GMO food are presented. Fourth, attitudes towards GMO foods, including likelihood of purchase, are analysed. Finally, the major implications of the study are reviewed and some areas for further research are noted.

Methods and Data

This study largely followed the approach and utilized a modified questionnaire that had been previously used in Ireland and the USA (Wolf *et al.*, 2001; Wolf and Domegan,

[1] Research assistance provided by Katie Canada, California Polytechnic State University, California, USA.

2002). This both facilitates international comparisons and provides a research instrument that has been validated in previous studies. The questionnaire is largely comprised of questions scaled around different degrees of frequency or different degrees of agreement. A Spanish translation was developed and pretested to ensure understanding. A few additional questions on attitudes were added.

This study was conducted in Cali, Colombia, an urban centre with a metropolitan population of around 2.5 million. Cali can reasonably be considered as a typical South American city in a region where over 70% of the total population is urban. As in the Irish and US studies, the questionnaire was randomly applied to people approaching or departing from points of food purchase. These included supermarkets and open air markets in six different neighbourhoods selected according to general indicators of economic status. A total of 150 questionnaires were conducted among food purchasers in March 2001 by a single experienced sociologist.

Females comprised 89.3% of respondents, 51.7% were between 25 and 44 years old, 68% were married or lived with partners and 69% were members of dual-income households. The respondents were almost evenly divided among those who work full time (34.7%), those employed part time (34.0%) and those not employed (31.3%). The sample was comparatively well educated with 33.5% having attended university. Some 71.8% of respondents have children under the age of 18 living at home.

Attitudes to Food Safety and Science

There is a high level of awareness among Cali consumers of possible food risks (Table 14.1). There is strong agreement among 63.3% that pesticides in food are dangerous and among 64% that food additives are dangerous. A majority of 52.7% strongly agrees that foods are adulterated with false ingredients. However, even though (or perhaps because) foot and mouth disease is endemic among cattle in some regions of Colombia, 65.3% of the sample disagrees or strongly disagrees that foot and mouth disease is a food risk for humans. Clearly the sample exercises discrimination about what it considers to be real food risks. It is sensitive to some potential food risks, like pesticide residues or food additives, but it is prepared to discount other factors, like foot and mouth disease, that could have been perceived as a food risk.

Overall, the Cali food purchasers have a positive view of science and technology. There is strong agreement among 68% of the sample that science improves the quality of life while 56% strongly agree that computers improve the quality of life. Thus, the sample would not appear to have a prior predisposition to be sceptical of scientific innovations such as GM food but rather might even be predisposed to associate new scientific discoveries with something positive.

There is also a fairly high level of confidence in the government assuring food safety. While 75.3% agree or strongly agree that the

Table 14.1. Consumer attitudes to food, safety, science, government and food producers, Cali, Colombia, 2001 ($n = 150$).

	Strongly agree	Agree	Disagree	Strongly disagree	Not sure
Pesticides dangerous to health	63.3	28.0	6.7	2.0	0.0
Foot and mouth disease a food risk	24.0	10.7	53.3	12.0	0.0
Food additives dangerous to health	64.0	29.3	6.0	0.7	0.0
False ingredients put in food	52.7	37.3	3.3	2.0	4.7
Science improves quality of life	68.0	26.0	3.3	2.0	0.7
Computers improve quality of life	56.0	30.7	10.7	2.7	0.0
Government assures food safety	38.0	37.3	15.3	6.7	2.7
Global producers assure food safety	49.3	24.7	17.3	4.0	4.7
Global food producers environmentally safe	30.9	27.5	30.2	2.0	9.4
Household food supply adequate	24.8	36.9	37.6	0.7	0.0
Price most important in choosing food	31.3	24.0	41.3	3.3	0.0

government assures food safety, only 22.0% disagree or strongly disagree. Similarly, 74.0% agree or strongly agree that food producers assure food safety, but only 21.3% disagree. There is less confidence in the environmental safety of food production, with 32.2% disagreeing or strongly disagreeing that food production is environmentally safe.

One major distinguishing characteristic of this sample is that nearly two-fifths, 38.3%, disagree or strongly disagree that there is always enough food to eat in their family. Likewise, a majority, 55.3%, agree or strongly agree that price is the most important factor in purchasing food. These two attitudes would support the hypothesis that for many people in low-income countries, they are unable to afford the quantity of food that they desire. This being the case, Colombian consumers may be less sensitive to potential but as yet unidentified food risks than consumers in high-income countries like those in Europe where food is abundantly available and quality issues come to the fore. These data suggest that in contrast to high-income countries, inadequate availability of food may be the most pressing food-related health issue for many people, thus decreasing the likelihood of resistance to transgenic food.

Knowledge of GM Food

There is a very low level of familiarity with GM food in Colombia. The vast majority of the sample, 77.6%, reports that they are not at all familiar with GM food. Only 5.4% indicate they are very familiar with transgenic food and 7.5% say they are somewhat familiar. These awareness levels are significantly lower than those found by Wolf *et al.* (2001) in their examination of familiarity with GM food in the USA and Ireland. Approximately half of the respondents in the USA and 40% of respondents in Ireland were at least somewhat familiar with GM food.

Television news has been the main source of information in Colombia, reaching 10% of the sample, while 6% had discussed it with acquaintances, 5.3% had heard about it over the radio, 4.0% had read about it in newspapers and 1.4% had read about it in magazines. Given the very low levels of familiarity, obviously many of those with some familiarity had had access to information about transgenic foods from more than one source.

Given the low levels of familiarity and access to information about GM food, it is probable that many of the attitudes towards GM food reported in this chapter may not be strongly held. The attitudes of the Colombian consumer could be subject to significant change in the light of additional exposure to information in the future.

Attitudes to GM Food

Attitudes towards GM foods among Colombian consumers are mixed. There is widespread agreement that some GM foods may be unsafe, but none the less only a minority would be unwilling to buy GM food.

Nearly three-quarters of Colombian consumers strongly agree or agree that some foods produced by genetic engineering may be unsafe (Table 14.2). Less than one-quarter would disagree. On the other hand, Colombian consumers are split quite evenly into three groups in terms of their willingness to purchase GM food. Some 33.6% would definitely or probably buy GM food, 32.9%

Table 14.2. Attitudes to GMO food, Cali, Colombia, 2001 ($n = 150$).

	Strongly agree	Agree	Disagree	Strongly disagree	Not Sure
GMO food unsafe	36.0	39.3	20.7	2.0	0.7
	Definitely	Probably	Maybe	Probably not	Definitely not
Willingness to buy GMO food	15.1	18.5	32.9	18.5	15.1

might buy it and 33.6% would probably or definitely not buy GM food. The Colombian probability of purchasing GM food is similar to that observed by Wolf *et al.* (2001) in the USA. However, it is higher than that observed for Ireland where only 17.2% would definitely or probably buy GM food.

Thus, nearly three-quarters of consumers perceive potential risks with GM food, but two-thirds would be willing to purchase GM foods. There are some possible explanations for this apparent inconsistency. In the first place, the widespread belief that some GM foods may be unsafe does not preclude the simultaneous belief that some GM foods may be safe. Given the previously reported high level of reported confidence in the food regulatory system, consumers may simply trust that some GM foods are safe, and those that are not would be excluded from the food supply by the regulatory authorities. Another explanation could be that consumers might be willing to absorb the risk of GM food if it met other important criteria for them.

The characteristics of the GM food would have an influence on consumers' likelihood to purchase GM food. Using a five-point scale of willingness to buy (definitely = 5; probably = 4; maybe = 3; probably not = 2; definitely not = 1), it can be seen that consumers are more willing to buy GM food if it has characteristics that they appreciate (Table 14.3). For example, consumers indicated that pesticides are dangerous to their health. The use of genetic modification to reduce the use of pesticides generated the highest purchase interest. Further, willingness to buy is significantly higher for characteristics that would be desired by consumers like improved nutrition or taste than for a characteristic like resistance to weedkillers that does not directly benefit consumers. Consistent with the finding reported above that Colombian consumers are aware of the risks of pesticide residue on food, this genetically modified pest resistance which would reduce the use of chemical pesticides also gives a higher willingness to buy. Further research might attempt to elucidate whether a lower cost of GM food would similarly elicit a greater willingness to purchase.

Attitudes of Colombian consumers to labelling may also have some relevance to these questions. A majority of Colombian consumers read food ingredient labels very or somewhat often (Table 14.4). This could indicate that they rely on the content and even the mere presence of food labels as a warrant of food safety. Moreover, 68% of consumers report that they think that mandatory labelling of GM food is very important and 22.7% think it is somewhat important. It is possible that this implies that with a system of labelling

Table 14.3. Likelihood to buy – attribute mean rating ($n = 150$).

	Attribute means	Paired *t*
To reduce the use of pesticides?	3.43	
To improve nutrition?	3.39	0.513**
For improved taste?	3.16	2.44**
To resist weed killers?	2.84	3.14**

**Significant difference at 0.05 level.

Table 14.4. Practices and attitudes towards food labelling, Cali, Colombia, 2001 ($n = 150$).

	Very often	Somewhat often	Not very often	Not at all
Read food labels for ingredients	38.1	26.5	22.4	12.9
	Very important	Somewhat important	Not very important	Not at all important
Importance of mandatory labelling of GMO food	68.0	22.7	4.7	4.7

most Colombians would be willing to purchase GM food even though they believe that some GM foods might be unsafe. Consequently, in order to be able to have the assurance that they can consume GM foods safely, Colombians think that mandatory labelling is very important.

This is not really contradicted by the fact that in practice far fewer consumers, 38%, often read food ingredient labels than think that mandatory labelling of GM foods is very important, 68%. To some extent, the mere presence of labels may be a sufficient indicator for many consumers that appropriate authorities are monitoring food safety. In addition, for items that are consumed regularly, people may not expect constant changes in ingredients and therefore do not need to read the labels of regularly purchased food on a frequent basis.

To understand better the attitudes of Colombian consumers to GM food, Table 14.5 shows mean willingness to buy scores for people holding different opinions. Thus, it is clear that among those who more strongly disagree that engineered foods are unsafe, that is among those who perceive less chance of risk from GM foods, willingness to buy is higher, 4.33, than it is among those who strongly agree that GM foods are not safe. Those strongly agreeing that GM foods are unsafe have a lower willingness to buy, 2.84. Perceptions of the risks of GM food thus have the expected relationship with willingness to buy. This relationship is consistent across the opinion categories and is statistically significant.

Furthermore, for those who strongly agree that low price is important in the food purchase decision, willingness to buy is higher, 3.41, than for those who strongly disagree that price is the main decision criterion for food purchase. Those who are less sensitive to price have a lower willingness to buy GM food, 2.40. This relationship of higher willingness to purchase GM food with higher sensitivity to food price is consistent across the categories of opinion with respect to price and is statistically significant. This would be consistent with the hypothesis that higher-income people, for whom the cost of food is less important, are more influenced by possible food quality characteristics and for this reason are less willing to purchase GM food. In contrast, for those consumers for whom food prices are a major criterion in food purchase, they may be more disposed to purchase GM food as long as it is cheaper. This suggests that poor consumers could benefit disproportionately from cheap GM food so long as it was indeed safe, but on the other hand if it really was not safe, then they could be more vulnerable to any risks associated with consuming GM food.

Similarly, among those who strongly disagree that the quality and variety of food in the family is good, that is, among those consumers in families where the quality and variety of food is less than desired, the willingness to buy GM food is high, 3.5. In contrast, among those families where they agree that the quality and variety of food is already good, willingness to buy GM food is low, 2.57. Although this relationship is not statistically significant for this sample, it does consistently suggest that the less adequate the quality of current food consumption, the more willing people are to buy GM food. The better the current quality of food, the less willing are people to buy GM food. This finding is again quite consistent with the previous result on the relationship between food price and willingness to purchase GM food.

Table 14.5. Mean willingness to buy genetically modified food by food attitude groups (5 = definitely willing; 1 = definitely not willing), Cali, Colombia, 2001 ($n = 150$).

	Strongly disagree	Disagree	Agree	Strongly agree	*F* statistic
Genetic engineered foods not safe	4.33	3.35	2.88	2.84	2.491*
Low price important to buy food	2.40	2.90	2.75	3.41	2.606**
Family food supply adequate	3.00	3.15	2.68	3.19	1.714
Family food quality good	3.50	3.18	3.10	2.57	1.856

*Significant difference at 0.01 level; **significant difference at 0.05 level.

There is, though, not a clear relationship between the adequacy of current diets in terms of quantity and willingness to buy GM food. It would have been hypothesized that consumers without an adequate quantity of food would have been more willing to purchase GM food, but there is no evidence for this.

Summary and Suggestions for Further Research

Although transgenic crops are being widely grown worldwide, consumer attitudes towards them have been found to vary substantially between Europe and the USA. This study of a sample of 150 food purchasers in Cali, Colombia, is believed to be one of the first studies to examine consumer attitudes towards GM food among consumers in tropical or low-income countries.

Consumers in Colombia are aware of possible food risks, with about two-thirds agreeing that residues of pesticides or food additives are dangerous. However, other factors that could have been perceived as a food risk, like foot and mouth disease, were not considered dangerous by nearly two-thirds of consumers, indicating that Colombian consumers do not simply accept as dangerous any hypothetical risk factor.

Well over half of consumers in Colombia appear to have positive views of science and technology and a surprisingly high level of three-quarters of consumers have confidence in government regulation of food safety. Holding these positive attitudes towards the benefits of science and the effectiveness of food safety regulations is likely more consistent with less concern about the risks of GM food.

Economic factors seem to affect the access to food of a significant number of Colombian consumers. Nearly two-fifths sometimes do not have enough food to eat in their family and for nearly one-half a low price is the most important factor in buying food. For consumers such as these, for whom the absolute quantity of food is a pressing concern, quality factors such as potential but unidentified food risks from GM foods may not play a major role in food purchase decisions.

About three-quarters of Colombian consumers agree that some GM foods may be unsafe. Nevertheless, some two-thirds of consumers would be willing to buy food with GM ingredients. This result is similar to that observed in the USA, but lower than the probability of purchase for consumers in Ireland. Concerns about safety do affect the willingness to buy GM food. There is a statistically significant relationship between perception of genetically engineered food as unsafe and the willingness to buy GM food: the stronger the safety concern, the lower the willingness to purchase.

Nevertheless, many consumers who perceive some safety risks in GM food would still be willing to buy it. Economic factors may be important in this regard. Those for whom low price is the most important factor in the food purchase decision are significantly more willing to buy GM food. Likewise, those for whom the current quality and variety of food is less than desired are also more willing to buy GM food. These findings suggest that for resource-constrained food consumers, ill-defined or uncertain risks would not necessarily be highly dissuasive of GM food consumption, especially if it were cheap. Thus, if GM food risks are indeed low or non-existent, then poor consumers would be most likely to reap the benefits of GM foods that reduce the price of food.

Finally, though highly suggestive, these results must still be taken as a very tentative picture of the attitudes of Colombian consumers to GM food. Familiarity with GM food is still very low and current attitudes could shift with increased familiarity.

Several further extensions to this initial research could be considered. It would be useful to more directly assess whether a lower cost of GM food would elicit a greater willingness to purchase. It could be useful to more purposively sample among consumers with a higher degree of familiarity with GM food to attempt to project what likely attitudes might be with increased familiarity in the future. The survey approach could be supplemented with a focus group approach to probe more into people's attitudes and to see how additional information might shape these attitudes. Further research is planned to contrast the results of this survey with those of surveys in high-income countries to compare and contrast the differences.

References

Altieri, M. (2001) *Genetic Engineering in Agriculture: Myths, Environmental Risks and Alternatives.* Food First Books, Oakland, California.

Bredahl, L., Grunert, K. and Frewer, L. (1998) Consumer attitudes and decision-making with regard to genetically engineered food products. *Journal of Consumer Policy* 21, 251–277.

Charles, D. (2001) *Lords of the Harvest: Biotech, Big Money and the Future of Food.* Perseus, New York.

Evans, L.T. (1998) *Feeding the Ten Billion.* Cambridge University Press, Cambridge.

Hopkins, K. (2001) The risks on the table. *Scientific American* 284(4), 60–61.

James, C. (2002) *Global Hectarage in GM Crops 2001.* International Service for the Acquisition of Agri-biotech Applications, Ithaca, New York.

Nelson, C.H. (2001) Risk perception, behavior, and consumer response to genetically modified organisms: toward understanding American and European public reaction. *American Behavioral Scientist* 44, 1371–1388.

Oxfam Policy Department (1999) *Genetically modified crops, world trade and food security.* Available at http://www.oxfam.org.uk/policy/papers/gmcrop.htm.

Sheehy, H., Legault, M. and Ireland, D. (1998) Consumer and biotechnology: a synopsis of survey and focus group research. *Journal of Consumer Policy* 21, 359–386.

Wolf, M.M. and Domegan, C. (2002) A comparison of consumer attitudes toward genetically modified food in Europe and the USA: a case study over time. In: Santaniello, V., Evenson, R.E. and Zilberman, D. (eds) *Market Development for Genetically Modified Food.* CAB International, Wallingford, UK, pp. 25–38.

Wolf, M.M., McDonnell, J. and Domegan, C. (2001) *A comparison of consumer attitudes toward genetically modified food in Ireland and the USA.* Paper presented at the International Conference on Agricultural Biotechnology Research, Ravello, Italy.

15 Consumer Acceptance and Development Perspectives of Functional Food in Germany

Heiko Dustmann and H. Weindlmaier
Centre of Life and Food Sciences Weihenstephan Professorship for Dairy and Food Industry Management, Technische Universität, München, Germany

Functional Food Market Facts

The association for consumer research (GfK) estimates that the turnover of functional food in the food retail trade amounts to €920 million (2000) in Germany.[1] This is proportionally slightly less than 1% of the total turnover in the food retail trade. According to a press release of 29 November 2001, A.C. Nielsen assigns 1.5% of the total food market to the functional food market. Functional foods are considerably more spread with a turnover of €3.5 billion in Japan and €11 billion in the USA (at the end of the 1990s) (Menrad *et al.*, 2000, pp. 154–157).

Nationally as well as internationally probiotic dairy products and functional drinks together make up the biggest part of the functional food market with more than 50% (Fig. 15.1) (A.C. Nielsen GmbH, 2001, p. 1).

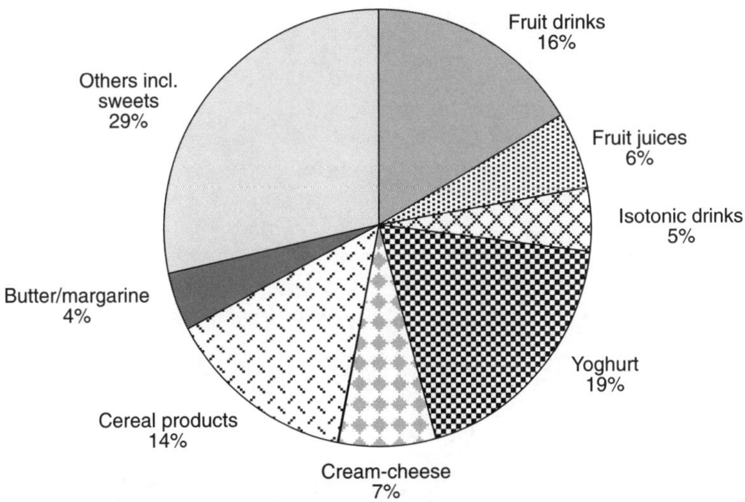

Fig. 15.1. Product groups of functional food in Germany in 2001. Source: Based on A.C. Nielsen GmbH (2001, appendix).

1 See GfK Panel Services Consumer Research GmbH, Birnbaum quoted in Soßna (2001, p. 22).

Consumer Trends

In order to classify development of potential consumer acceptance for functional food in Germany, consumer trends in the total system of food demand and supply were analysed as the basis for the further steps of the project (Fig. 15.2). The trends towards health and convenience seem to be particularly decisive for the successful positioning of functional food.

The functional components add to the food promise an additional positive effect on health. A health-orientated composition of food is substituted by functional foods incorporating different healthy components; though flops in this field showed that consideration of the consumer trends of health and convenience are not sufficient for the market success of functional food.[2] It is necessary to pay special attention to, for example, sensory qualities in contrast with prescription drugs or over-the-counter business.

Consumer Motivation to Buy Functional Food

In order to bring the current consumer trends and consumer motivation to buy into line, it is necessary to analyse the trends in detail. The definite consumer habit resulting from the trends plays the decisive part. A written survey as well as ten nationwide group discussions (with 102 participants in total) were carried out in 2001 by the Institute for the Management of Dairy Companies to investigate the potential acceptance of functional food.

The members of the group discussions who also participated in the written survey exclusively consist of housekeeping persons of all Nielsen-fields. For the selection of participants two facts were considered most important: on the one hand, a rather representative distribution of age and on the other hand, an appropriate distribution of urban and rural areas. A selection of the informative results for the development of acceptance of these products is presented here.

The participants of the group discussions associate with 'functional food' various products, especially probiotic yoghurts, physical fitness, health, activity and added value. In the discussion of the question: 'What do you make of functional food?' there were different answers, for example:

- 'If I get a balanced diet, I don't need functional food.'
- 'Unsureness about health value.'

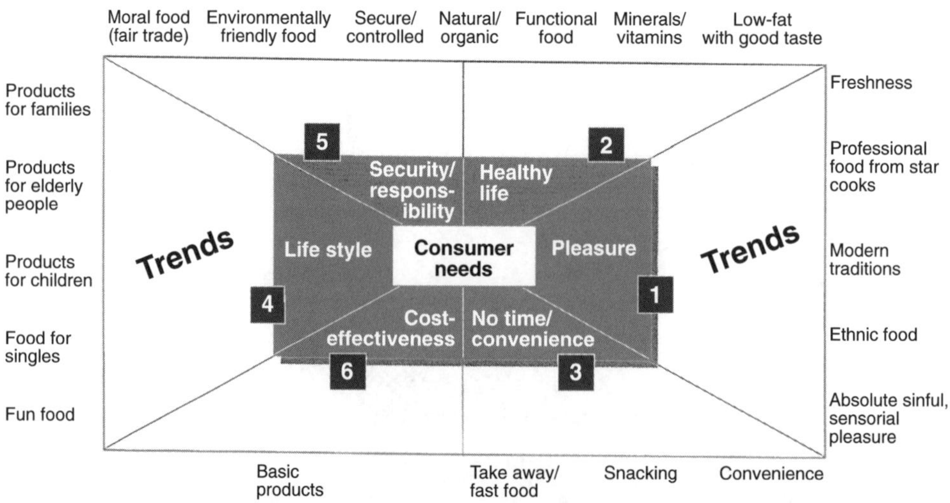

Fig. 15.2. Most important consumer trends observed in the German food market. Source: Weindlmaier *et al.* (2001, p. 68).

[2] See Biester (2001, p. 33). A popular flopped functional food brand is 'Aviva' (Novartis).

- 'In support of a varied nutrition are these products reasonable, especially when it has to go quickly.'
- 'It tastes. But I can't say that I feel better after eating these products.'
- 'A little bit of the content of the advertising must be true. There is no chance. You have to trust in the products as you can't control the effectiveness by yourself.'
- 'Only wheeling and dealing.'
- 'I like tasting new products, if I believe in them.'

Figure 15.3 presents the most important aspects of consumer acceptance of functional food. Generally, good taste, high quality, reasonable prices, freshness, and a proven health value are the most important demand drivers for functional food.

With the aid of tasting some products it was possible to substantiate the comments of the participants. The evaluation of the different product groups varied significantly (Table 15.1).

Perspectives of Functional Food

In order to determine the future perspectives of functional food a Delphi-Study was carried out. Sixty-one experts from the food industry, the food trade, public and science were asked in two stages about subjects related to the whole value-added chain.

The experts of the Delphi-Study expect a continuous growth of functional food during the next 10 years. Within the next 5 years a growth of 3% is forecasted (Fig. 15.4).[3]

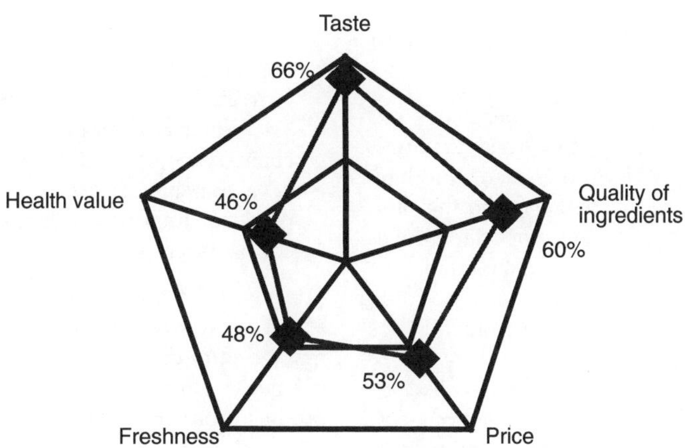

Fig. 15.3. Most important demand drivers for functional food.

Table 15.1. Evaluation/comments by the majority of consumers.

Probiotic yoghurt	♦ Efficient completion of the range of traditional yoghurt. ♦ A surcharge of about 10% is acceptable.
Functional bread	♦ Healthy effects are accepted. However, the diversity of bread offered in Germany already stands for wellness. Necessity of bread with functional ingredients is not given.
ACE drinks	♦ Not really perceived as a healthy food – no alternative to fresh pressed juice or fruits.
Functional fat	♦ Criticized with respect to taste, however, belief in health value.
Functional confectionery	♦ Product is evaluated to be contradictory (health versus sweet). but still seen as an alternative to normal sweets for children.

3 Median of expert opinion from the second inquiry of the Delphi-Study.

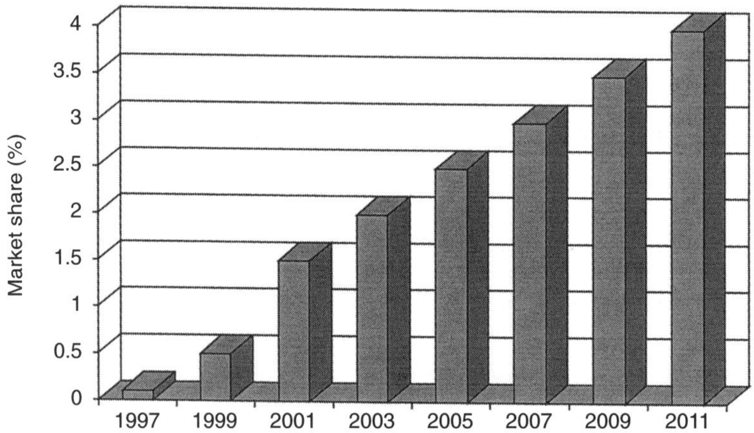

Fig. 15.4. Growth of functional food during the next 10 years.

Further Results of the Delphi-Study

Further results of the Delphi-Study can be summarized as follows:

- Actual trends in consumer behaviour such as pleasure seeking, the trend to fitness, the preference for convenient products, the polarization of markets, and the obsolescence of society are important drivers for the demand for functional food.
- Heavy users of functional food are expected to be athletes and women in specific stages of life.
- Additional functional properties like the prevention of cancer and cardiovascular diseases and the compensation of malnutrition are important. Research efforts are necessary to incorporate some of the properties of functional food into the raw materials, especially into plants.
- For a further growth of the market the positioning as premium brands, marketing orientation on specific target groups and intensive promotional activities are important.
- Legal barriers for advertising the health-related added value, shortcomings in R&D in the German food economy and the inadequate perception of the value added to functional food by the consumer are the main restrictions for a fast market growth.
- The different actors in the value-added chain will not equally participate in the development of further functional food products. Important will be a closer vertical cooperation.

The results of the Delphi-Study and the consumer survey allow the conclusion that the market entry of a functional food product will positively affect the total demand for the product group (Fig. 15.5).

Conclusion

Functional food presents chances for all actors in the value-added chain, if the consumer is correctly addressed. The right target group, clear and simple information about the health effects, tastefulness and an emotional communication will be the basics for the acceptance of functional food.

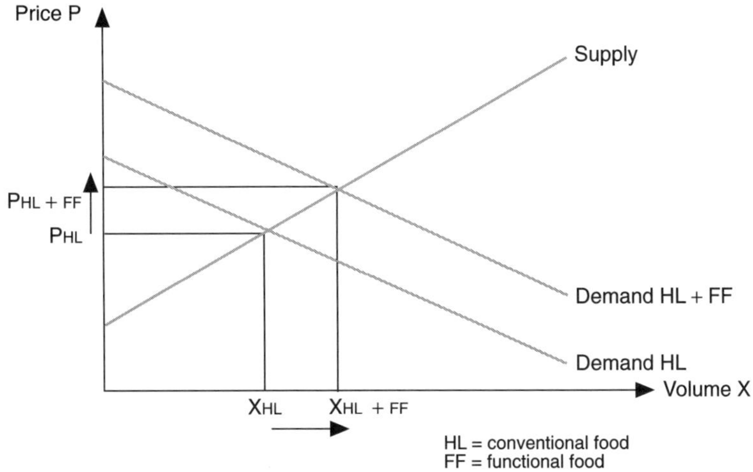

Fig. 15.5. Demand development at market entry of functional food into the product group. Source: Based on own data collection of the Delphi-Study and group discussions.

References

A.C. Nielsen GmbH (2001) Functional Food weiter im Aufwind. A.C. Nielsen press report, 29 November. Frankfurt am Main.

Biester, S. (2001) Verhaltene Stimmung. Der Markt für Functional Food kann bisher nicht halten was vielversprechende Prognosen glauben machen. *Lebensmittel Zeitung* 53, 33.

Menrad, K., Reiss, T., Hüsing, B., Menrad, M., Beer-Borst, S. and Zenger, A. (2000) *Functional Food*. TA Publikationen 37/2000. Zentrum für Technikfolgen-Abschätzung (Hrsg.), Bern, Switzerland.

Soβna, R. (2001) 'Health food': an opportunity in stagnating markets. *European Dairy Magazine* 11, 22.

Weindlmaier, H., Fallscheer, T. and Dustmann, H. (2001) Dem Trend auf der Spur. Weiβe Linie: Perspektiven und Erfolgspotentiale. *Milch-Marketing* 18, 66–71.

16 Factors Explaining Opposition to GMOs in France and the Rest of Europe

Sylvie Bonny
INRA (French National Institute of Agricultural Research), UMR d'Economie Publique INRA-INAPG, BP 1, 78850 Grignon, France

Introduction

The agricultural applications of biotechnology were generally seen as highly promising in the 1980s. Yet, at the end of the 1990s and in the early 2000s a strong movement of opposition to genetically modified organisms (GMOs) has developed throughout the world, particularly in some countries. How can its rapid growth be explained? Asking this question does not mean that such opposition is unjustified or, conversely, that it is legitimate but expressed with great differences in emphasis according to the country. The aim of this article is to analyse various factors explaining this opposition in France, a country in which it is particularly strong. This seems especially important in view of the current deadlock. The French case is fairly representative of various European countries in this respect: even if differences exist, depending on cultural characteristics and economic situations, a number of factors of opposition are found throughout. We studied the processes and mechanisms of the development of the opposition movement. We also examined recurring topics in discussions and debates on the subject, in the discourse of opponents and in articles on GMOs in the media. We furthermore monitored and observed initiatives by the various actors, and drew on the results of several surveys, which gave insights into the reasons for the opposition. The aim of this chapter is thus to contribute to a better understanding of this opposition movement, its determinants and its processes. Over the past few years extensive literature has been published on GMO issues. Thus, this text does not address certain topics already studied in depth, such as the segmentation of opinions according to various categories of consumers or the economic and political effects of this opposition (e.g. agenda setting, decision making, public policy, regulatory aspects concerning GMOs, etc.).

Growing Suspicion and Concern About GMOs Throughout the World

Various recent opinion polls reveal fairly strong and increasing rejection of GMOs throughout the world. Indeed, the opposition movement intensified after 1998; prior to that, opinions were on the whole more favourable (Hoban, 1997). To be sure, the results of such surveys have to be interpreted with caution, for they have various limits. These limits are above all: the risk of artefacts when the respondent has to choose, in a short space of time and out of context, an answer in a series of items proposed on a complex subject; the influence of the formulation of the questions and their interaction on the answers obtained; the impact of the context or of recent events; and, lastly, the risk of superficiality of the approach compared to more in-depth interviews. Moreover, a stated opinion may differ from an effective opinion, and from behaviour in a real situation where other factors play a part. In spite of all, polls prove to be a useful source of information, to be used with other methods; they have the advantage of providing indicators on vast samples representing the group under study. Bearing these limits in mind, we hereafter present the data of some recent surveys of perceptions of GM foods.

International level

At the international level, a survey carried out in early 2002 showed that concerns about GMOs have spread around the world, even to developing countries and the USA (Table 16.1). In February–March 2002, the Ipsos-Reid Group, an opinion survey institute, carried out a poll dealing with some food issues in 12 countries (500 people questioned per country, 1000 in the USA). One question asked regarding the level of concern about the safety of GM foods (Ipsos-Reid, 2002). Across the 12 countries, an average of 76% of those polled said they were worried about GM food safety. Almost half (48%) 'strongly agreed' and 28% 'somewhat

Table 16.1. Concern about GM food in various countries throughout the world by country and socio-demographic characteristic (Ipsos-Reid, 2002): 'Please tell me if you agree strongly, agree somewhat, disagree somewhat or disagree strongly with the statement: I am concerned about the safety of genetically modified foods?' (% of answers).

	Agree strongly	Agree somewhat	Total agree
Japan	54	37	91
South Korea	55	27	82
Germany	62	18	80
Brazil[a]	58	21	79
France	54	23	77
Canada	46	29	75
USA	44	28	72
India[a]	44	28	72
South Africa[a]	41	30	71
Russia[a]	40	31	71
UK	45	22	67
China[a]	34	29	63
Overall	48	28	76

[a]Brazil, China, India, Russia and South Africa data represent urban-only samples.

	Overall strong agreement[a]	Gender		Age			Relative income		
		Men	Women	18–34	35–54	55+	Lower	Middle	Higher
North America	44	37	51	41	47	42	49	43	40
Europe	50	45	55	46	54	51	53	50	50
Asia-Pacific	51	50	51	44	52	59	47	50	59
Overall	48	44	52	45	51	49	49	48	49

Europe includes France, Germany, urban Russia, UK.
Asia-Pacific includes urban China, urban India, Japan, South Korea.
Income groupings are nationally relative, and represent the bottom, middle and top income strata within each country.
[a]Percentage of respondents who agree strongly.

agreed' when they were asked if they agreed or disagreed, strongly or moderately, with the statement 'I am concerned about the safety of genetically modified foods'. Only one in five disagreed 'somewhat' or 'strongly' with the statement. From country to country, the proportion of people who are concerned ranges from a low of 63% (in urban China) to a high of 91% (in Japan). The percentage who 'strongly agreed' is the highest (more than 54%) in Germany, Brazil, South Korea, France and Japan. In many countries women are more likely than men to be concerned about this safety aspect (52% versus 44%, respectively) (Table 16.1). According to the chosen delimitation of age groups and large geographical zones, the gender difference appears quite pronounced in North America, less so in Europe, but hardly perceptible in the other countries surveyed. Old people seem to be more disquieted, particularly in Asia, but elsewhere age seems to be less a factor of differentiation as far as GM food safety is concerned. In North America people from lower-income households are more worried than the wealthier about the safety of GM food. It is the opposite in Asia-Pacific, perhaps because people with lower income are less educated in such zones and less aware of this issue. In addition, in China and India, consumers from lower-income households may have more urgent concerns and so be less likely to pay attention to these questions.

USA

In the USA, the International Food Information Council (IFIC) commissioned surveys on a sample of about 1000 people in 1997, 1999, 2000, 2001 and 2002 (IFIC, 2002). It enables a follow up of opinion several years in a row because the same questions were asked several years in a row. Various questions on attitudes toward food biotechnology were formulated by referring mainly to its advantages. Therefore the survey results can be used more as regards the trends than in absolute value compared to others which have more neutral formulation. Americans seem to be favourable to biotechnologies. However, the feeling that biotechnology will be beneficial diminished between 1997 and 2000, particularly in 1999, but tends to be quite stable later (Table 16.2a).

In January 2002, the Pew Initiative on Food and Biotechnology commissioned a survey dealing with environmental issues of genetic engineering (Pew, 2002). People were asked 'Overall, which do you feel is greater – the environmental risks or the environmental benefits of using biotechnology to genetically modify plants, animals, fish or trees?' This question was asked twice: on the one hand, at the beginning of the survey, on the other hand, at its end (Table 16.2b). The rest of the questionnaire mainly addressed several other points dealing with more

Table 16.2a. Benefits expected from biotechnology in the USA (IFIC, 2002): 'Do you feel that biotechnology will provide benefits for you or your family within the next five years?' (% of answers).

	March 1997	Feb. 1999	Oct. 1999	May 2000	Jan. 2001	Sept. 2001	Aug. 2002
Yes	78	75	63	59	64	61	61
No	14	15	21	25	22	17	18
Don't know/ refused	8	10	16	16	14	21	21

Table 16.2b. Perception of environmental risks and benefits of genetic engineering (Pew, 2002) 'Overall, which do you feel is greater – the environmental risks or the environmental benefits of using biotechnology to genetically modify plants, animals, fish or trees?' (% of answers).

Which do you feel are greater?	Risks	Benefits	About the same	Not sure
Pre-statement (first question)	40	33	19	9
Post-statement (last question)	38	38	21	3

US nationwide survey of 1214 adults conducted by Zogby International in January 2002.

detailed and precise aspects of environmental risks or benefits of GM. Thus, at the end, the respondents were more aware of GMOs; however, there could be a bias in the answer to the last question according to the characteristics of the examples mentioned in the rest of the interview. In this list of possible environmental risks and benefits of GMOs, the interviewed people had to grade each of them according to the importance of effect.

> The specific environmental risks on the poll were: drifting genes, creating 'superweeds', increasing pest resistance, affecting non-target organisms, reducing biodiversity, or changing the ecosystem. Benefits listed were: engineering plants to clean up toxic waste, reducing soil erosion, reducing run-off, needing less water to grow crops, saving endangered or threatened species, reducing the need to log in native forests, or reducing pesticide use. Asked to rank these 13 items in terms of personal importance, the environmental benefits scored significantly higher than any of the risks listed, with the exception of the non-target organism issue nationally ... Prior to reading a series of informational statements about the possible benefits and risks of biotechnology, respondents nationwide were more likely to say that the risks of biotechnology outweighed the benefits (40% to 33%), while 19% thought the benefits and risks were about the same, and 9% were unsure. However, after being read a series of questions about specific environmental risks and benefits (without specifically identifying which were risks or benefits), respondents were exactly evenly divided, with 38% saying the risks outweigh the benefits and another 38% saying the benefits outweigh the risks. (Pew, 2002)

European Union

In the European Union, Eurobarometer surveys reveal high suspicion towards biotechnology. The most recent Eurobarometer survey on this topic was carried out in spring 2001 on about 16,000 people in the EU[1]. It enables a comparison between the different EU countries because the same questionnaire has been used in the 15 EU countries. The questionnaire, filled in during face-to-face interviews, addressed various issues related to technological and scientific progress (European Commission, 2001). Among them, some questions dealing with biotechnology show a high level of mistrust of GMOs. Previously, in the 1990s, other Eurobarometer surveys were specifically devoted to biotechnology and tackled many aspects of opinion about it (European Commission, 2000; Gaskell *et al.*, 2000). From the 2001 Eurobarometer, the results of only one question are presented here: the feeling of danger about food based on GMOs (Table 16.3).

A majority of Europeans (56.4%) believe that food based on GMOs is dangerous, as opposed to 17.1% who do not; however, this is an open question for more than a quarter of Europeans (26.5% of 'don't know'). Variations by country are significant: in The Netherlands, Finland, the UK and Sweden less than 46.5% of people are wary about the danger of GM food. However, in Greece, France and Luxembourg, more than two-thirds of the inhabitants believe the same. This greater level of variation by country than by usual socio-demographic variables can be linked to the importance of cultural aspects, as well as to the differences in public debate, government intervention, history of economic development and industrial situation between the various European countries (Zechendorf, 1998; De Cheveigné *et al.*, 2002; Springer *et al.*, 2002). However, variations per socio-demographic characteristics are smaller than those per country. The feeling of danger is a little lower among managers, students, high-income people, educated people (i.e. those

1 More precisely, a total of 16,029 people was questioned between 10 May and 15 June 2001. In each EU Member State a representative sample of the national population aged 15 and over was taken, with an average of some 1000 people per country, except in Germany (1000 in the new Länder and 1000 in the former Länder), in the UK (1000 in Great Britain and 300 in Northern Ireland) and in Luxembourg (600). This opinion poll, managed and organized by the EC Directorate-General for Press and Communication, Public Opinion Sector, was carried out at the request of the Directorate-General for Research. It was conducted under the general coordination of EORG, the European Opinion Research Group, a consortium of market study and public opinion agencies.

Table 16.3. Opinion of GMOs in the EU, by country and by socio-demographic group (European Commission, 2001): 'Do you think it is true or false that food based on GMOs is dangerous?' (% of answers).

Country	True	False	Don't know
Netherlands	37.9	31.7	30.4
Finland	43.0	34.3	22.8
UK	44.8	23.9	31.3
Sweden	46.6	25.0	28.4
Denmark	48.,9	25.8	25.4
Belgium	51.4	21.3	27.2
Germany (E)	53.8	15.1	31.0
Germany	53.8	15.8	30.4
Germany (W)	53.8	16.0	30.2
Total EU 15	56.4	17.1	26.5
Portugal	57.2	10.5	32.3
Ireland	58.3	13.4	28.3
Italy	59.8	14.6	25.6
Spain	60.8	14.2	25.0
Austria	64.4	15.0	20.6
Luxembourg	66.6	10.9	22.5
France	67.6	12.4	20.0
Greece	88.8	3.2	8.0

Germany (E), new Länder; Germany (W), former Länder.
Countries are ranked by increasing level of perception of GMOs as being dangerous.

Socio-demographic variables		True	False	Don't know
Gender	Male	54.0	20.3	25.7
	Female	58.7	14.1	27.2
Age (years)	15–24	54.6	20.8	24.6
	25–39	56.4	19.3	24.3
	40–54	58.4	16.9	24.7
	55+	55.9	13.5	30.6
Education level	⩽15	58.0	11.7	30.4
(age when left school)	16–19	57.9	17.5	24.6
	20+	53.2	20.8	26.0
	Still studying	53.0	23.4	23.7
Area	Rural/village	59.0	15.3	25.7
	Small town	55.6	17.9	26.5
	Large town	55.0	17.9	27.1
	Don't know	49.9	16.9	33.1
Profession	Self-employed	61.1	15.6	23.3
	Managers	48.8	24.1	27.1
	Employees	57.1	19.1	23.8
	Manual workers	58.0	17.5	24.5
	House-persons	59.3	12.5	28.3
	Unemployed	58.4	15.5	26.1
	Retired	55.7	13.2	31.1
	Students	52.3	23.6	24.1
Income level	– –	57.7	13.7	28.6
	–	56.6	16.6	26.8
	+	56.6	18.8	24.5
	+ +	53.1	22.5	24.4
	Don't know	57.3	15.5	27.2
EU 15	EU 15	56.4	17.1	26.5

who have studied beyond age 20) and men: among them, less than 54% think that it is true that 'food based on GMOs is dangerous'. However, among self-employed, house-persons, rural area or village inhabitants, and women, more than 58.7% are worried and the feeling of danger is higher. Thus, people who are a little more vulnerable or fragile appear to be a little more worried about the potential risks of GMOs.

The Focus on Potential Risks and the Extensive Publicity Given to Them

Thus in France, opposition to GMOs and concern about their risks appear to be among the strongest, particularly when one compares the opinions in France with those expressed in most other European countries. What factors explain this increasing hostility in public opinion? A number of processes commonly used and a number of products currently on the market present drawbacks and offer only illusory benefits. They are, however, well-established and generate only a small degree of rejection. In the same way, in technical fields numerous process innovations are hardly noticed or are known only in limited circles. However, the controversy over transgenic plants drew a wide audience and received extensive concern, especially in the late 1990s. The intensity of the GMO debate heightened and opinions became increasingly radical. One important factor is the frequently critical, or even negative, information spread about them, in particular by opponents and the media (see below). Of course, the influence that the information received from the media has on people's opinions is not linear and direct: people are not merely passive spectators and 'recipients' of information, particularly if significant risks are involved. However, the media play a significant role in particular when several actors are in contest to advance their views.

> The media are active mediators through (i) their focus on particular issues and neglect of others, (ii) the process of legitimising certain points of view – i.e. who gets to speak, and (iii) their significant ability to resonate with the public mood. They compete effectively on the field because of their control over timing, their ability to have their position treated as credible (at least parts of the media), and their ability to command an agenda and debate. In all of these elements they have frequently been more successful than the risk experts and managers. (Petts, 2002)

In the GMO field, the media played a significant role as has been shown (Petts *et al.*, 2000; Ipsos Corporation unpublished study, 2002; Frewer *et al.*, 2002; Petts, 2002).

In addition, GMOs surfaced in force at about the same time as the public's confidence in institutions and certain technological advances had been shaken by several safety affairs, in particular the issues of contaminated blood, mad cow disease, asbestos, etc. To make an extremely brief summary of these affairs, one can say that in the case of 'contaminated blood', through blood transfusions patients received blood products contaminated by the AIDS virus when in fact the state of knowledge at the time could have allowed this risky practice to be limited. In the case of 'mad cow disease', despite presumptions of risks, stringent measures on cattle feed and meat imports were sometimes taken with much delay – or were not complied with – primarily to protect economic interests in the sector. About asbestos, although its risks had been known for a long time, it continued to be used, especially to protect the interests of this industry which was an influential player in the official body responsible for evaluating and managing risks. These events led to definite distrust of firms and public authorities and increased the public's attention to critical voices, and so the principle of precaution became an omnipresent reference.

The strong influence of associations that focus on risks

In France GMOs have been strongly opposed by various associations. Initially these consisted essentially of ecologist associations (Greenpeace, Friends of the Earth, etc.) and groups of various tendencies (e.g. Ecoropa, the Natural Law Party), as well as supporters of the Green political parties and organic agriculture associations. This movement progressively expanded from environmentalist circles towards groups active in the economic domain including, for example, a farmer's union – the Confédération Paysanne – anti-globalization organizations (ATTAC, the Association for the Taxation of Financial Transactions for the Aid of Citizens), LETS (Local Exchange Trading Systems), etc. Finally, small circles of associations were created for the very purpose of fighting against GMOs, for example OGM Danger!, OGM Dangers, Terre sacrée, etc.

The impact of these associations has been strong, owing to the dynamism of their action which gave them extensive publicity: numerous strongly worded press releases, the repeated mass dissemination of alerts and warnings, petitions, leaflets, standard letters to send to elected representatives, agro-food firms or food retailing companies, lawsuits, demonstrations, and so on. In particular, these groups took advantage of the new communication technologies: multi-transmission of information via electronic mailing lists, forums of discussion on the internet, very well-documented and frequently updated websites used extensively by many as sources of information, etc. The endless reuse of certain information (sometimes very partial or biased) gave it credibility due to multiple repetition that ended up making it seem reliable (since it was frequently mentioned, it was corroborated). In addition, the influence of groups that had taken a stand against GM extended way beyond their own supporters to many sympathizers or people close to them.

The mobilization of the staff and members of many associations on this issue was intense, not only because they felt strongly about it but also because it helped to establish their audiences and legitimacy, especially in the case of diverse associations that were formerly in a tiny minority. For example, its anti-GMO action was instrumental in strengthening Greenpeace-France which had been in serious financial straits and was experiencing a relative drop in its membership compared to other north European countries. Greenpeace now has sound legitimacy and is invited to many debates and conferences. More generally, the GMO issue has enabled various groups or associations to enhance their renown, recognition and resources, and to acquire a degree of legitimacy by presenting themselves as defenders of consumers and of their health and interests, but also of the environment and of the interests of developing countries or of future generations. Since it has proved to be so fruitful, this encourages them to pursue their militancy in this field and to devote more resources to it.

Many actors are involved more or less directly in the GMO field but their respective influence varies widely. For the entire EU, the Eurobarometer survey in late 1999 showed that the actors who were judged most often by respondents as 'doing a good job for society' as regards GMOs were primarily consumers' unions, doctors, the media and environmentalist groups. By contrast, industry is the only actor judged most often as not doing 'a good job' for society in this respect (Table 16.4). The countries most hostile to industry's biotechnology activities are Sweden, Greece, France, Denmark, Ireland, Italy, Austria, Luxembourg and the UK. This sheds light on certain determinants of opposition to GMOs and on the respective impact of the actors involved. Industry seems to have little credit and its arguments are therefore taken into consideration relatively little or are even discredited. By contrast, other actors that are often opposed to GMOs – consumer unions, environmentalist associations and the media – have more legitimacy and are therefore taken into account and quoted more often.

Behaviour of other actors involved in publicizing information on risks

The publicity given to various associations' denunciation of GMOs has been noteworthy, particularly in the case of the media which have played a significant part in making GMOs widely known and in highlighting their potential dangers, especially from 1999 when many journalists became increasingly opposed to GMOs. Whereas previously – especially in the early 1980s when there were few articles on the subject – the media presented biotechnology as a promising inno-

Table 16.4. How the activity of the actors involved in biotechnology in the EU and in France is perceived (European Commission, 2000) (% of answers).

	Tend to agree		Tend to disagree		Don't know	
Do you think they are doing a good job for society:	EU	France	EU	France	EU	France
Consumer organizations checking products of biotechnology[a]	70	73	12	14	19	12
Medical doctors keeping an eye on the health implications of biotechnology	69	69	11	15	20	16
Newspapers and magazines reporting on biotechnology	59	56	18	25	23	19
Shops making sure our food is safe	59	48	21	35	20	17
Environmental groups campaigning against biotechnology[b]	58	66	18	18	24	16
Farmers deciding which types of crop to grow	55	53	20	27	25	20
Ethics committees looking at the moral aspects of biotechnology	53	55	18	23	29	22
Our government making regulations on biotechnology	45	40	29	39	26	21
The churches giving their points of view on biotechnology	33	20	31	44	35	36
Industry developing new products with biotechnology[c]	30	25	38	51	32	25

[a]The countries most favourable to the consumerist associations are The Netherlands, Finland, Greece, Denmark and Austria.
[b]The countries most favourable to the ecologist associations are Greece, Austria, Denmark, France and Finland
[c]The countries most favourable to industrial activity in biotech are Finland, The Netherlands, ex-East Germany, Belgium and Portugal.
Countries which are the most opposed to industrial activity: Sweden, Greece, France, Denmark and Ireland.

vation, in 1999 and 2000 increasingly critical opinions were often expressed. A number of journalists focused on risks and expressed standpoints opposed to GMOs, sometimes entering into opposition movements themselves (Durant and Lindsey, 2000; Kassardjian, 2002). This can be explained by various factors. Initially, the subject of biotechnology was treated by scientific journalists who were relatively in favour of it. Later, when the topic became more politico-economic, it was also covered by other journalists, for example those who had worked on the issues of contaminated blood, mad cow disease, etc., and who drew parallels between these issues. Another explanation lies in the characteristics of the journalistic profession and the strong competition within the media sector. Shocking headlines revealing hidden dangers and dramatic presentation of issues guarantee wider audiences and have more impact than more moderate, qualified articles; hence, this tendency to overstate and try to outmatch one another.

Furthermore, the communication methods of associations opposed to GMOs often guaranteed them a strong impact in the media. These associations focused on spectacular actions announced in advance. Pictures of activists chained to or climbing on to strategic or symbolic places, photographs of large protest banners, destruction of transgenic crops, and so on, had every chance of receiving extensive media coverage due to their characteristics and attractiveness. This is precisely one of the aims of this type of action (Ruckus Society, no date). Likewise, their press communiqués were particularly lively, stimulating and clear, and their websites well documented.

On the other hand, the firms involved have often maintained a more traditional type of communication, strongly influenced by their usual clientele – the agricultural sector, not the public at large. Moreover, until 1997–1998 they often underestimated suspicion of GMOs, considering it to be the product of irrational and somewhat residual fears that would progressively disappear as more information became available. They often seemed to consider that the rejection of GMOs was mainly the result of poor knowledge of biotechnology and that the public only needed to be educated. But their promotion of the advantages of GMOs did not convince the public.

As for the public research organizations, on the whole they did relatively little public relations work on the subject in France, especially compared to the associations involved. Institutional communication often remained focused on the presentation of important results obtained by research teams. No statements were issued to clarify the matter when facts or controversies on specific points concerning GMOs were mentioned in the media (which was very often). As a result, explanations and interpretations disseminated very widely among the general public were often those from associations opposed to GMOs. Researchers from public research organizations were interviewed, but they were often quoted too selectively (or partially) in articles that mainly reflected the viewpoints of associations opposed to GMOs. In addition, the views expressed by scientists tend to be complex while those expressed by opponents are very loud and clear: 'GMOs are dangerous, we must ban them'. We note that the opinions of researchers in the life sciences on genetic engineering vary, depending essentially on their specific discipline, but that the vast majority of those working in molecular biology

> believe that recombinant DNA techniques constitute powerful and safe means for the modification of organisms and can contribute substantially in enhancing quality of life by improving agriculture, health care, and the environment. ... We ... express our support for the use of recombinant DNA as a potent tool for the achievement of a productive and sustainable agricultural system. (AgBioWorld, 2002)

To be sure, the point of view of researchers working in the environmental sciences is often somewhat more reserved, and many scientists are concerned about patents or other economic aspects, but generally researchers working in the life sciences think that genetic engineering is a useful tool. However, evaluation methods in public organizations urge them to publish in highly specialized scientific journals far more than in magazines for the general public or popularized science magazines, and to participate in scientific conferences rather than in debates with the general public. In fact, the latter forms of publicizing results are even frequently discredited in the scientific world. Even if researchers have participated in public debates, in total these have reached only a very small audience. Thus, on the whole little but silence can be heard from public research. So, views against genetic engineering voiced by certain scientists are quite frequently cited. These scientists are judged as saying aloud what others dare only to think (for fear of losing their contracts with private firms, or dissuading public financing). Public research has, moreover, published relatively few books or statements for the general public on GMOs. It has participated in many fairly specialized scientific conferences on this theme, but these have received little attention outside scientific circles. In contrast, the book *Tais-toi et mange! L'agriculteur, le scientifique et le consommateur* [*Shut-up and eat! The farmer, the scientist and the consumer*] (Paillotin and Rousset, 1999), written by the president of INRA, was judged simply on the basis of its title as legitimizing reservations.

Various consumers' unions also became strongly involved in the GMO controversy, without being opposed from the outset. They stressed the need to take into consideration risks and the principle of precaution. In late 2000, delegates at the International Consumers' Organization conference in Durban called for a moratorium:

> governments and international institutions should require full pre-market evaluation and social and safety impact assessments of GM foods and the products of other new food technologies to ensure that they are safe, environmentally sustainable and acceptable to consumers, and impose a moratorium on the cultivation and marketing of new GM foods until this is done. (Consumers International, 2000)

In the EU the European Consumers' Organization (BEUC) emphasizes that the recent years have shown that GM food will never be accepted without consumer choice. A press release from July 2002 entitled 'GMOs or No GMOs: the choice should be ours!' reminds of its demand 'to ensure that the key consumer rights to information and choice are met' and claims 'only clear labelling will ensure that consumers can choose whether or not to buy GM foods'.

The question of labelling of GM products helped to radicalize the debate. In 1998 Greenpeace launched an 'Info-conso network' with the slogan 'no GMOs on my plate', listing products and brands according to whether or not they contained GM ingredients, and stigmatizing those that did. It urged consumers to ask producers or distributors to adopt measures necessary to 'preserve Europe and the food chain from contamination from GMOs against consumers' will'. This movement was strongly relayed. To avoid a loss of market share, one by one many agro-food or mass distribution groups committed themselves to excluding GMOs in France, in Europe and sometimes also in the USA.

In this context in which many influential actors (the media, associations) denounced the risks of GMOs, hostility towards them seemed to many to 'stand to reason', simply on the basis of information received or because that was the standpoint of the ideological movement to which they felt closest.

Many risks or negative effects are suspected in a very wide field

We have established a typology of these risks, fears and reasons for refusal, on the basis of the subjects mentioned repeatedly in debates, articles and arguments against GMOs (Table 16.5). This focus on the risks of a technical innovation is nothing new. At the time of their introduction many innovations were violently opposed, e.g. industrial mechanization, the railway, the potato, etc. (Salomon, 1984). But in the case of GMOs this opposition has been particularly strong and widespread, so that in Europe this innovation has comparatively few advocates or supporters.

A Risk–Benefit Assessment of GMOs Perceived as Very Unbalanced

One of the causes of European opposition to GMOs is that their advantages in food production are often considered to be weak or non-existent, while their risks are considered to be substantial (Gaskell *et al.*, 2000; European Commission, 2000, 2001; De Cheveigné *et al.*, 2002).

Advantages of GMOs judged weak by many

Opponents of GMOs presented them as a technology with high potential risks and with no advantages except for the firms that developed them. They strongly emphasized two arguments:

1. GMOs comprise many risks, from which no one can escape since they concern daily food and the immediate environment. They also comprise other, more global, potential dangers and risks for farmers in developing countries and for biodiversity, which legitimizes opposition and actions against them and even makes this opposition essentially ethical.
2. Their possible benefits will go primarily to the firms that produce them and not to society as a whole or to consumers. On the contrary, consumers' and society's safety is sacrificed.

It is not surprising that, thus presented, GMOs were met with suspicion, especially since these arguments of distrust of GMOs were perceived as credible in a context where agricultural productivism is strongly called into question and suspicion reigned in the aftermath of several public health affairs. On the other hand, arguments which tried to present the potential advantages of genetic engineering were often rejected because they were perceived as hypocritical.

In the USA, where opinions were more favourable to GMOs, people often thought that opposition in Europe came from its relative backwardness in this field, and that arguments on risks concealed a form of protectionism aimed at avoiding the dismantling or buy-out of the European seed industry. Another frequent US interpretation is that GMO refusal aimed to protect the European market from US grain importations. But, while this fear might sometimes have had an influence, it did not stem mainly from

Table 16.5. Motives put forward for GMO rejection: risks, fears and reasons for refusal (typology developed by the author on the basis of the themes repeatedly treated in debates, articles and declarations made by opponents).

Types of risk	Fears and perceived risks
Troublesome, violent gene	Transgenesis = transgression of the barrier between species
transfer process	Risk engendered by troubling the 'order of the genome', which may appear only later
	insufficient knowledge of the genome to authorize such tinkering with the transfer of foreign genes (living organisms are not just 'building blocks')
Health,	Allergies, long-term toxicity
for example	Insufficient safety tests: 'consumers = guinea pigs'
• Bt maize	Gene coding for Bt toxin → consuming continuously secreted insecticide toxins
• glyphosate-tolerant soya	Gene coding for the enzyme which degrades glyphosate → GMOs accumulate products of degradation
Environmental	Gene flow towards related wild species → 'superweeds', invasive plants, accelerated decrease in biodiversity
Agro-economic	Gene flow towards nearby crops of the same species → impure harvests, 'contaminated'
	Problem of volunteer plants in the following crop (rapeseed)
	Risk of a drop in Bt or glyphosate efficiency, interesting molecules for use in other agricultural sectors
Economic	Of little interest to consumers, 'product imposed' by the multinationals
	Increasingly dependent agriculture (farmers must buy seeds every year)
	Difficulty for developing countries to access such technology (patents)
	→ hypocrisy of saying 'Genetic engineering is necessary to feed humanity'
	Appropriation of genetic resources by a few large multinationals
	GMOs = symbol of privatization of all resources, now even genetic resources
	'Imperialist' technology because coexistence with non-transgenic production is difficult (gene flow)
Agricultural and food	Reinforcing of the industrialized model, the limits of which have already been critically portrayed
production model	Consumer perception: 'They're playing with our health to make more money' (cf. BSE and contaminated blood).
More socio-political motives	Innovation neither asked for nor desired, but set up solely for the profits of some multinational firms
(value systems and beliefs)	No respect for consumer free choice due to the presence of GMOs in many additives and fortuitous 'contamination' of grain through gene flow
	Media showing scientists (or associates) opposed to GMOs } → opinion: 'They're hiding something from us'
	Vacillation in the positions taken by public authorities }
	Perception 'Everything is messed with more and more' → the desire to return to true nature (growing interest in organic products)
	GMOs symbolize development towards a type of society which is perceived negatively
	'Such progress, why bother?' (a certain loss of faith in science and progress)

economic protectionism since the French and other European people's rejection affected the biotechnology activities of their own countries just as much.

GMOs thus lacked supporters and allies in many European countries, including France. Moreover, in the late 1990s the public authorities adopted a hesitant attitude in this respect, often backtracking and procrastinating, which heightened confusion and perplexity (De Cheveigné *et al.*, 2002). Thus, faced with strong denunciation by various associations, extensively relayed by the media, there were few actors to present GMOs in a favourable light: firms were judged as having little credibility and public research organizations made few public and official statements on the subject; the few scientists or their allies interviewed by the media were in some cases against GMOs; and, lastly, the authorities seemed confused and hesitant. The European and French situation is, in this respect, very different from that in the USA where GMOs enjoyed extensive support.

Yet an abundant scientific literature has been published on the potential benefits of GMOs, generally evaluated as being greater than the foreseen risks: more efficient agricultural production with reduced losses; increase in productive capacities in difficult conditions; improvement in various qualitative characteristics; diversification of uses of plants with the possibility to produce diverse molecules, etc. We mention only a few references here on this topic which alone necessitated a considerable amount of synthesis work (Conway and Toenniessen, 1999; Borlaug, 2000; Interacademies, 2000; AgBioWorld, 2002; ASPB, nd; ACN, 2002; SOT, 2002). Biotechnology opens the possibility of a new path for technological development, based more on living processes and on information (knowledge) than on chemical inputs. But in Western Europe, in a context of agricultural overproduction and extensive calling into question of agricultural productivism, these aspects hardly aroused much public interest. Moreover, the very first products commercialized (transgenic soybeans and maize) seemed to consumers to have little interest, especially in Europe. As Boy puts it

> there is a missing fundamental element in this progress which constitutes the king-pin between science and society: the utility function ... One of the basic springboards towards acceptance of innovation is the risk/utility equation ... If an invention arouses incomprehension as to its usefulness while presenting a potential risk factor it is doomed to a motivated rejection. (Boy, 1999).

In other words, even if GMOs can have advantages for society as a whole and for all actors, the actors situated downstream from production – who today have considerable weight – judged them as being of negligible and often even of no interest compared to their potential and unknown risks. Nothing justified the use of GMOs, perceived as serving only the interests of the firms involved; any risk-taking seemed unjustified.

Some words on risk perception

Concern as regards GMOs cannot simply be imputed to a lack of knowledge in biology, as many actors arguing for better education of the public have done (Miller and Conko, 2000). In addition, the public at large cannot be accused of irrationality, as research conducted in several European countries shows (Marris *et al.*, 2001). Various studies have enabled us to better understand risk perception. Experts evaluate it in relation to two components: the probability of an undesirable event actually happening, and the seriousness of its consequences. The public, on the other hand, takes into account a set of other factors in its assessment of risks, as many studies have shown (Slovic, 1987; Morgan, 1993; Slovic *et al.*, 1995; Powell, 1998; Siegrist, 2000). These factors are in particular:

1. The knowledge of and familiarity with the specific risk: household accidents and automobile accidents generate less worry than the potential dangers of GMOs. The invisible or uncontrollable is especially prone to provoke anxiety.

2. The delay before the appearance of bad consequences: some important risks (such as heavy smoking or sun-tanning without precaution) are quite often taken deliberately with lack of concern for the consequences because they will appear only in the distant future.

3. The semblance of catastrophe: an accident affecting several people at the same time and place usually has more impact than individual accidents spread across time and space, even if the total number of people affected is much lower.
4. The possibility for those exposed to risk to control the risk: the feeling of mastery is the essential point.
5. The voluntary or involuntary nature of the event. One is more angry about being exposed to an inescapable risk than to a risk from which one can escape (or choose for oneself); one tolerates deliberate risk best.
6. Risk-related advantages for the person exposed to or taking the risk: a risk which brings profits to the person responsible for its creation but not to the person exposed to it induces a high level of indignation. People are much more shocked by the impact of accidents to children because of the 'innocence' of the victims.
7. Scientific uncertainty and controversy: poorly understood risks make everyone nervous. In the case of controversy the public suspects those who minimize the risk of having vested interests in the affair or of being obliged to take their position by people who want to avoid a crisis (as for BSE in the UK at the end of the 1980s).
8. Confidence in institutions.

> The perceived risk of biotechnology will be significantly influenced by trust in the system that produces it … Components of the relationship that builds trust, or distrust, is the extent to which an individual feels affiliated with the system, agrees with the distribution of decision-making power, and shares the values enabled by the system. (Espey, 1998)

Thus, some individual practices which represent a true danger, such as heavy smoking or driving a car, arouse less worry than genetic engineering which is less well known, unobservable, difficult to control and can lead to risk exposure which is not a question of personal choice. So, acceptability depends on numerous factors in relation to risk perception. In addition, this acceptability is highly contingent on the consumer assessment of the expected benefits justifying risk taking and offsetting potential bad effects. But the considered risks of GMOs have been extended to a very wide field, including many socioeconomic or political aspects. Hence consumer suspicion is not rooted solely in a lack of knowledge of production and risk. The underlying issues are much more complex:

> The argument that scientific literacy, or more knowledge about a technology, will increase support assumes that there is no reasonable basis for opposition. It assumes that technology is objectively a desirable thing and that opposition must stem from ignorance of its true benefits and costs. This approach not only fails to receive empirical support, but adhering to it, conceptually, severely jeopardizes policy development and communication strategies … Proponents of technology often argue that public perceptions of risk are irrational. Yet, public risk calculations are rational, given the socio-ethical perspective from which they are derived. In fact, expert risk assessments also stem from a specific socio-ethical perspective. The difference between expert and public risk perceptions should be seen as the difference in the socio-ethical perspective which defines the calculation, not as the difference in rational ability. (Espey, 1998)

Diverse Opposition to and Concerns About the Functioning of Society and its Evolution Crystallized Around GMOs

Limited trust in the institutions and firms involved

Despite parliamentary debates in 1992 at the time of the transposition of European directives concerning the dissemination of GMOs, and various articles in the media on biotechnology, discussion on the subject remained limited in the early 1990s to a fairly small circle. It started to spread in the public at large mainly from late 1996 when the very first imports of transgenic oilseeds from the USA arrived in Europe and animated debate surrounded authorization of Bt maize from the firm Novartis. At that stage public opinion was strongly marked by various affairs, especially 'contaminated blood' and BSE, which led to strong distrust and caused people to think that firms and public authorities sometimes disregard certain health risks in order to protect certain economic or political interests.

In the period 1998–2000, debate on GMOs (authorization, importation, labelling, impact, etc.) was situated in a context strongly influenced by food safety issues (BSE, listeriosis, etc.) that had been widely publicized. As a result, GMOs were perceived as an additional indication of negligence when it came to health hazards. The precautionary principle therefore became extensively invoked and adduced.

One of the arguments often put forward by the promoters of biotechnology to justify its development is that it is necessary for feeding the world's population, particularly in the coming decades. Although it is valid in several respects, this argument has frequently been perceived as highly hypocritical when used by multinationals. Indeed, these corporations frequently adopted a policy of patenting and prohibition on the free reuse of saved seeds by farmers, that is, commercial policies that could strongly limit poor farming communities' access to biotechnology. Furthermore, non-governmental organizations stress that genetic engineering is likely to increase the risk of food dependence on major agro-exporting countries. Hence, mistrust regarding the policies of the public authorities and firms involved in the commercialization of GMOs increased sharply. It was, moreover, fuelled by the many turnarounds and instances of procrastination which could give the impression that 'they're hiding something from us' or that too many unknowns still existed (Table 16.6). In July 2000 a majority (58%) of respondents said they tended to disagree with the opinion that 'the public authorities can be trusted to make good decisions on GMOs' while 40% tended to agree (IFOP and Libération, 2000).

GMOs – symbol of negatively perceived trends

Biotechnology is often seen as an ultimate reinforcement of highly industrialized agriculture that has been the object of much criticism in the past few years (Bonny, 2000a). It is blamed for deterioration in the quality of foods, damage to the environment, an accelerated reduction in the number of farms, etc. This mistrust generated by the modernization of agriculture appears in a survey carried out

Table 16.6. Trust in French public authorities to protect people in several areas, and the belief that the truth is told on related dangers (Charron *et al.*, 2000).

	Respondents having confidence (%)			Respondents thinking the truth is told (%)		
Area	No (not at all or not really)	More or less	Yes (strongly or somewhat)	No (not at all or not really)	More or less	Yes (strongly or somewhat)
Nitrates and pesticides	53.3	30.2	13.5	55.6	26	16.1
Genetically modified crops	49.3	28.6	18.4	62.7	23	10.8
Young people's smoking	49.0	26.8	23.4	28.3	24.5	46.5
Lake, river and sea pollution	47.4	32.7	19.2	52.7	31.1	15.8
Genetic manipulations	46.1	28.0	20.7	57.3	26.9	12.9
Radioactive waste	45.4	31.5	21.9	64.4	21.1	13.8
Chemical waste	44.8	33.1	20.4	60.8	25.7	12.2
Air pollution	43.5	34.8	21.1	45.3	31.8	22.3
Alcoholism	39.8	30.4	29.0	25.5	21.4	52.5
Nuclear plants	38.7	27.2	32.7	55.2	26.9	17.2
Food products	37.8	31.9	30.2	47.9	27.8	24.0
Chemical plants	35.0	36.4	25.2	54.4	30.6	12.3
AIDS	28.7	27.0	43.8	24.9	20.9	24.9
Road accidents	28.0	30.4	41.2	18.2	23.3	58.3
Water from the tap	26.5	32.1	40.9	36.2	29.2	33.8

BVA poll carried out on 16–31 October 2000, representative sample of 1000 people.

in late 2000 and early 2001. In late 2001 over one-third of respondents considered that 'agricultural use of scientific and technological innovations' is 'a bad thing for the consumer', while a quarter thought that 'it is a good thing' (Table 16.7a) (UNCAA-SIGMA, 2001; Union InVivo 2002). The Eurobarometer survey carried out in 2001 (European Commission, 2001) also shows a relative scepticism in France about the impacts of science and technology on agricultural and food production. The French are the most sceptical of the Europeans particularly when compared with northern European countries (Table 16.7b).

For some people, especially many activists, biotechnology also symbolizes the negative aspects of globalization and economic liberalism: destruction of local cultures and economies, growing trend of commodifying everything including genetic resources, and aggravated competition often perceived as disloyal due to the rivalry created between economies with different levels of development. So, certain surveys reveal that economic motives have become a significant cause of opposition to GMOs – at least for some groups – (Table 16.8). Arguments put forward by active opponents show that they often perceive this struggle as a form of opposition to extreme

Table 16.7a. Opinion about the agricultural use of scientific and technological progress (UNCAA-SIGMA, 2001; Union InVivo, 2001) (% of answers).

'What do you think of the agricultural use of scientific and technological progress?' Would you say that it is...	December 2000	November 2001
A good thing for the consumer?	34	25
A bad thing for the consumer?	31	35
A neither good nor bad thing?	32	32
Don't know.	3	8

UNCAA-SIGMA-SOFRES poll carried out in late 2000 (from 27 December 2000 to 3 January 2001) and late 2001 (in November 2001); representative sample of 1000 people.

Table 16.7b. Opinion about the impacts of science and technology on agricultural and food production (European Commission, 2001): 'Do you think it is true or false that science and technology will improve farming and food production?' (% of answers).

	True	False	Don't know
France	49.2	31.6	19.2
Luxembourg	51.6	28.9	19.5
Italy	52.0	23.1	25.0
Austria	53.1	25.2	21.7
Spain	55.4	21.4	23.2
Portugal	58.0	14.7	27.3
Germany (former Länder)	58.5	19.9	21.6
Belgium	58.6	23.1	18.3
European Union EU 15	59.0	20.7	20.3
Germany total	60.1	18.5	21.4
Ireland	62.4	13.8	23.8
UK	65.5	16.6	17.9
Germany (new Länder)	66.1	13.5	20.4
Greece	69.5	14.9	15.6
The Netherlands	75.7	12.0	12.4
Sweden	75.7	12.4	11.9
Denmark	78.0	12.5	9.5
Finland	78.3	9.7	11.9

Table 16.8. Reasons for the opposition to GMOs in 1999–2000, according to surveys aimed at explaining purchasing behaviour towards such products by consumers, in experimental economics (non-representative sample) (Ruffieux and Robin, 2001) (% of answers).

Reasons mentioned in the questionnaire: Producers and food distributors sell products ...	Tend to agree	Tend to disagree	Don't know
Which are above all profitable for themselves	90	5	5
Which are not safe for my health	64	22	14
Which are too far removed from nature	61	22	17
Which are not environmentally sound	53	27	20
Which are not in line with my moral standards and my principles	39	42	19

economic liberalism. Militancy in this respect is in a sense a sort of metamorphosis of anti-capitalist militancy, or at least of protest against its excesses. Since the collapse of the communist ideal has made direct opposition to capitalism more difficult today, it seems to have found new forms of expression including, in particular, criticism of globalization, certain aspects of consumption, technical developments, etc.

For the general public, GMOs are perceived above all as hardly useful, non-natural and risky. This suspicion, along with limited trust in the institutions and firms concerned, often leads to the suggestion that greater participation of citizens in scientific and technological choices would be desirable and useful. Some people believe that it would help to solve the current deadlock regarding acceptance. Others believe in the need for a renewal in democracy, as this extract from an editorial vehemently illustrates:

> People no longer automatically accept that scientific development is necessarily beneficial to humanity. Particularly because that progress has become inextricably tied up with money, hijacked by companies greedy for profit ... In addition, our decision-makers have developed a bad habit of mortgaging our collective futures without first asking us, the people. The basis of the democratic pact has thus been altered. As a result, people have become more and more suspicious. They are increasingly unwilling to give the powers-that-be the authority to play with our collective futures by rubber-stamping scientific innovations that are risky and insufficiently tested. A new spirit of distrust is abroad among the sorcerer's apprentices of neo-scientism ... Shouldn't we all have a say in defining what is acceptable risk, and not just leave it to the 'experts'? (Ramonet, 2000).

Conclusion

It has been quite often said that the rejection of GMOs was the result of poor knowledge in biology and that the public mainly needed to be educated. However, it is necessary to analyse the causes of GMO refusal more deeply and further: such was the goal of this chapter, which presented various factors, actors and mechanisms of this movement. Opposition to GMOs stems from the many potential risks particularly highlighted by associations and many media, and from a stigmatization of their possible advantages. By presenting themselves as defenders of consumers' interests and health, the opposition rallied a substantial proportion of the Western public who saw no advantages in GMOs. GMOs thus seem to have become a symbol for many negative aspects of global economic development when in fact they are by no means the only forms or embodiment of that development. In this respect they differ from many other innovations that also strongly represent general economic development but the advantages of which are judged more clearly apparent by those who have access to them, and which are therefore the focus of little opposition. The quite low trust in companies performing gene technology and selling GM seeds – mainly agro-chemical companies – has an important effect on the benefit and risks perceived and thus on the acceptance of GMOs. On the opposite side, organizations or media putting forward GMOs' risks receive more trust and their risk alerts are therefore valued more and listened to.

A large part of denunciations and economic criticisms of GMOs are not in fact on specific dimensions of these products but concern techno-economic aspects affecting many goods and sectors. In other words, the economic criticism of GMOs could apply to many other products (which are spared the same opprobrium). In fact, they should target not GMOs themselves, but rather the context and economic conditions of their production and use. Thus, for example, the large *concentration* of firms in large multinational groups exists in many sectors, as does *the commodifying of* new activities; underprivileged populations in developing countries are exposed to and will probably continue to be exposed to *difficulties of access to many goods* requiring resources to obtain them or the infrastructure to produce them; *patents* have existed for a long time for many goods that are sometimes vital. As for the *potential impact on the environment*, it is considerable for multiple human activities; moreover, biotechnology can be considered as being able to contribute to greater sustainability, and not the opposite. Yet these issues are raised most forcefully for GMOs – as if they were the only subject to warrant them, perhaps because of the particular place of food and agriculture in society. They therefore seem to be acting as a *scapegoat*, in a sense. Unlike many other products with identical characteristics, GMOs are accused, even when they are not really directly concerned. For example, it is not genetic engineering itself that imposes patents for technical reasons; it is current economic conditions that lead to the use of patents; however GMOS are often accused of being responsible for the patents on living organisms. *Finally, GMOs are suspected due to their very essence, and not in relation to the way in which they are used.* And yet in fact, the impact of techniques depends on the way and conditions in which they are used, the purpose given to them, the orientation of their applications, etc. (Bonny, 2000a,b); they are accused because they are perceived as having little utility.

In this context a change of attitude towards GMOs seems difficult to achieve at present in France and the rest of Western Europe due to the strength of the current opposition (Bonny, 2003). On the contrary, the opposition movement is spreading around the world. A change of attitude would require that GMOs no longer be considered the symbol of various unpopular trends but rather for themselves, in relation to their potential and the objectives to be set for them. However, the history of techniques shows that many innovations, after strong initial rejection, were subsequently widely diffused but with considerable improvement, especially as regards risk reduction, improved convenience of use and usefulness. Changes in the general socioeconomic context as well could perhaps play an essential role by allowing GMOs to be perceived in a different light. For example, other risks, such as global change and its impacts, could move into the foreground and make biotechnology seem to be a possible solution. Transgenic plants are still in their early stages and various subsequent developments could reduce their potential risks or highlight more positive aspects of this technique or its products. But could this reversal take place when some have made GMOs a scapegoat that has to be eliminated because it symbolizes trends perceived as negative? Another solution may be the development of other applications of biotechnology and life-science research. The development of biotech and genomics applications could lead to new prospects for plant breeding and farming and so, perhaps, make foreign gene transfer less necessary. However that may be, the debate on this subject has had a considerable impact by inducing many questions about technical progress, scientific expertise, trust in private companies and public authorities, participation of the public, uncertainty, acceptable risk and choice of technological development paths.

References

ACN (American College of Nutrition) (2002) The future of food and nutrition with biotechnology. *Journal of American College of Nutrition* 21 (suppl. 3).

AgBioWorld (2002) *Scientists in support of agricultural biotechnology*. Declaration in support of agricultural biotechnology signed by more than 3330 scientists around the world until early December 2001. Available at http://www.agbioworld.org/

ASPB (American Society of Plant Biologists) (nd) *Statement on Genetic Modification of Plants Using Biotechnology*. ASPB, Rockville, Maryland. Available at http://www.aspb.org/

Bonny, S. (2000a) Consumer concerns about industrialized agriculture and food safety: importance, origin and possible solutions. *Annales de Zootechnie: An International Interdisciplinary Journal in General and Comparative Animal Science* 49, 273–290.

Bonny, S. (2000b) Will biotechnology lead to more sustainable agriculture? In: Lesser, W.H. (ed.) *Transitions in Agbiotech: Economics of Strategy and Policy*. University of Connecticut, Food Marketing Policy Center, Storrs, Connecticut, pp. 435–453. Available at http://agecon.lib.umn.edu/cgi-bin/pdf_view.pl?paperid=1986&ftype=.pdf

Bonny S. (2003) Why are most Europeans opposed to GMOs? *EJB Electronic Journal of Biotechnology* 6, (1).

Borlaug, N.E. (2000) Ending world hunger: the promise of biotechnology and the threat of antiscience zealotry. *Plant Physiology* 124, 487–490.

Boy, D. (1999) *Le progrès en procès*. Presses de la Renaissance, Paris.

Charron S., Mansoux, H. and Brenot, J. (2000) *Perception des risques et de la sécurité: résultats du sondage d'octobre 2000* (Baromètre IPSN). Note SEGR 00/112. IPSN, Clamart.

Consumers International (2000) *Consumers, social justice and the world market*. Statement from Consumers International's 16th World Congress, Durban, November 2000. Available at http://www.consumersinternational.org/homepage.asp

Conway, G. and Toenniessen, G. (1999) Feeding the world in the twenty-first century. *Nature* 402 (suppl. 2), C55–C58.

De Cheveigné, S., Boy, D. and Galloux, J.-C. (2002) *Les biotechnologies en débat: pour une démocratie scientifique*. Balland, Paris.

Durant, J. and Lindsey, N. (2000) *The Great GM Food Debate – a Survey of Media Coverage in the First Half of 1999*. Parliamentary Office of Science and Technology, Report 138, London. Available at www.parliament.uk/post/home.htm

Espey, J. (1998) *Socioethical Implications of Biotechnology*. Industry Canada, Office of Consumer Affairs, Ottawa, Canada.

European Commission (2000) *Europeans and biotechnology. Eurobarometer 52(1)*. European Commission, Brussels.

European Commission (2001) *Europeans, Science and Technology. Eurobarometer 55(2)*. European Commission, Brussels.

Frewer, L.J., Miles, S. and Marsh, R. (2002) The media and genetically modified foods: evidence in support of the social amplification of risk. *Risk Analysis* 22, 701–711.

Gaskell, G., Allum, N., Bauer, M., Durant, J., Allansdottir, A., Bonfadelli, H., Boy, D., de Cheveigné, S., Fjaestad, B., Gutteling, J.M., Hampel, J., Jelsoe, E., Jesuino, J.C., Kohring, M., Kronberger, N., Midden, C., Nielsen, T.H., Przestalski, A., Rusanen, T., Sakellaris, G., Torgersen, H., Twardowski, T. and Wagner, W. (2000) Biotechnology and the European public. *Nature Biotechnology* 18, 935–938

Hoban, T.J. (1997) Consumer acceptance of biotechnology: an international perspective. *Nature Biotechnology* 15, 232–234.

IFIC (International Food Information Council) (2002) *US Consumer Attitudes Toward Food Biotechnology*. Washington: IFIC, press release, September 2002.

IFOP and Libération (2000) Les Français et les risques alimentaires. *Libération*, 3 August.

Interacademies (2000) *Transgenic Plants and World Agriculture*. Report prepared under the auspices of seven Academies of Science (Brazil, China, India, Mexico, Third World, UK, USA). National Academic Press, Washington, DC.

Ipsos-Reid (2002) Genetically modified foods and food labeling. *Ipsos World Monitor*, 2nd Quarter, 28–37. Available at http://www.ipsos-reid.com/index.cfm.

Kassardjian, E. (2002) Appropriation de concepts en situation d'éducation non formelle, Cas d'une exposition scientifique sur les OGM. Thesis, Université Claude Bernard, Lyon, France.

Marris, C., Wynne, B., Simmons, P. and Weldon, S. (2001) *Public Perceptions of Agricultural Biotechnologies in Europe*. Final report of the PABE Research Project commissioned by the EC. Available at http://www.pabe.net.

Miller, H.I. and Conko, G. (2000) The science of biotech meets the politics of global regulation. *Issues in Science and Technology* 17, 47–54.

Morgan, G. (1993) Risk analysis and management. *Scientific American* 269, 32–41.

Paillotin, G. and Rousset, D. (1999) *Tais-toi et mange! L'agriculteur, le scientifique et le consommateur.* Bayard Centurion, Paris.

Petts, J. (2002) Science, society and risk: bridging the gap? Centre for Environmental Research and Training, University of Birmingham, inaugural lecture, 30 May.

Petts, J., Horlick-Jones, T. and Murdock, G. (2000) *Social Amplification of Risk: The Media and the Public.* Contract Research Report no. 329/2001. HSE Books, Health and Safety Executive, Sudbury, UK. Available at http://www.hse.gov.uk/research/crr_pdf/2001/crr01329.pdf.

Pew (2002) *Environmental Savior or Saboteur? Debating the Impacts of Genetic Engineering.* Pew Initiative on Food and Biotechnology, Washington, DC. Available at http://pewagbiotech.org/newsroom/releases/020402.php3.

Powell, D. (1998) Impacts of biotechnology, environment, food safety: communications. Paper presented at the Agriculture Risk Management Conference, Hull, Canada. University of Guelph, Guelph, Canada.

Ramonet, I. (2000) Les peurs de l'an 2000. Le Monde Diplomatique 561, p 1.

Ruckus Society (nd) *The Ruckus Society Media Manual.* Ruckus Society, Berkeley, California. Available at http://www.ruckus.org/man/media_manual.html.

Ruffieux, B. and Robin, S. (2001) *Analyse économique de la disposition à payer des consommateurs pour des produits garantis sans utilisation d'OGM et choix du signal distinctif pertinent.* Rapport du contrat 'Pertinence et faisabilité d'une filière 'sans OGM''. IREPD-ENSGI, Grenoble.

Salomon, J.-J. (1984) *Prométhée empêtré: la résistance au changement technique.* Anthropos, Paris (Pergamon, Paris, 1982).

Siegrist, M. (2000) The influence of trust and perceptions of risks and benefits on the acceptance of gene technology. *Risk Analysis* 20, 195–203.

Slovic, P. (1987) Perception of risk. *Science*, 236, 180–285.

Slovic, P., Malmfors, T., Krewski, D., Mertz, C., Neil N. and Bartlett, S (1995) Intuitive toxicology. II. Expert and lay judgments of chemical risks in Canada. *Risk Analysis* 15, 661–675.

SOT (Society of Toxicology) (2002) *The safety of genetically modified foods produced through biotechnology.* Position paper, SOT, Reston, Virginia.

Springer, A., Mattas, K., Papastefanou G. and Tsioumanis, A. (2002) Comparing consumer attitudes towards genetically modified food in Europe. In: *Proceedings of the Xth European Congress of Agricultural Economists* (CD-ROM), Zaragoza, Spain, pp. 053–073.

UNCAA-SIGMA (2001) Baromètre UNCAA/SIGMA sur les français et l'agriculture (sondage SOFRES). Media release of 10 January. UNCAA, Paris. Available at http://www.tns-sofres.com/etudes/pol/120101_agri_r.htm

Union InVivo (2001) *Enquête SOFRES sur les Français et l'agriculture.* Press release, December, Union InVivo, Paris.

Zechendorf, B. (1998) Agricultural biotechnology: why do Europeans have difficulty accepting it? *AgBioForum* 1, 8–13.

17 Introducing Novel Protein Foods in the EU: Economic and Environmental Impacts[1]

Xueqin Zhu, Ekko van Ierland and Justus Wesseler
Environmental Economics and Natural Resources Group, Wageningen University, Hollandsweg 1, 6706 KN, Wageningen, The Netherlands

Introduction

Animal protein production, in particular pork production, has important environmental impacts, and animal health problems are a major concern of consumers. These considerations may result in shifting consumers' preferences for proteins from animal proteins to novel protein foods (NPFs). To analyse the impacts of enhancing the demand for NPFs is one of the purposes of the research programme PROFETAS[2] in The Netherlands. This programme was initiated to study whether or not a substantial shift from animal to plant protein foods is 'environmentally more sustainable, technologically feasible and socially desirable'. Novel protein foods are plant protein-based food products, which are developed by modern technology (including biotechnology) and designed on the basis of consumers' preferences for flavour and texture. Examples are 'vegetable burgers', high-quality soybean products or protein products made of peas. Baggerman and Hamstra (1995) suggest that NPFs can reduce environmental pressure because the conversion of plant proteins into meat proteins is biochemically and environmentally inefficient.

There is extensive literature (e.g. Carlsson-Kanyama, 1998; Mattsson, 1999; Kramer, 2000) on the relationship between food consumption and the environment. These studies focused on the environmental pressure of the different types of food, and the environmental assessment of food production and consumption by life-cycle assessment or other environmental assessment methods.

Some agricultural and environmental applied general equilibrium (AGE) models investigate the impacts of agricultural policies and environmental policies. Examples are the GTAP model (Hertel, 1997), the ECAM model (Folmer *et al.*, 1995) and the MERGE model (Manne and Richels, 1995). The GTAP model (global trade analysis project model) uses the general equilibrium modelling framework combined with a huge database on international trade. It is a very dis-aggregate model and it is continuously updated by a team of researchers. Version 5 of the GTAP model includes 57 sectors and 66 regions (GTAP, 2002). The ECAM model (EC agricultural model) distinguishes two sectors (farm sector and non-farm sector), two consumers (farmer and non-farmer) and two commodities (agricultural commodity and non-agricultural

1 The research is financed by the Netherlands Organization for Scientific Research (NWO). The usual disclaimer applies.

2 PROtein Foods, Environment, Technology And Society, see http://www.profetas.nl/ for details.

commodity) for each country. We mention the MERGE model (a model for evaluating regional and global effects of greenhouse gas reduction policies), because it integrates the AGE framework with climate change and damage assessment sub-models to assess climate change policy proposals in a multi-regional context.

The purpose of this chapter is to study some of the potential economic and environmental consequences of a shift from animal protein foods to NPFs in the European Union (EU). In order to investigate the consequences of a shift from animal protein foods to NPFs, we apply the AGE framework and include the environmental aspects in the utility function of consumers and in the production function of the producers. The contribution of this chapter is to present an environmental AGE model that can capture the environmental concerns of the consumers, and that is applied for examining the impacts of the enhanced introduction of NPFs.

The environmental concern of the consumers is embodied in the utility function of our AGE model. The consumer's utility depends not only on the consumption of the rival goods but also on the environmental quality, as a non-rival public good. The introduction of NFPs to society is simulated in the model by an exogenous shift in consumer demand, i.e. by increasing the expenditure share of NPFs in the protein budget (δ) to partially replace the consumption of pork. We use an increasing expenditure share of NPFs because we simulate a voluntary shift to NPFs, which is the central hypothesis in the research programme. This shift might be considered to be the result of consumers' orientation to 'green products' and to the safety of the plant protein products. The substitution between pork and NPFs is represented by the substitution elasticity (σ) in the utility function. In the application of the model, the expenditure share of NPFs in the total protein budget of the consumers (δ) is increased from 0% in the base run to 30% after the enhanced introduction of NPFs in the simulation run. The substitution *elasticity* between pork and NPFs (σ) is chosen to be 0.8, considering the consumers' concerns with health and the tendency to the new products on the one hand, and the present diet habits on the other hand. For the environment, we temporarily only consider the atmospheric emissions of CO_2 as an environmental indicator for several pollutants and environmental effects related to the use of energy in the model application. The nitrogen and phosphate emissions from the manure of pork production could also be included, but they are not yet considered in the application because of data problems. The consumers' concern for environmental quality is represented by the willingness to pay for the environment. To be specific, by the utility elasticity for environmental quality (ε) in the utility functions. Since the value for ε is difficult to obtain, we analyse the impacts of NPFs by means of sensitivity analysis for ε over a relatively wide range of values (0.05–0.20). The new runs for the different values of ε construct different scenarios. The comparison between the results of the base run and those of the scenarios provides insights into the potential economic impacts of a shift towards the consumption of NPFs, considering consumers' concern for the environment.

The chapter is organized as follows: the next section includes the theoretical background, and the motivation for the choice of the AGE format. It gives the specification of the AGE model considering environmental pollution, where the environment is viewed as an input of the production and the consumers have to pay for their consumption of the environmental good. The following section includes the model application and the sensitivity analysis of the utility elasticity for environmental quality. In this section, some simulation results are presented and a brief interpretation of the results of the model application is also given. The final section gives the preliminary conclusions of the impacts of NPFs based on the application of the model, and some discussions of the model.

The Specification of the AGE Model

The postulates of the general equilibrium model are that consumers maximize their utility subject to their budget constraint and producers maximize their profits subject to the technological constraint. In an AGE model, all the goods, services and factors in the economy are called commodities. Industries produce outputs using factor inputs and/or intermediate inputs. Products produced in industries can be used as intermediate inputs for production of other products, or consumed by households as final goods or exported to the rest of the world. In equilibrium the demand for all commodities cannot exceed their supply.

Our central research aim is to analyse the potential impacts of an increase in demand for NPFs. We choose the AGE model because AGE modelling is considered the best choice to anticipate *adjustments* in the economy and it has been used to evaluate a wide range of policy issues, including changes in direct taxes, trade policy, income redistribution and public investments (Ginsburgh and Keyzer, 1997). As the problems of agricultural production increasingly became issues of efficiency in allocation, there was a growing need for such an embedding of the agricultural and food sector in a wider setting (e.g. Keyzer, 1989; GTAP, 2002).

Mathematically we can represent the general equilibrium models in several formats. There are five alternative formats according to Gunning and Keyzer (1995) and Ginsburgh and Keyzer (1997): (i) excess demand format; (ii) Negishi format; (iii) full format; (iv) open economy format; and (v) CGE format. Each format is best suited for specific purposes. It should, however, be stressed that all formats describe the same model and lead to the same equilibrium solution. For this study, we have chosen the Negishi format because it provides a direct link to welfare analysis. It starts with a welfare programme, which is subsequently decentralized through commodity and agent-specific signals (e.g. prices).

According to Ginsburgh and Keyzer (1997), choosing the Negishi format implies that only primal forms can be used. Following the *primal approach*, we represent the production technology (production set) by a functional form with a finite number of parameters. The parameters have to be chosen so that the production set satisfies the condition of *strict convexity*. In applied general equilibrium applications, the global properties of functional forms become important (Perroni and Rutherford, 1996). For the uniqueness of the equilibrium, the Cobb-Douglas functional form for production and strictly concave utility function (e.g. CES or Cobb-Douglas utility function) can be chosen (Ginsburgh and Keyzer, 1997).

Based on the theoretical structure of the AGE model with environmental concerns presented in Appendix 17.I, we have specified the model for this study. In this section we describe the characteristics of the applied model, and specify the functional forms of the model.

The characteristics of the model

In the AGE model applied in this chapter, the world is divided into two regions: the EU and the rest of the world (ROW). Thus, we have two representative consumers, i = EU and ROW. The flow of the commodities in these two regions is shown in Fig. 17.1. Six products are distinguished: pork, other food, non-food, NPFs, peas and feed. The first four goods – pork, other food, non-food and NPFs – are the consumption goods of the consumers. Peas are both direct consumption goods and intermediate goods for production of NPFs. Feed is an intermediate input of pork production. For the production of pork, the factor inputs (labour, capital and land) and intermediate input (feed) are used, while for the production of other food and feed, only the factor inputs are used. NPFs are produced by capital, labour and an intermediate good of peas. The non-food product only uses the factors capital and labour. Feed and peas are both produced as intermediate goods in agriculture by the factor inputs labour, capital and land. The environment is specified in two ways. Firstly, the use of environmental services is included as an input for production. Secondly, the utility of each consumer is related to the consumption of private goods and services, and to the level of an environmental quality indicator. Thus there are nine commodities (pork, other food, non-food, NPFs, peas, feed, labour, capital and land) and one non-rival good (expressed by an environmental quality indicator) in the model. All the goods and

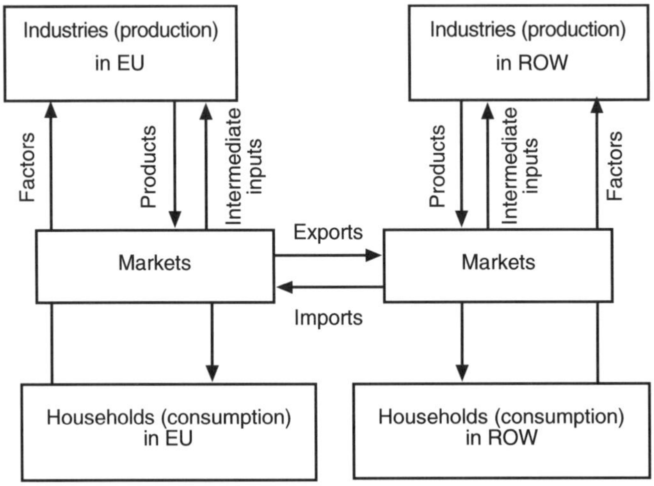

Fig. 17.1. The flows of the two-region AGE model.

services can be exported or imported based on the comparative advantages of each region under free trade. In our application the factors of production are immobile between the two regions. For simplicity, the model is comparatively static.

Objective function and utility functions

The objective function of the welfare programme in the Negishi format is:

$$W = \mathrm{Max}[\alpha_{EU} \cdot \log U_{EU} + \alpha_{ROW} \cdot \log U_{ROW}] \quad (1)$$

where W is the total welfare, U_{EU} and U_{ROW} are the utility of the EU and ROW, and α_{EU} and α_{ROW} are the Negishi weights of the EU and ROW, respectively. A list of symbols is given in Appendix 17.II. For the equilibrium solution of the model, the Negishi weights have to be found such that the budget constraints hold. Analytically in the sequential joint maximization (SJM) method, the Negishi weights are the respective shares in total income in the economy when Cobb-Douglas utility functions and production functions are chosen (Ermoliev *et al.*, 1996; Rutherford, 1999).

The utility function in our model is a nested function of three levels. The substitution structure of the consumption of goods is shown in Fig. 17.2. At Level 1, it is a Cobb-Douglas function with substitution between the consumption of rival goods and a non-rival good (environmental quality). At Level 2, it is also a Cobb-Douglas function with substitution between proteins, other food, non-food and peas for the consumption of rival goods. At Level 3, it is a CES function with substitution between pork and NPFs for the consumption of proteins.

The demand function (Shoven and Whalley, 1992) for pork and NPFs will then be[3]:

$$C_{EU,pork} = \frac{(1-\delta)E_{pr,EU}}{p_{pork}^{\sigma} \cdot [(1-\delta)p_{pork}^{(1-\sigma)} + \delta p_{NPFs}^{(1-\sigma)}]}$$

$$C_{EU,NPFs} = \frac{\delta E_{pr,EU}}{p_{NPFs}^{\sigma} \cdot [(1-\delta)p_{pork}^{(1-\sigma)} + \delta p_{NPFs}^{(1-\sigma)}]}$$

where $C_{EU,NPFs}$ is the consumption of NPFs, $C_{EU,pork}$ is the consumption of pork in the EU, σ is the elasticity of substitution between pork and NPFs, δ is the expenditure share of NPFs in protein budget, $E_{pr,EU}$ is the expenditure of the consumers on protein consumption in the EU (the protein

[3] The demand function of pork and NPFs are based on the CES utility function in Level 3 for the protein consumption: $U(\text{protein}) = [(1-\delta) \cdot C_{EU,pork}^{\frac{(\sigma-1)}{\sigma}} + \delta \cdot C_{EU,NPFs}^{\frac{(\sigma-1)}{\sigma}}]^{\frac{\sigma}{\sigma-1}}$.

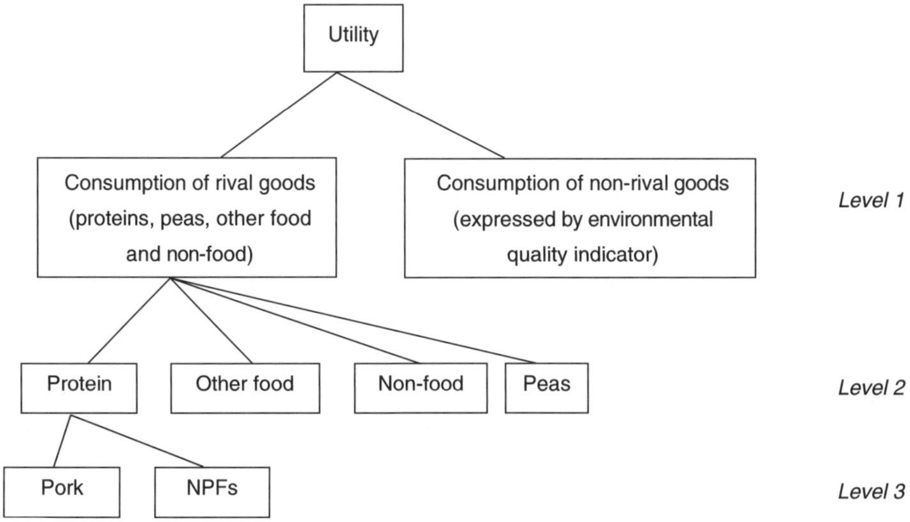

Fig. 17.2. Nesting structure in utility function in the EU.

budget), p_{pork} and p_{NPFs} are prices of pork and NPFs, respectively. Therefore, according to the substitution effects and expenditure share of the two proteins, the following relation exists[4]:

$$C_{\text{EU,NPFs}} = \frac{\delta}{1-\delta}\left(\frac{p_{pork}}{p_{NPFs}}\right)^{\sigma} \cdot C_{\text{EU,pork}}. \tag{2}$$

The protein consumption in the EU $C_{\text{EU,pr}}$ and in the ROW $C_{\text{ROW,pr}}$ is as follows:

$$C_{\text{EU,pr}} = C_{\text{EU,pork}} + C_{\text{EU,NPFs}} \tag{3}$$

$$C_{\text{ROW,pr}} = C_{\text{ROW,pork}} \tag{4}$$

where $C_{\text{ROW,pork}}$ is the pork consumption in the ROW.

For the environment use, we consider the simple case in which environmental services are used as input in the production process. The use of environmental inputs decreases the utility of the consumers by reducing environmental quality that we express in the model by means of an environmental quality indicator. In this manner environmental quality is affected by the use of the environmental services in production and by preferences of the consumers for the non-rival good 'environmental quality'. The utility function $u_i(x_i,g_i)$ is continuous, concave, increasing in (x_i,g_i) and satisfies: $u_i(0,g_i)=0$, where x is the vector of consumption goods, g is the non-rival consumption of environmental quality and i is the consumer. This results in the following utility functions with C_s as the consumption of rival good s, s = proteins (pork + NPFs), other food, non-food and peas, and g as the non-rival consumption of an environmental good (expressed by the environmental quality indicator):

$$U_i = g_i^{\epsilon_i}(f(C_{si}))^{1-\epsilon_i} = g_i^{\epsilon_i}\left(\prod_s C_{si}^{\beta_{si}}\right)^{1-\epsilon_i} \tag{5}$$

where i indicates the consumer (i = EU and ROW), ϵ is the utility elasticity for environmental quality, β_s are the utility elasticities for consumption of rival goods without considering the environment, and $\sum_s \beta_{si} = 1$. The utility functions used in the applied model are given in Appendix 17.III.

[4] If σ =1.0, this relationship between pork demand and NPFs demand does not hold any more. Then the consumption of both goods is dependent on the protein balance function and the utility function. The consumer will only consume the cheaper one.

Production functions

A production function describes the technical relationship between the inputs and outputs of a production process represented by a mathematical function. The production of pork, or animal protein products (processed pork), and NPFs can be described by the production chains because the agriculture process is very different from the industrial production. The two representative chains are shown in Fig. 17.3a and b.

Along the chains, many inputs and outputs (including the environmental emissions) are involved. It is impossible to include all the inputs along the chains in the production function of the pork and NPFs production, and simplification is necessary. As we have noticed, the production processes not only use production factors as inputs but also generate the emissions from production. For technical reasons *pollution* in our model is not viewed as a negative externality but as the use of a natural resource. The production inputs of pork include labour, capital, land, the intermediate good 'feed' and an environmental input (e.g. emission). For the production of all the goods, an environmental input is also used. The Cobb-Douglas production function for production of good j with environmental input can in general be presented as follows:

$$Y_j = EM_j^{\xi j} F(LB_j, LD_j, K_j, FD_j, PI_j)^{1-\xi j}$$

where j is the production good (j = pork, NPFs, other food, non-food, peas and feed), Y is the production, EM is the environmental input, ξ is the exponent of the emission in the production function indicating the cost share of the emission EM for production, $0 < \xi < 1$, LB refers to labour input, LD land input, K capital input, FD the feed input and PI the pea input. One can consider EM as the use of 'environmental services', which reflects that the firm must release its emissions to the environment. We can think of the firm as requiring EM emission permits in order to produce (Copeland and Taylor, 2003). Therefore when environmental services are treated in the production function in this way, an emission permit system reflecting the annual endowment of environmental services for each region is necessary for the modelling. Thus the following relationship holds

$$\sum_j EM_{ij} \leq \overline{EM_i} \qquad (6)$$

where EM_{ij} is the use of environmental services in region i for good j, and $\overline{EM_i}$ is the number of emission permits in region i.

The production function for good j is then:

$$Y_j = EM_i^{\xi i}[(LB_j)^{\eta_{1j}}(LD_j^{\eta_{2j}}(K_j)^{\eta_{3j}}\left(FD_j\right)^{\eta_{4j}}(PI_j)^{\eta_{5j}}]^{1-\xi j} \qquad (7)$$

where η_1, η_2, η_3, η_4 and η_5 is the cost share of each input (LB, LD, K, FD, PI) for production without considering the cost of emissions, with $\eta_1+\eta_2+\eta_3+\eta_4+\eta_5=1$.

For the parameters of the production functions, we use information from other studies. For example, the feed costs amount to 60% of the total production costs in the Netherlands (Jogeneel, 2000). For the EU an average of 45% of the feed costs is used in the pork production function. The technological parameters in the production functions of the EU and the ROW are 1.0 and 0.6

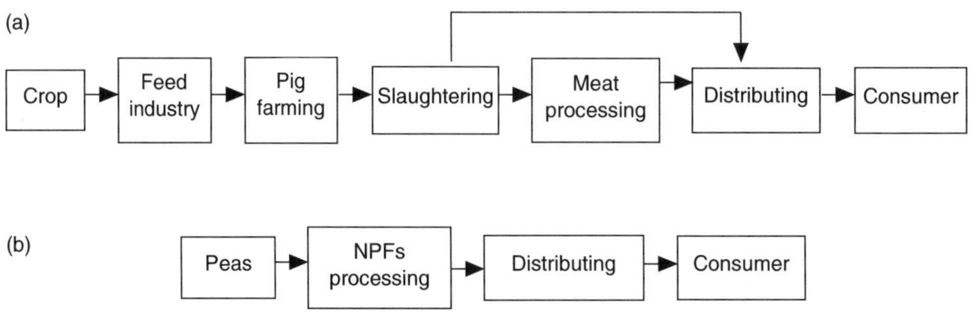

Fig. 17.3. Production and consumption chains of pork (a) and novel protein foods (b).

respectively[5]. The production functions in this manner *grosso modo* reflect the production technology for the region that we distinguish in our study. The production functions and balance equations are reported in Appendix 17.III.

Environmental quality

The balance equation for environmental goods (e.g. clean air), which are inputs to the production process, is assumed to be determined by the initial stock and production inputs as shown in equation (A2) of Appendix 17.I. But the initial stock of 'environmental services' is hard to know and the link between output and the use of environmental goods in the production process is hard to establish. With the emission permit system, we established the relationship in equation (6) for producers. For consumers, the environment is valued in terms of environmental service which is constrained by equation (A2)' of Appendix 17.I. For our applied model, we only consider a one-dimensional ($e = 1$) environmental service g, which reflects a number of environmental issues that are related to the energy use and the release of pollutants like NO_x and SO_x or greenhouse gases. As a proxy for energy use and related emissions we use the level of CO_2 emissions in the respective regions. Then we can define the 'environmental quality indicator' to be determined by the level of emissions. If the emissions are above a critical level, the environmental quality indicator will decrease. We next use the environmental quality indicator as the non-rival consumption of environmental goods in the utility function. Of course, the model can be easily expanded to include more dimensions of environmental goods g, by explicitly modelling emissions of nitrogen oxides and other pollutants as long as the data are available.

Obviously, the environmental quality that consumers face in region i is determined by the total use of the environmental services of all the producers in region i. In the present model version we approximate this relation by means of a linear function in the use of the environmental services:

$$g_i = \overline{\Psi}_i - \sum_{j=1}^{n} EM_{ij} \qquad (\phi_i) \tag{8}$$

where Ψ_i is the intercept and $\sum_{j=1}^{n} EM_{ij}$ is the total emissions of all the producers in region i. This relationship shows that the higher the emissions the lower the environmental quality. Since consumers will enjoy and pay for this environmental quality, it can be seen as a product produced by a certain environmental sector.

Budget constraints

Under constant return to scale, profits are zero so that income is the value of initial endowments, which are employed in the production. According to the endowments of production factors and emission permits the income is:

$$h_i = rl_i \cdot LD_i + w_i \cdot LB_i + rk_i \cdot K_i + p_{mi} \cdot EM_i$$

where rl is the price of land, w is the wage, rk is the price of capital and p_m is the price of emission permit. It should be equal to the total revenue of the production sectors and the 'environmental sector':

$$h_i = \sum_j p_j \cdot Y_{ij} + \phi_i g_i \tag{9}$$

where p_j is the price scalar of good j, the first item of the right-hand side $\sum_j p_j \cdot Y_{ij}$ is the revenue of the production sectors, and the second item $\phi_i g_i$ is the revenue of the 'environmental sector' which maintains certain environmental quality demanded by the consumer.

Budget constraints say that the expenditure of the consumer should be equal to his/her income. Now that the non-rival environmental quality is one of consumption goods, the consumer has to pay for his/her consumption. Just like the producer has to pay for the emission permit for production, the consumers who simply enjoy the presence of the

[5] These technological parameters are chosen to the best of our knowledge but require further research. The model specification in GAMS (General Algebraic Modeling System) is available on request from the authors and the impact of different parameter values can be easily established.

resource or environmental quality should pay to the 'environmental sector' for the environmental services. The budget constraint of the consumer now looks like:

$$\sum_s p_s \cdot C_{si} + \phi_i g_i = h_i \qquad (10)$$

where p_s is the price scalar of good s, s = proteins (pork + NPFs), other food non-food, and peas, C_{si} is the consumption of good s in region i, $\sum_s p_s \cdot C_{si}$ is the total expenditure on the consumption of all rival goods and $\phi_i g_i$ is the payment by the consumers for the environmental quality g, and h is income.

In this welfare programme, where both the consumers and producers have to pay for the environmental use, the Lindahl equilibrium is reached (Ginsburgh and Keyzer, 1997).

The Model Application and Results

The data and scenarios

The base run and scenarios

We have applied the model to develop the base run, a scenario for the enhanced consumption for NPFs and some scenarios of sensitivity analysis.

BASE RUN There are no NPFs, the environmental concern is indicated by the utility elasticity for environmental quality ε, which is assumed to be 0.05 for both regions.

NPF SCENARIO For the simulation of the new scenarios, we assume that the substitution elasticity of the NPFs for pork is $\sigma = 0.8$ and we simulate a situation where the expenditure share of NPFs in the protein budget is increased to 30% (δ = 0.3) after enhanced introduction of NPFs. We do not assume NPFs as *perfect* substitutes of pork (σ =1) because we think in the short run it is impossible to replace all the animal protein products by NPFs. In this scenario, we use the same value for the utility elasticity for the environment ($\epsilon = 0.05$) as in the base run.

SENSITIVITY ANALYSIS As a consumer-driving economy, the sensitivity of the results to the parameters in the utility functions is a very interesting issue. We carry out the sensitivity analysis for the value of parameter ε. The values of 0.1 and 0.2 for ε are simulated under two cases of four runs where (i) different values for the EU and ROW and (ii) similar values for the EU and ROW are used, respectively. The results of all these scenarios are compared with the results of the base run. The comparison gives an impression of some potential impacts of the enhanced introduction of the NPFs in the EU on the economy and the environment.

The data

As stated, the model is applied to the economy with two regions: the EU and ROW. The data for labour, land and capital are based on the database of the FAO (2002) and the Penn World Table (Penn World, 2002). The labour force in 1998 in the EU is 252 million and 3323 million in the ROW. The total land area in 1998 in the EU is 313,000 ha and 12,149,000 ha in the ROW. Non-residential capital stock per worker in the EU is approximately €30,000 per worker and €5000 per worker in the ROW according to the Penn World Table. The total capital stock in the EU is €7560 billion and in the ROW it is €16,615 billion. The data for emissions is based on the little Green Data Book (World Bank, 2000). The EU contributes about 12% of the global CO_2 emissions (3000 million tonnes in the EU and 22,000 million tonnes in the ROW in 1998). As we have already mentioned, emission permits should be given when the emissions are taken as an input for the production function. In the model run, we initially allocate emission permits to the EU and ROW according to the emission levels of 1998. The initial endowments are shown in Table 17.1. These data are used for the model applications.

The results

The results for the base run

When there are no NPFs, and $\epsilon = 0.05$, we run the model as the base case. The results for the 'base run' are reported in Table 17.2. Firstly, for production the table shows that the EU is basically the major producer of pork

Table 17.1. Factor endowments of labour, land, capital and CO_2 emission permits.

	Labour (millions)	Land (ha × 1000)	Capital (billion €)	Emissions (million tonnes)
EU	252	313	7,560	3,000
ROW	3,323	12,149	16,615	22,000

Table 17.2. Baseline: production, consumption, trade, emissions and income.

	Production					Consumption				
	Pork	Other food	Non-food	Peas	Feed	Pork	Other food	Non-food	Peas	Feed
EU	304	0	1283	0	0	94	668	1218	3	301
ROW	39	2422	3163	43	340	249	1754	3229	40	39
Total	343	2422	4446	43	340	343	2422	4447	43	340
	Trade (+, export; −, import)					Emissions		Income per worker		Utility (welfare)
EU	+210	−669	+	−3	−301	1162		12.4		779
ROW	−210	+669	−	+3	+301	8188		2.5		2140
Total	0	0	0	0	0	9350				(7.39)

and non-food. It exports pork and non-food to the rest of the world and imports other food, peas and feed from the rest of the world. Secondly, for the use of environmental services, the entry 'emissions' in Table 17.2 shows that the EU emits 12% of the global emissions, which is consistent with the endowment of environmental services that we used. Pork is, in our analysis, the most polluting product with the highest environment input in the production function. Pork is more expensive, because its production needs more factor inputs, including feed as an extra intermediate input. Finally, Table 17.2 shows that income per worker in the EU is five times higher than the rest of the world.

The results for the NPF scenario

By introducing an exogenous increase in the consumption of NPFs in the EU by increasing the expenditure share of NPFs in protein budget, with the same environmental concern in the two regions as the base run ($\varepsilon = 0.05$), a new equilibrium will be reached. The results are reported in Table 17.3.

Comparing Tables 17.2 and 3, we observe the implications of the enhanced introduction of NPFs in the EU to the economy. The budget share of 30% for NPFs results in a reduction of consumption of pork in the EU by 28%. Pork production in the EU will be decreased by 5% (15 units) from 304 to 289 units. The reduction in consumption of pork is more than the decrease of the pork production in the EU because the EU will benefit from exporting pork to the ROW. The international trade of pork is increased by 5% from 210 to 221 units.

Since the production of NPFs is less polluting than that of pork production, the total emissions will decrease by about 0.8% in the EU. As for the ROW, the emissions are decreased by 0.2% because they produce less pork by importing from the EU. The total emissions are reduced by about 0.2% because the emissions of the EU are much less than those of the ROW.

For income related to the remuneration of factors, we observe that income for the EU falls slightly because the production of NPFs needs simpler processing than pork and thus

Table 17.3. NPFs scenario: production, consumption, trade, emissions and income.

	Production						Consumption					
	Pork	NPFs	Other food	Non-food	Peas	Feed	Pork	NPFs	Other food	Non-food	Peas + input	Feed input
EU	289	52	0	1281	0	0	68	52	662	1205	3 + 66	284
ROW	30	0	2421	3164	109	313	251	0	1759	3240	40	29
Total	319	52	2421	4445	109	313	319	52	2421	4445	109	313

	Trade (+, export; −, import)						Emissions	Income per worker	Utility (welfare)
EU	+221	0	−662	+77	−69	−283	1153	12	794
ROW	−221	0	+662	−77	+69	+283	8170	2.5	2148
Total	0	0	0	0	0	0	9323		(7.4)

less primary input. Therefore the factors are less in demand than before the enhanced introduction of NPFs, and prices of factors are lower. Given the fixed volume of factors, the remuneration will be lower.

The utility is increased slightly because in our model the utility depends on both the consumption of rival goods and the environmental quality indicator. The environmental quality indicator is linear and declining in the level of emissions. The consumers have to make a trade-off between more consumption of the rival goods with lower environmental quality and better environmental quality with less consumption. More consumption of the rival goods means more pollution but more pollution implies lower environmental quality. In this manner the preference of consumers for environmental quality gives feedback to consumption of rival and non-rival goods and then to production.

Sensitivity analysis

As the preferences of consumers for environmental quality will have feedback on production and consumption in a competitive model, the interesting question is how the consumers value this environmental quality. We carried out some sensitivity analysis for the valuation of the consumer for the environmental quality, because little information is available on the role of the environment in the utility function of the consumers. In the above two applications of the model, a modest value of 0.05 for the utility elasticity for the environment is used for both regions. This means that the consumers are willing to pay 5% of their expenditure for a good environment. But in reality different people have different willingness to pay for the non-rival consumption of environmental goods. Therefore, it will be interesting to see how the attitude of the consumers will influence their consumption bundle. In the first case, we consider the different environmental concerns in different regions. The market for environmentally friendly goods is located mainly in the member countries of the OECD, where during the last few years consumers have started to articulate strong environmental concerns. These concerns have been translated into both individual purchasing decisions and government regulations (Bharucha, 1997). In the second case, we will increase the value of ε from 0.05 to 0.1 and 0.2 for both regions. Therefore, we will carry out sensitivity analysis under these two cases of the four runs which are shown in Table 17.4.

Table 17.4. The runs for sensitivity analysis for ϵ.

Model runs	Values of ϵ
Case 1	
Run 1	$\epsilon_{EU} = 0.1$, $\epsilon_{ROW} = 0.05$
Run 2	$\epsilon_{EU} = 0.2$, $\epsilon_{ROW} = 0.05$
Case 2	
Run 3	$\epsilon_{EU} = 0.1$, $\epsilon_{ROW} = 0.1$
Run 4	$\epsilon_{EU} = 0.2$, $\epsilon_{ROW} = 0.2$

In Case 1, we fix the value of the utility elasticity for the environment ε in the ROW at 0.05, and increase the value for the EU from 0.05 to 0.1 and 0.2. See Tables 17.A1 and 17.A2 in Appendix 17.IV for the detailed results. With the increased value for the EU, pork production in the EU will decrease. If the value increases to 0.2, pork production in the EU will be hampered severely and at the given technology as described by the present production functions will eventually disappear, because pork is the most polluting product. The price of pork decreases because it is demanded less. Non-food production will increase in the EU because the EU has a comparative advantage and it is less polluting than the other products. As a result, the export of non-food to the ROW will increase and the price of non-food falls because more production takes place. The emissions in the EU decrease with the increase of consumer's valuation of the environmental goods in the EU, because the EU switches to produce more non-food and less pork. In contrast, pork production in the ROW will increase as a result of the increase in the value of the environmental goods in the EU, because the EU will reduce production and export of pork. Since pork and non-food become cheaper with the increase of ε, the ROW is also better off. The emissions in the ROW increase, however, because the ROW has to produce the polluting product 'pork' for its own consumption and export to the EU.

In Case 2, we have increased the value of ε for both regions from 0.05 to 0.1 and 0.2. The simulation results are reported in Tables 17.A3 and 17.A4 in Appendix 17.IV. The results show that pork production for both regions decreases and the emissions decrease greatly.

Conclusions and Discussion

In this chapter we have sketched some important aspects and possible implications of an enhanced demand for NPFs, by means of an AGE model. Although we are aware that the model is far from perfect and that it is formalized at a high level of aggregation, we think it is worthwhile to discuss some of the characteristics, the assumptions and the results of the analysis. The model considers both the utility from the consumption of goods and the disutility from environmental pollution. The emissions from production give feedback on utility and on the bundle of rival and non-rival consumption, and then indirectly on production. For a value of 0.05 for the utility elasticity for the environment, the enhanced introduction of NPFs decreases the emissions from pork production in the EU and decreases the total emissions slightly. The EU will consume less pork by consuming some NPFs and will export more pork than before. Moreover, pork production in the ROW will decrease slightly because slightly more pork can be imported from the EU. Thus, the introduction of NPFs decreases in this setting the emissions in the ROW slightly. As a result, the total emissions in the world will *decrease* slightly too.

Nevertheless, the model results are sensitive to the value of the utility elasticity for the environment. If the EU has a higher utility elasticity for the environment than the ROW (0.1 versus 0.05) pork production in the EU will decrease more strongly and the export of pork to the rest of the world will decrease. As a result, the rest of the world has to increase pork production for their high demand for pork. The emissions in the ROW will *increase* by 1.7% from 8188 to 8328 units. If the utility elasticity for the environment in the EU increases to 0.2, then a stronger trend will occur. The EU will stop producing pork and will import some pork from the ROW. Then the emissions in the ROW will *increase* by 2.7% (to 8413 units). To summarize, if only consumers in the EU increase their environmental concern, the introduction of NPFs does not reduce the emissions in the ROW. But switching to produce more NPFs and less pork in the EU is helpful to reduce the unevenness of the income distribution by improving the income share of the ROW.

If the two regions have the same concern for the environment, the increase in the value of ε will limit pork production in both regions and limit the emissions globally. The model strongly suggests that the enhanced introduction of NPFs is meaningful for global environmental improvement by emission reduction, only if both regions increase their preferences for environmental quality.

This chapter presents an AGE model that captures the environmental concerns in the utility function. The model presented in this chapter shows how the economy can be modelled by general equilibrium modelling when facing some changes in preferences. Despite its simplicity, it illustrates some of the most important fundamental environmental economic mechanisms that might occur as a result of the enhanced introduction of NPFs based on the classification of the goods and their production functions of our model. The model provides a useful framework for further empirical studies on the role of biotechnology in the economy and for studies on the environmental concerns of consumers. The inclusion of agricultural elements, like land use, water use and agricultural chemicals use, effects of the common agricultural policy (CAP) and other environmental issues (such as environmental policy measures) are important aspects for expansion and application of the model. At the theoretical level, embodying the dynamic properties of the environment and introducing explicit environmental feedback on production and consumption in the AGE model is an interesting challenge.

References

Baggerman, T. and Hamstra, A. (1995) Motives and perspectives from consumption of NPFs instead of meat [Motieven en perspectieven voor het eten van NPFs in plaats van vless]. DTO-werkdocument VN9, DTO, Delft.

Bharucha, V. (1997) The impact of environmental standards and regulations set in foreign markets on India's export. In: Jha, V., Hewison, G. and Underhill, M. (eds) *Trade, Environment and Sustainable Development A South Asian Perspective*. Macmillan Press, Basingstoke, pp. 123–124.

Carlsson-Kanyama, A. (1998) Climate change and dietary choices – how can emissions of greenhouse gases from food consumption be reduced? *Food Policy* 23, 277–293.

Copeland, B.R. and Taylor, M.S. (2003) *International Trade and the Environment: Theory, Evidence and Policy*. Princeton University Press, Princeton, New Jersey.

Ermoliev, Y., Fischer, G., and Norkin, V. (1996) Convergence of the sequential joint maximization method for the applied equilibrium problems. Working paper. IIASA (International Institute for Applid Systems Analysis), Laxenburg.

FAO (2002) Land and Population from Data Collections. Available at http://www.fao.org/.

Folmer, C., Keyzer, M.A. and Merbis, M.A. (1995) *The Common Agricultural Policy beyond the MacSharry Reform*. Elsevier Science B.V., Amsterdam.

Ginsburgh, V. and Keyzer, M.A. (1997) *The Structure of Applied General Equilibrium Models*. The MIT Press, London.

GTAP (2002) GTAP 5 documentation. Available at http://www.gtap.agecon.purdue.edu/

Gunning, J.W. and Keyzer, M.A. (1995) Applied general equilibrium models for policy analysis. In: Behrman, J. and Srinivasan, T.N. (eds) *Handbook of Development Economics*. Elsevier Science B.V., Amsteram, pp. 2025–2107.

Hertel, T.W. (1997) *Global Trade Analysis: Modeling and Applications*. Cambridge University Press, Cambridge.

Jongeneel, R.A. (2000) *EU's Grains, Oilseeds, Livestock and Feed Related Market Complex: Welfare Measurement, Modelling and Policy Analysis*. Social Science Department, Wageningen, Wageningen University.

Keyzer, M.A. (1989) Some views on agricultural sector modeling. In: Bauer, S. and Henrichemeyer, W. (eds) *Agricultural Sector Modeling: Proceedings of the 16th Symposium of the European Association of Agricultural Economists (EAAE)*. Wissenschaftsverlag Vauk Kiel KG, Kiel, pp. 23–30.

Kramer, K.J. (2000) *Food Matters: On Reducing Energy Use and Greenhouse Gas Emissions from Household Food Consumption*. Groningen University, Groningen.

Manne, A. and Richels, R. (1995) MERGE: a model for evaluating regional and global effects of EGH reduction policies. *Energy Policy* 23, 17–34.

Mattsson, B. (1999) *Environmental Life Cycle Assessment (LCA) of Agricultural Food Production*. Department of Agricultural Engineering, Swedish University of Agricultural Sciences, Alnarp.

Penn World (2002) Penn World Table 5.6. Available at http://pwt.econ.upenn.edu/

Perroni, C. and Rutherford, T.F. (1996) A comparison of the performance of flexible functional forms for use in applied general equilibrium analysis Available at http://www.gams.com/solvers/mpsge/domain.htm

Rutherford, T.F. (1999) Sequential joint maximization. In: Weyant, J. (ed.) *Energy and Environmental Policy Modeling*. Kluwer Academic Publishers, Dordrecht.

Shoven, J.B. and Whalley, J. (1992) *Applying General Equilibrium*. Cambridge University Press, Cambridge.

Appendix 17.I. The Theoretical AGE Model with Environmental Concerns – The Negishi Format

Consider an economy consisting of m consumers, indexed by i, $i = 1, 2, \ldots, m$ and n producers, indexed by j, $j = 1, 2, \ldots, n$. There are r commodities (goods and factors), indexed by k, $k = 1, 2, \ldots, r$. Environmental goods indexed by g ($g = 1, 2, \ldots, e$) are involved in the economy for consumption and production. The welfare programme in the Negishi format, which allocates the resources in the economy optimally (Ginsburgh and Keyzer, 1997), is as follows[6]

$$\max_{x_i, g_i \geq 0, y_j, \forall i,j} \sum_i \alpha_i u_i(x_i, g_i) \qquad \text{(A1)}$$

subject to the balances of rival commodities and environmental goods:

$$\sum_i x_i + x_g \leq \sum_j y_j + \sum_i \omega_i \qquad (p) \qquad \text{(A2)}$$

$$g_i \leq x_g. \qquad (\phi_i) \qquad \text{(A2)}'$$

Production technology: $y_j \in Y_j$ (A3)

With welfare weights α_i, such that

$$px_i + \phi_i g_i = p\omega_i + \sum_j \theta_{ij} \Pi_j(p) \quad (\lambda_i) \qquad \text{(A4)}$$

and

$$\alpha_i = \frac{1}{\lambda_i}. \qquad \text{(A5)}$$

In this model, equation (A1) is the objective function of the model, where u_i is the utility function of each individual i ($i = 1, 2, \ldots, m$), x is the vector of consumption goods with k dimension and g is the vector of consumption of non-rival environmental goods with e dimension. The objective of this welfare programme is to maximize the total welfare, which is a weighted sum of the utility of all the m consumers in the economy, the Negishi weight of consumer i is given by α_i.

The equations in (A2) are the balance equations for each commodity k ($k = 1, \ldots, r$) and each environmental good g ($g = 1, 2, \ldots, e$). In this equation, x_g is the vector of consumption of environmental goods with e dimensions, y_j is the vector of the net output of a producer j with $k + e$ dimension if each producer produces only one good, and ω_i is the vector of initial endowments (including the environmental goods) of consumer i with $k + e$ dimensions. Positive y_j indicates the output of a production process and negative y_j indicates the input of the production process. A vector of Lagrange multipliers associated with the balance constraints, i.e. a vector of the shadow prices of each commodity or environmental good is indicated by p in the bracket. The commodity can be a final product, a production factor or an intermediate good. This equation states that the consumption of a commodity or environmental good must be smaller than or equal to its production plus its initial endowments.

Equation (A2)′ is the balance equation of non-rival environmental consumption goods, where a consumer's individual consumption should not exceed the common consumption of all the consumers. This also makes it possible to obtain explicit Lagrange multipliers for the value that each consumer attributes to the environmental consumption x_g. The vector of the Lagrange multiplier f_i in the bracket with e dimension is the price vector that each consumer has to pay for the consumption of environmental goods.

Equation (A3) shows that the production plan must belong to some feasible set, or is constrained by the production technology. Y_j is the production set of firm j reflecting its feasible technology.

Equation (A4) states that the expenditure of the consumer must be equal to his/her income, where the left-hand side shows the total expenditure and the right-hand side shows the income of the consumer. The total expenditure includes the total expenditure on the consumption of all rival goods, px_i, and the payment for the environment, $\phi_i g_i$. The income of consumer i includes the value of his/her initial endowments $p\omega_i$ and his/her total profit, received from firm j ($j = 1, 2, \ldots, n$). θ_{ij} is the profit share of consumer i in firm j, $\Pi_j(p)$ is the profit of firm (producer) j.

[6] In this appendix we follow the original notation of Ginsburgh and Keyzer (1997).

Equation (A5) shows how welfare weights are related to the budget constraints in this welfare programme. The Lagrange multiplier associated with the budget constraint of consumer i is indicated by λ_i, its inverse is the welfare weight that is attributed to consumer i such that the equilibrium of the economy exists. The allocation resulting from the equation system from equations (A1) to (A5) is called the Lindahl equilibrium.

Appendix 17.II. A List of the Symbols

Variables

C = consumption
E = expenditure
EM = emission
$\overline{EM}$ = emission permits (exogenous)
FD = feed input in the production of pork
h = income
g = vector of consumption of environmental goods in general model, or environmental quality indicator in applied model
K = capital
$\overline{K}$ = capital endowment (exogenous)
LB = labour
$\overline{LB}$ = labour endowment (exogenous)
LD = land
$\overline{LD}$ = land endowment (exogenous)
PI = peas input for the production of NPFs
u or U = utility of the consumer
W = total welfare (Negishi welfare)
X = net export
x = vector of consumption goods
Y = production sets or production quantity
y = vector of net production of goods.

Parameters

α = welfare weights
β = parameter in the utility function
γ = parameters in Cobb-Douglas production function
δ = expenditure share of NPFs in protein budget
ϵ = utility elasticity for the environment (in utility function)
η = parameter in the utility function
θ = profit share
ξ = cost share of the emission input in the production
Π = profit
σ = substitution elasticity
ψ = environmental standard
ω = vector of initial endowments.

Shadow prices

λ = Lagrange multipliers associated with the budget constraint of the consumer
ϕ = shadow price of the environmental goods
p = shadow price vector of commodities
p_j = shadow price (scalar) of production good j
p_s = shadow price (scalar) of consumption good s
p_m = shadow price of emission permit
rk = shadow prices of capital
rl = shadow prices of land
w = shadow prices of labour.

Subscripts

g = environmental goods, $g = 1, 2, \ldots, e$
i = consumers, $i = 1, 2, \ldots, m$ for theoretical model, and i = EU and ROW for applied model
j = goods or products, $j = 1, 2, \ldots, n$ for general model, and j = pork, other food, non-food, NPFs, peas and feed for the applied model
k = commodities, $k = 1, 2, \ldots, r$ for general model
s = consumption goods in applied model, s = proteins (pork + NPFs), peas, other foods and non-food
EU = the European Union
ROW = the rest of the world.

Appendix 17.III. Utility and Production Functions, and Balance Equations of the Applied Model

Utility functions

$$U_{EU} = (g_{EU})^{0.05} \cdot (C_{EU,pr}^{0.12} \cdot C_{EU,otf}^{0.299} \cdot C_{EU,peas}^{0.001} \cdot C_{EU,nf}^{0.58})^{0.95}$$

$$U_{ROW} = (g_{ROW})^{0.05} \cdot (C_{ROW,pork}^{0.12} \cdot C_{ROW,otf}^{0.295} \cdot C_{ROW,peas}^{0.05} \cdot C_{ROW,nf}^{0.58})^{0.95}$$

Production functions

Here Y indicates the production quantity, LB the labour input, LD the land input, K the capital input, FD the feed input and EM the emission input.

Pork

$$Y_{EU,pork} = EM_{EU,pork}^{0.05}[LB_{EU,pork}^{0.2} \cdot LD_{EU,pork}^{0.15} \cdot K_{EU,pork}^{0.20} \cdot FD_{EU,pork}^{0.45}]^{0.95}$$

$$Y_{ROW,pork} = 0.6EM_{ROW,pork}^{0.05}[LB_{ROW,pork}^{0.2} \cdot LD_{ROW,pork}^{0.15} \cdot K_{ROW,pork}^{0.20} \cdot FD_{ROW,pork}^{0.45}]^{0.95}.$$

Other food

$$Y_{EU,otf} = EM_{EU,otf}^{0.04}[LB_{EU,otf}^{0.30} \cdot LD_{EU,otf}^{0.35} \cdot K_{EU,otf}^{0.35}]^{0.96}$$

$$Y_{ROW,otf} = 0.6EM_{ROW,otf}^{0.04}[LB_{ROW,otf}^{0.3} \cdot LD_{ROW,otf}^{0.4} \cdot K_{ROW,otf}^{0.3}]^{0.96}.$$

Non-food

$$Y_{EU,nf} = EM_{EU,nf}^{0.02}[LB_{EU,nf}^{0.45} \cdot K_{EU,nf}^{0.55}]^{0.98}$$

$$Y_{ROW,nf} = 0.6EM_{ROW,nf}^{0.02}[LB_{ROW,nf}^{0.45} \cdot K_{ROW,nf}^{0.55}]^{0.98}.$$

Feed or peas

Here feed is the yield of feed crops. The following production functions are used for feed crops and peas:

$$Y_{EU,feed} = EM_{EU,ROW}^{0.03}[LB_{EU,feed}^{0.3} \cdot K_{EU,feed}^{0.20} \cdot LD_{EU,feed}^{0.4}]^{0.97}$$

$$Y_{ROW,feed} = 0.6EM_{ROW,feed}^{0.03}[LB_{ROW,feed}^{0.4} \cdot K_{ROW,feed}^{0.3} \cdot LD_{ROW,feed}^{0.3}]^{0.97}$$

$$Y_{EU,peas} = EM_{EU,peas}^{0.03}[LB_{EU,peas}^{0.4} \cdot K_{EU,peas}^{0.3} \cdot LD_{EU,peas}^{0.3}]^{0.97}$$

$$Y_{ROW,peas} = 0.6EM_{ROW,peas}^{0.03}[LB_{ROW,peas}^{0.4} \cdot K_{ROW,peas}^{0.3} \cdot LD_{ROW,peas}^{0.3}]^{0.97}.$$

NPFs in EU

$$Y_{EU,NPFs} = EM_{EU,NPFs}^{0.015}[LB_{EU,NPFs}^{0.1} \cdot K_{EU,NPFs}^{0.2} \cdot LD_{EU,NPFs}^{0.7}]^{0.985}.$$

Balance equations

Feed balance

The production of feed crops Y_{feed} is equal to the intermediate input for pork production FD_{pork} plus its net export X_{feed}.

$$Y_{feed} = FD_{Pork} + X_{feed}$$

Balance of peas

Peas are produced for direct use and production of NPFs, and NPFs are only produced in the EU[7]

$$Y_{EU,peas} = C_{EU,peas} + PI_{EU,peas} + X_{EU,peas}$$

$$Y_{ROW,peas} = C_{ROW,peas} + X_{ROW,peas}$$

where C is the direct consumption of peas, PI is the intermediate input of peas for production of NPFs and X is the net export of peas.

Balance of pork, other food and non-food

The production of a good in one region Y_{ij} equals the consumption of a good C_{ij} plus its net export X_{ij}.

$$Y_{ij} = C_{ij} + X_{ij}$$

where j = pork, other food, non-food, but $j \neq$ feed, peas.

[7] We assume that NPFs are particularly developed in the European market and that in the short run they will mainly be produced within Europe.

Balance of NPFs

The production of the NPFs equals its consumption.

$$Y_{\text{EU,NPFs}} = C_{\text{EU,NPFS}}.$$

Trade balance

$\sum_i X_{ij} = 0$, for j = pork, other food, non-food, feed and peas, but $j \neq$ NPFs.

Balance of factors

$$\sum_j LB_{ij} \leq \overline{LB_i}$$

$$\sum_j LD_{ij} \leq \overline{LD_i}$$

$$\sum_j K_{ij} \leq \overline{K_i}$$

where j includes pork, other food, non-food, feed, peas and NPFs, LB is the labour usage, LD is the land usage and K is the capital usage for production. $\overline{LB_i}$, $\overline{LD_i}$ and $\overline{K_i}$ are the factor endowments of region i.

Appendix IV. Results for Sensitivity Analysis

Table 17.A1. Production, consumption, trade, emissions and income ($\epsilon_{EU} = 0.1$, $\epsilon_{ROW} = 0.05$).

	Production						Consumption					
	Pork	NPFs	Other food	Non-food	Peas	Feed	Pork	NPFs	Other food	Non-food	Peas + input	Feed input
EU	223	49	0	1334	0	0	64	49	619	1133	3 + 62	207
ROW	95	0	2400	3103	105	301	254	0	1781	3304	40 + 0	94
Total	318	0	2400	4437	105	301	318	49	2400	4437	105	301

	Trade						Emissions	Income per worker	Utility (welfare)
EU	159	0	−618	201	−65	−207	669	10.7	802
ROW	−159	0	618	−201	65	207	8328	2.4	2181
Total	0	0	0	0	0	0	8997		(7.43)

Table 17.A2. Production, consumption, trade, emissions and income ($\epsilon_{EU} = 0.2$, $\epsilon_{ROW} = 0.05$).

	Production						Consumption					
	Pork	NPFs	Other food	Non-food	Peas	Feed	Pork	NPFs	Other food	Non-food	Peas + input	Feed input
EU	0	35	0	1410	0	180	46	35	445	827	2 + 45	0
ROW	319	0	2354	3001	90	138	273	0	1909	3584	43 + 0	318
Total	319	35	2354	4411	90	318	319	35	2354	4411	90	318

	Trade						Emissions	Income per worker	Utility (welfare)
EU	−46	0	−445	583	−47	180	311	7.0	719
ROW	46	0	445	−583	47	−180	8413	2.3	2345
Total	0	0	0	0	0	0	8724		(7.54)

Table 17.A3. Production, consumption, trade, emissions and income ($\epsilon_{EU} = 0.1$, $\epsilon_{ROW} = 0.1$).

	Production						Consumption					
	Pork	NPFs	Other food	Non-food	Peas	Feed	Pork	NPFs	Other food	Non-food	Peas + input	Feed input
EU	280	51	0	1268	0	0	66	51	648	1193	3 + 65	280
ROW	28	0	2371	3131	107	308	242	0	1723	3206	39 + 0	28
Total	308	51	2371	4399	107	308	308	51	2371	4399	107	308

	Trade						Emissions	Income per worker	Utility (welfare)
EU	214	0	−648	75	−68	−280	702	12	837
ROW	−214	0	648	−75	68	280	4883	2.5	2385
Total	0	0	0	0	0	0	5585		(7.50)

Table 17.A4. Production, consumption, trade, emissions and income ($\epsilon_{EU} = 0.2$, $\epsilon_{ROW} = 0.2$).

	Production						Consumption					
	Pork	NPFs	Other food	Non-food	Peas	Feed	Pork	NPFs	Other food	Non-food	Peas + input	Feed input
EU	369	50	0	1250	0	0	63	50	631	1177	3 + 64	274
ROW	27	0	2307	3088	105	302	233	0	1676	3161	38 + 0	28
Total	296	50	2307	4338	105	302	296	50	2307	4338	105	302

	Trade						Emissions	Income per worker	Utility (welfare)
EU	205	−631	73	−274	−66	−274	380	12	945
ROW	−205	631	−73	274	66	274	2525	2.5	2994
Total	0	0	0	0	0	0	2905		(7.69)

18 Consumer Attitudes Towards GM Foods: The Modelling of Preference Changes

Chantal Pohl Nielsen,[1] Karen Thierfelder[2] and Sherman Robinson[3]
[1]*Danish Institute of Agricultural and Fisheries Economics, Rolighedsvej 25, 1958 Frederiksberg C, Denmark;* [2]*US Naval Academy, 589 McNair Road, Annapolis, MD 21402, USA;* [3]*International Food Policy Research Institute, 2033 K Street NW, Washington, DC 20006, USA*

Introduction

The sharp reaction against the use of genetically modified organisms (GMOs) in food production by some consumers has already initiated the creation of differentiated marketing systems for genetically modified (GM) and conventional maize and soybeans in the USA. Consumer attitudes will be an important determinant for the profitability and hence the viability of markets for non-GM varieties in the longer term. For producers it is a matter of assessing the benefits and costs of gaining access to niche markets for non-GM crops relative to the benefits of lower production costs associated with cultivating GM crops. Furthermore, many consumers are not only critical of the use of genetic engineering techniques in the production of bulk commodities such as soybeans and cereal grains; they are also concerned about GM ingredients in animal feed and processed foods. To the extent that consumers are in fact willing to pay the additional costs of having these preferences, identity preservation systems will develop so that these demands can also be satisfied.

These divergent consumer attitudes toward GM foods and the increasing demand to be informed about production processes through identity preservation systems etc. will have consequences for the structure and pattern of world food trade. Regardless of whether a country is a net exporter or net importer of agricultural and food products, it will be affected to some extent by the changing consumer attitudes toward GMOs in the developed world. Some countries are highly dependent on exporting particular primary agricultural products to GM-critical regions. Depending on the strength of opposition towards GM products in such regions, the costs of segregating production, and the relative productivity difference between GM and non-GM production, such countries may benefit from the establishment of segregated agricultural markets for GM and non-GM products. In principle these countries may choose to grow GM crops for the domestic market and for exports to countries where consumers are indifferent as to GMO content, and to supply GMO-free products to countries where consumers are willing to pay a premium for this characteristic. Such a market development would be analogous to the niche markets for organic foods. Other countries are net importers and can benefit from the widespread adoption of GM technology. To the extent that consumers in those countries are not opposed to GM products, they will benefit from lower world market prices.

This chapter analyses the price, production and trade consequences of changing consumer preferences regarding the use of GMOs in food production. The analytical framework used is an empirical global general equilibrium model, in which the two primary GM crops, soybeans and maize, are specified as either GM or non-GM. This GM vs. non-GM split is maintained throughout the entire processing chain: GM livestock and GM food processing industries use only GM intermediate inputs; likewise non-GM livestock and non-GM food-processing industries use only non-GM intermediate inputs. This approach is an extension of the authors' earlier work, where only the primary crop markets were segregated (see Nielsen *et al.*, 2000).

The following section provides a concise overview of the current status of GM crops in food production and briefly discusses selected issues related to the segregation of GM and non-GM marketing systems. Section three presents the empirical analysis of changing consumer preferences regarding GMOs. Consumer reactions against GMOs may be interpreted to mean several things. Hence the empirical analysis illustrates two different approaches and how they have been implemented in the model. This too is an extension of the authors' earlier work: preference changes may not only be taken to mean reduced price sensitivity – in this analysis we also investigate what it means to interpret preference changes as a structural shift. The empirical results are examined in the fourth section, and a final section provides some concluding remarks.

Genetic Engineering in Food Production

Current status

Genetic engineering techniques and their applications have developed rapidly since the introduction of the first GM plants in the 1980s. In 1999, GM crops occupied 40 million hectares of land – making up 3.4% of the world's total agricultural area and representing a considerable expansion from less than 3 million hectares in 1996.[1] Cultivation of transgenic crops has so far been most widespread in the production of soybeans and maize, accounting for 54% and 28% of total commercial transgenic crop production in 1999, respectively. Cotton and rapeseed each made up 9% of transgenic crop production in 1999, with the remaining GM crops being tobacco, tomato, and potato (James, 1997, 1998, 1999). To date, genetic engineering in agriculture has mainly been used to modify crops so that they have improved *agronomic* traits such as tolerance to specific chemical herbicides and resistance to pests and diseases. Development of plants with enhanced agronomic traits aims at increasing farmer profitability, typically by reducing input requirements and hence costs. Genetic modification can also be used to improve the final *quality* characteristics of a product for the benefit of the consumer, food-processing industry or livestock producer. Such traits may include enhanced nutritional content, improved durability and better processing characteristics.

The USA holds almost three-quarters of the total crop area devoted to GM crops. Other major GM-producers are Argentina, Canada and China. At the national level, the largest shares of genetically engineered crops in 1999 were found in Argentina (approximately 90% of the soybean crop) Canada (62% of the rapeseed crop) and the USA (55% of cotton, 50% of soybean and 33% of maize) (James, 1999). The US Department of Agriculture (USDA), figures for the USA are similar in magnitude (Economic Research Service (ERS), 2000a): it is estimated that 40% of maize and 60% of soybean areas harvested in 1999 were genetically modified. Continued expansion in the use of transgenic crops will depend in part on the benefits obtained by farmers cultivating transgenic instead of conventional crops relative to the higher cost for transgenic seeds.[2] So far the

[1] Calculations are based on the FAOSTAT statistical database accessible at www.fao.org.

[2] As long as private companies uphold patents on their transgenic seeds they will be able to extract monopoly rents through price premiums or technology fees.

improvements have been not so much in increased yields per hectare of the crops, but rather by reducing costs of production (OECD, 1999). Empirical data on the economic benefits of transgenic crops are still very limited, however. The effects vary from year to year and depend on a range of factors such as crop type, location, magnitude of pest attacks, disease occurrence and weed intensity.

Future market structures

As indicated in the previous section, adoption of GM crop varieties has been extremely rapid in North America and in some large developing countries such as Argentina and China. Lacking systematic evidence of the economic benefits of cultivating GM crops relative to conventional varieties, this rapid adoption must be taken to reflect real or anticipated benefits for farmers. Furthermore, it may seem that the strong reaction against GMOs by consumers in, for example, Western Europe and Japan has not been fully anticipated. The lack of consumer acceptance in these countries has made the market for GM crops more uncertain. As shall be seen below, several of the large producers of GM-potential crops are highly dependent on exporting to GM-critical countries and hence there are important commercial interests in maintaining access to these markets. As the use of genetic engineering moves into its 'second stage' and increasingly provides quality-enhanced foods (e.g. better nutritional content, improved durability, etc.) consumer attitudes toward GMOs might change. However, as long as environmental and food safety issues remain uncertain there will be some consumers that wish to avoid GMOs altogether.

The USDA (ERS, 2000b) reports that the present demand for non-GM maize and soybeans in the USA is very limited. Markets for non-GM crops have, however, developed in response to GM labelling requirements in the European Union and to serve a handful of niche markets domestically, in Europe and in Japan. This demand for non-GM varieties could very well expand rapidly. Maize and soybeans are used as ingredients in a wide range of processed foods as well as in animal feed products. Given that consumers in many developed countries are generally becoming more aware of the production processes that lie behind their food products, they are also increasingly beginning to formulate demands about how these processes should take place. This includes whether or not the use of genetic engineering techniques is deemed acceptable or desirable. Hence livestock producers and food-processing industries also need to consider the consequences of their input choices both in terms of domestic and foreign demand.

Current methods of testing a food product for possible GMO content are not completely reliable. Heating GM maize, for example, eliminates the GM proteins. This renders current testing methods unsatisfactory if the information demanded by consumers is whether or not genetic engineering techniques have been used at any stage in the production process. Therefore, in order to provide consumers with the choice of purchasing guaranteed non-GM foods, the principles of identity preservation (IP) must be followed in the food-marketing systems. IP systems are well known from existing specialty markets (e.g. high-oil maize), but are also applied to a greater or lesser extent for almost all traded agricultural products. Existing grading systems based on the type, length, colour, weight, water content, share of broken or damaged grains, etc. can be thought of as basic IP systems. After classification has taken place, the subsequent handling, storage and processing systems must ensure that the identity of the product is retained throughout the supply chain – as far and as detailed as the final user or the regulatory authorities require (see Buckwell *et al.*, 1999, for a more detailed discussion of the economics of IP systems).

Identity preservation adds administration and marketing costs at all stages along the supply chain, and they can be considerable. How these costs are shared between the farmer, supplier, processor, distributor and consumer depends on how price sensitive demand is at each stage. The lower the demand elasticity, the larger the share of the extra costs that must be borne by the purchaser. To the extent that demand for non-GM varieties is strong enough to support the price differential that will arise between GM and non-GM foods, product and processing methods will be segregated into GM and non-GM varieties to serve these differentiated demands, and private suppliers of marketing services to enable producers to segregate their products will develop. The larger the segregated section of the market, the lower these costs need be as economies of scale are realized. In summary, the economic consequences of changing consumer attitudes toward GMOs in food production will depend crucially on the relation between three aspects: (i) the nature and extent of the preference change (willingness to pay); (ii) the costs of preserving the identity of products in completely segregated markets; and (iii) the size of the relative productivity difference between GM and non-GM production methods.

Empirical Analysis of Consumer Preferences

GM-potential foods in world production and trade

The data used in the empirical analysis described below are from version 4 of the Global Trade Analysis Project (GTAP) database, which is estimated for 1995 (McDougall *et al.*, 1998). As discussed above, the main crops that have been genetically modified to date are soybeans and maize. The sectoral aggregation of the database for use in this analysis therefore comprises a cereal grains sector (which includes maize but not wheat and rice) and an oilseeds sector (which includes soybeans) to reflect these two GM-potential crops. The livestock, meat and dairy, vegetable oils and fats, and other processed food sectors are also singled out, since they are important demanders of cereal grains and oilseeds as intermediate inputs in production.

In terms of the importance of the two GM-potential crops in total primary agriculture, Table 18.1 shows that the cereal grains sector accounts for almost 20% of agricultural production in the USA, approximately 11% in South America, but less than 7% in all other regions. Oilseed production accounts for 6–7% of agricultural production in the Cairns group[3], low-income Asia, the USA and sub-Saharan Africa, while its share is even smaller in Western Europe, high-income Asia and the South American countries outside the Cairns group.

As discussed above, maize and soybeans, the two crops that have benefited most from genetic engineering techniques to date, are widely used both directly as animal feed and as ingredients in processed feed and food products. With the exception of some final consumption of oilseeds in high-income Asia, cereal grains and oilseeds are generally not used for final consumption in the three high-income regions, high-income Asia, the USA

Table 18.1. Agricultural production structures, 1995.

	Cairns group	High-income Asia	Low-income Asia	USA	Rest of South America	Western Europe	Sub-Saharan Africa	Rest of world
Cereal grains	4.9	1.8	4.9	19.1	10.7	5.4	2.9	6.7
Oilseeds	6.1	0.4	7.2	6.2	1.8	1.9	6.4	2.4
Wheat	4.6	0.8	6.2	4.8	2.0	5.1	4.3	7.8
Other crops	50.4	67.1	52.4	24.2	53.6	33.6	69.1	43.7
Livestock	34.0	29.9	29.3	45.7	31.9	54.0	17.3	39.3
Total agriculture	100.0	100.0	100.0	100.0	100.0	100.0	100.0	100.0

Source: Multi-region GMO model database derived from GTAP version 4 data.

[3] Consists of Argentina, Australia, Bolivia, Brazil, Canada, Chile, Colombia, Costa Rica, Guatemala, Indonesia, Malaysia, New Zealand, Paraguay, the Philippines, South Africa, Thailand and Uruguay.

and Western Europe. In the Cairns group and in the developing regions, cereal grains and oilseeds are to a larger extent used for final consumption, although these shares vary substantially across regions. In terms of intermediate use, there are some interesting differences regarding the use of these two crops as direct feed for livestock or as inputs into the food- and feed-processing industries (the latter including manufacturers of prepared feeds, e.g. soybean meal, feed pellets, etc.). Within the Cairns and high-income Asia groups, by far the largest share of these crops is used in food and feed processing rather than directly as feed. This is also the case for oilseed use in both Western Europe and the USA. For cereal grains, however, three-quarters of the total use in the USA is directly for livestock feed, whereas for Western Europe there is more of a 50–50 split for use of cereal grains directly as livestock feed and as inputs into the processing industries. These use patterns illustrate how important it is for GMO-critical consumers to be able to trace GM cereal grains and oilseeds throughout both the livestock and other food-producing chains.

The importance of trade in these two commodities varies across the regions. Table 18.2 shows that the value of oilseed exports relative to total value of production is significant for the Cairns group, the USA and the rest of South America. Cereal grain exports are also moderately large in value terms for the first two regions, but otherwise most of the production value of these two crops is captured on the domestic markets. For the Cairns group, the rest of South America, the USA and sub-Saharan Africa, the impact of genetic engineering would be much larger if these techniques were applicable to the crops contained in the much larger aggregate 'other crops' sector. On the import side, the value of oilseed imports into Western Europe amounts to almost 40% of the total value of oilseed absorption. High-income Asia is also heavily dependent on imports of oilseeds and to a lesser extent cereal grains. Table 18.2 also shows the importance of trade in livestock and processed food products by region. These trade dependencies are generally lower than those just described.

Table 18.2. Trade dependence: agricultural and food products, 1995.

	Cairns group	High-income Asia	Low-income Asia	USA	Rest of South America	Western Europe	Sub-Saharan Africa	Rest of world
Value of exports in % of total production value								
Cereal grains	9.7	0.2	0.7	16.0	0.7	3.7	4.3	0.7
Oilseeds	15.7	4.1	2.7	28.7	32.4	1.8	5.8	11.2
Wheat	28.5	0.0	0.3	39.2	6.6	6.8	0.1	1.5
Other crops	15.4	0.7	3.5	18.9	29.2	4.7	20.0	6.6
Livestock	7.3	0.2	1.5	2.4	2.9	1.2	2.4	1.7
Vegetable oils and fats	32.8	4.8	3.2	7.2	4.0	4.3	10.3	6.7
Meat and dairy	10.2	0.4	12.6	4.9	1.5	3.1	11.3	1.7
Other processed foods	12.6	0.7	10.3	5.2	10.9	6.2	15.7	4.1
Value of imports in % of total absorption value								
Cereal grains	7.2	18.3	5.5	0.9	14.8	5.0	7.2	10.3
Oilseeds	6.5	71.1	0.9	2.4	55.2	38.2	0.4	10.6
Wheat	11.9	17.1	10.4	3.4	51.4	3.7	15.5	17.7
Other crops	5.5	6.5	2.3	17.8	5.7	18.3	1.4	8.0
Livestock	0.9	5.4	1.5	2.1	1.6	2.3	0.4	2.4
Vegetable oils and fats	3.1	19.0	17.2	5.0	15.3	4.1	14.5	23.1
Meat and dairy	2.0	9.9	6.4	1.8	8.9	1.5	35.1	10.4
Other processed foods	4.6	4.2	3.5	4.6	5.9	3.6	15.8	10.3

Source: Multi-region GMO model database derived from GTAP version 4 data.

Compared with trade in other crops, meat, dairy and other processed foods, the value of world trade in cereal grains and oilseeds is modest, as the net trade values in Table 18.3 show. The USA is by far the dominant exporter of both crops followed by the Cairns group. Western Europe is the main importer of oilseeds and high-income Asia is the main importer of cereal grains and the second largest importer of oilseeds. In terms of processed food trade, countries in the Cairns group and Western Europe are large exporters of meat and dairy products and other processed foods, of which high-income Asia is a major importer.

In terms of bilateral trade flows, it may be noted that processed food exports from the Cairns group are mainly destined for high-income Asia and Western Europe. As mentioned above, the USA is a substantial exporter of both cereal grains and oilseeds. Half of the former exports are sold in high-income Asia. For US oilseed exports, 34% are sold in Western Europe and 37% in high-income Asia. A large share of meat and dairy and other processed food exports from the USA goes to the high-income markets of Asia. Cereal grain imports into high-income Asia are very narrowly sourced in the sense that more than 90% of this region's cereal grain imports come from the USA, whilst its imports of processed food products come from both the Cairns group and the USA. For Western Europe, almost 45% of its oilseed imports come from the USA and another almost 35% from the Cairns group. As men-

Table 18.3. Net trade flows and composition of world trade, 1995.

	Cairns group	High-income Asia	Low-income Asia	USA	Rest of South America	Western Europe	Sub-Saharan Africa	Rest of world	Total
Net trade (US$ billion)									
Cereal grains	0.31	−4.51	−0.88	8.03	−1.20	0.13	−0.06	−1.81	0
Oilseeds	1.64	−2.86	0.46	4.57	−0.52	−3.55	0.23	0.05	0
Wheat	2.71	−1.62	−2.40	5.35	−1.00	1.23	−0.55	−3.72	0
Other crops	13.37	−11.59	2.09	0.62	9.23	−20.16	8.42	−1.97	0
Livestock	5.84	−4.47	-0.11	0.50	0.32	−1.62	0.22	−0.68	0
Vegetable oils and fats	7.18	−1.16	−3.02	0.50	−0.61	0.16	−0.14	−2.91	0
Meat and dairy	9.81	−11.09	0.82	5.48	−1.93	5.53	−0.69	−7.93	0
Other processed foods	18.09	−24.09	6.03	1.17	3.11	11.01	−0.46	−14.85	0
Value of exports in % of value of world trade									
Cereal grains	11.29	0.10	1.06	75.88	0.55	9.29	0.71	1.13	100
Oilseeds	26.48	0.48	6.89	49.83	4.18	2.43	2.52	7.20	100
Wheat	31.88	0.01	0.64	48.20	0.86	15.68	0.03	2.69	100
Other crops	28.05	1.83	8.78	16.29	15.01	7.47	12.44	10.13	100
Livestock	40.41	1.27	8.95	17.73	4.06	15.57	1.59	10.41	100
Vegetable oils and fats	55.86	2.16	3.50	11.37	1.30	18.22	1.67	5.91	100
Meat and dairy	34.65	1.05	4.68	24.33	1.01	29.84	0.45	3.98	100
Other processed foods	27.44	3.70	8.90	16.39	6.54	27.83	2.30	6.89	100
Value of imports in % of world trade									
Cereal grains	8.50	40.82	8.97	3.42	11.42	8.14	1.27	17.46	100
Oilseeds	9.65	29.88	2.20	2.85	9.54	38.99	0.19	6.71	100
Wheat	8.45	13.99	21.40	2.00	9.49	5.10	4.74	34.82	100
Other crops	9.48	17.94	5.88	15.43	2.20	35.47	0.75	12.86	100
Livestock	4.79	28.59	9.62	14.66	2.12	25.43	0.26	14.53	100
Vegetable oils and fats	4.10	10.50	25.26	7.81	5.69	17.04	2.71	26.90	100
Meat and dairy	6.74	32.60	2.36	8.75	6.51	14.10	2.40	26.53	100
Other processed foods	10.63	26.08	3.30	15.31	3.65	17.60	2.73	20.69	100

Source: Multi-region GMO model database derived from GTAP version 4 data.

tioned above, Western Europe is a large exporter of meat, dairy and other processed foods. About one-fifth of these exports are destined for the high-income Asian markets. (Tables containing these data are to be found in Nielsen *et al.*, 2001.)

Global CGE model with segregated food markets

The modeling framework used in this analysis is a multi-region computable general equilibrium (CGE) model consisting of eight regions, which are inter-connected through bilateral trade flows: the Cairns group, high-income Asia, low-income Asia, the USA, the rest of South America, Western Europe, sub-Saharan Africa and the rest of the world.[4] For the purpose of describing the model,[5] it is useful to distinguish between the individual regional models and the multi-region model system as a whole, which determines how the individual regional models interact. When the model is actually used, the *within* region and *between* region relationships are of course solved simultaneously. Each regional CGE model is a relatively standard trade-focused CGE model, with 12 sectors: five of which are primary agriculture, three are food-processing industries and the remaining four are manufacturers and services. Each regional model has five factors of production: skilled and unskilled labour, capital, land and natural resources. For each sector, output supply is specified as a constant elasticity of substitution (CES) function over value-added, and intermediate inputs are demanded in fixed proportions (a Leontief specification). Profit-maximization behaviour by producers is assumed, implying that each factor is demanded so that marginal revenue product equals marginal cost, given that all factors are free to adjust. Each regional economy contains domestic market distortions in the form of sectorally differentiated indirect consumption and export taxes, as well as household income taxes. There is a single representative household in each economy, which demands commodities according to fixed expenditure shares, maximizing a Cobb-Douglas utility function.

As in other CGE models, it is only relative prices that are determined – the absolute price level is set exogenously. In this model, the aggregate consumer price index in each sub-region acts as the numeraire. A convenient consequence of this specification is that solution wages and incomes are in real terms. The solution exchange rates in each region are also in real terms and can be seen as equilibrium price-level-deflated exchange rates, using the country consumer price indices as deflators. The international numeraire is defined by fixing the exchange rate for North America. World prices are converted into domestic currency using the exchange rate, including any tax or tariff components. Cross-trade price consistency is imposed, so that the world price of country A's exports to country B are the same as the world price of country B's imports from country A.

Sectoral export-supply and import-demand functions are specified for each region. As is common in other CGE models, the multi-regional model used in this analysis specifies that goods produced in different countries are imperfect substitutes. On the supply side, sectoral output is a constant elasticity of transformation (CET) aggregation of total supply to all export markets and supply to the domestic market. The allocation between export and domestic markets is determined by the maximization of total sales revenue. On the demand side the assumption of product differentiation is combined with the almost ideal demand system (AIDS) to determine the input aggregation equation. Although not used in this application, this specification allows for non-unitary income elasticities of demand for imports and pairwise substitution elasticities that vary across countries (unlike the more typical CES specification). The macro closure of the model is relatively simple. First of all, aggregate real investment and government consumption are assumed to be fixed. Secondly, since the trade balances in each

4 Note that the bilateral trade figures that link these regions are net of trade within the region, and that in the model intra-regional trade is treated as another source of domestic demand.

5 The description of the standard version of the model draws in part on Lewis *et al.* (1999).

region also are assumed to be fixed with the real exchange rates adjusting to equilibrate aggregate exports and imports, the macro closure of the model is achieved by allowing domestic savings for each region to adjust to achieve macro equilibrium.

Data base adjustments and model extensions

IDENTITY PRESERVATION: SEGREGATING INTERMEDIATE USERS OF GM AND NON-GM CROPS In order to operate with segregated GM and non-GM sectors in the extended model, the base data must also reflect this segregation. First of all, the base data are adjusted by splitting the cereal grain and oilseed sectors into GM and non-GM varieties.[6] It is assumed that all regions in the model initially produce some of both GM and non-GM varieties of cereal grains and oilseeds. Specifically, the assumed shares are as shown in Table 18.4, adapted from estimates provided in James (1999) and (ERS 2000a)[7]. In the Cairns group, for example, 40% of total cereal grain production is assumed to be of the GM variety, whilst 60% of oilseed production is assumed to be GM. The structures of production in terms of the composition of intermediate input and factor use in the GM and non-GM varieties are initially assumed to be identical. The destination structures of exports are also initially assumed to be the same, and this determines the resulting import composition by ensuring bilateral trade flow consistency.

The next step is to identify the sectors that use cereal grains and oilseeds as intermediate inputs as GM and non-GM sectors to reflect the concept of identity preservation. The GM/non-GM split is applied to the following sectors: livestock, vegetable oils and fats, meat and dairy, and other processed foods. In the base data the GM/non-GM split for these four sectors is determined residually, based on the share of GM inputs of cereal grains and oilseeds in total (GM plus non-GM) inputs of cereal grains and oilseeds for each sector. These shares are then used to split the data into GM and non-GM varieties of the four processing sectors. At this stage, the described procedure leaves all agricultural and food sectors using some of both GM and non-GM inputs. The input–output table is then adjusted so that GM sectors only use GM inputs and non-GM sectors only use non-GM inputs. As mentioned above, in the description of the model, intermediate demand is

Table 18.4. Assumed initial shares of GM varieties in cereal grain and oilseed production.

	Cairns group	High-income Asia	Low-income Asia	USA	Rest of South America	Western Europe	Sub-Saharan Africa	Rest of world
GM cereal grains as a % of total (GM + non-GM) cereal grain production	40	20	40	40	40	15	15	15
GM oilseeds as a % of total (GM + non-GM) oilseed production	60	35	60	60	60	15	15	15

6 As will be discussed later, the distinguishing characteristic between these two varieties is the level of productivity. Furthermore, although acknowledging the fact that there may be environmental risks and hence externality costs associated with GM crops, they are impossible to estimate at this time and this chapter makes no attempt to incorporate such effects in the empirical analysis.

7 It is due to technical data limitations that the shares of GM varieties in total crop production in high-income Asia are as high as they are. The important point here is the relative size of the shares between regions, not so much the levels within each region. Producers in high-income Asia and Western Europe are assumed to be reluctant with respect to adopting GM crops due to negative consumer attitudes. GM adoption in sub-Saharan Africa and the rest of world is assumed to be restricted by other factors such as lack of access to the technology. The other regions are assumed to adopt GM crop varieties enthusiastically and in fact be able to do so.

characterized by a Leontief specification, i.e. the input-output coefficients are fixed. Given the adjustments of the input–output data described above, this means that in the model, intermediate use in the GM sectors is restricted to only GM inputs and intermediate use in the non-GM sectors is restricted to only non-GM inputs. This is an important difference compared with the authors' earlier work (Nielsen *et al.*, 2000), where intermediate users of oilseeds and cereal grains had a choice between GM and non-GM varieties.

ENDOGENOUS FINAL DEMAND CHOICE BETWEEN GM AND NON-GM FOODS In the model consumers have a choice between GM and non-GM varieties. Final demand for each composite good (i.e. GM plus non-GM) is held fixed as a share of total demand, while introducing a choice between GM and non-GM varieties. In this way, all the initial expenditure shares remain fixed, but for six of the food product categories (oilseeds, cereal grains, livestock, vegetable oils and fats, meat and dairy, and other processed foods), a choice has been introduced between GM and non-GM varieties. All other expenditure shares remain fixed.

The choice between GM and non-GM varieties is determined by a CES function (here shown for an arbitrary food product f in country k):

$$C(f,k) = a(f,k) \cdot [\alpha_G(f,k) \cdot C(gmf,k)^{-\rho_G(f,k)} + (1-\alpha_G(f,k)) \cdot C(ngf,k)^{-\rho_G(f,k)}]^{-1/\rho_G(f,k)} \quad (1)$$

where $C(f,k)$ is the share of food product f in total final consumption in country k, gmf is the GM variety and ngf is the conventional variety. The parameter $a(f,k)$ is the CES demand shift parameter. The exponent is defined by the elasticity of substitution between GM and non-GM varieties, $\sigma_G(f,k)$: $\rho_G(f,k) = [1/\sigma_G(f,k)] - 1$. The CES share coefficients are $\alpha_G(f,k)$ and $(1-\alpha_G(f,k))$. In the model, the following first-order conditions are included for the demand for the six food types that come in GM and non-GM varieties – one set of equations for each of the six food categories (here shown for one arbitrary food product f):

$$\frac{C(gmf,k)}{C(ng,k)} = \left[\frac{PC(ngf,k)}{PC(gmf,k)} \cdot \frac{\alpha_G(gmf,k)}{(1-\alpha_G(gmf,k))}\right]^{1/(1+\rho_G(f,k))} \quad (2)$$

The adding-up constraints are included as follows, ensuring that expenditure on each composite good (i.e. GM plus non-GM) remains fixed:

$$C0(gmf,k) + C0(ngf,k) = C(gmf,k) + C(ngf,k) \quad (3)$$

where $C0(gmf,k)$ and $C0(ngf,k)$ are the base level share coefficients for the GM and non-GM varieties, respectively.

Design of experiments

GM and non-GM production technologies

As mentioned above, the distinguishing characteristic between the GM and non-GM maize and soybean sectors is the level of productivity. The GM cereal grain and oilseed sectors are assumed to benefit from increased productivity in terms of primary factor use as well as a reduction in chemical use.[8] The available estimates of agronomic and hence economic benefits to producers from cultivating GM crops are very scattered and highly diverse (see e.g. OECD, 1999 for an overview of available estimates). Nelson *et al.* (1999), for example, suggest that glyphosate-resistant soybeans may generate a total production cost reduction of 5%, and their scenarios have GM maize increasing yields by between 1.8% and 8.1%. For present purposes, the GM-adopting cereal grains and oilseed sectors are assumed to make more productive use of the primary factors of production as compared with the non-GM sectors. In other words, the same level of output can be obtained using fewer primary factors of production, or a higher level of output can be obtained using the same level of production factors. In our scenarios, the GM oilseed and GM cereal grain sectors in all regions are assumed to have a 10% higher

8 Note that this is an asymmetric shock and that it will therefore have different effects in different regions because of different cost structures: the shares of primary factor costs and chemical costs in total production costs are different.

level of factor productivity as compared with their non-GM (conventional) counterparts. Furthermore, there seems to be evidence that cultivating GM varieties substantially reduces the use of chemical pesticides and herbicides (see e.g. Pray *et al.*, 2000). Hence the use of chemicals in the GM oilseed and GM cereal grain production is reduced by 30% to illustrate this cost-saving effect.

Consumer preferences

There are many ways to model changes formally in consumer preferences. This chapter illustrates how two such ways can be implemented in a computable general equilibrium model. This is done by shifting and altering the curvature of the indifference curve between GM and non-GM commodities. Each alternative has a different interpretation of what consumers might mean when they say they disapprove of GM foods. The starting point for the consumer preference experiments is that food products come in two varieties, distinguished by their method of production: GM and non-GM. As described above, the model has the representative consumer demanding these two product varieties in terms of a CES function, meaning that the two varieties are imperfect substitutes with a region-specific elasticity of substitution, $\sigma_G(f,k)$, between the two. Three different consumer response scenarios are examined (summarized in Table 18.5). In the base case consumers in all countries are relatively indifferent with respect to the introduction of GM techniques in food production, and so they find GM and non-GM food varieties highly substitutable. It is assumed that the elasticity of substitution between GM and non-GM varieties is high and equal in all regions. Specifically, $\sigma_G(f,k) = 5.0$ for all for all regions k.

The next two experiments then attempt to reflect the fact that citizens in Western Europe and high-income Asia dislike the idea of GM foods. In the second experiment this is illustrated by lowering the elasticities of substitution between the GM and non-GM varieties for consumers in these two regions $\sigma_G(f,k) = 0.5$ for k = Western Europe and high-income Asia). Consumers in these regions are simply assumed to be less sensitive to a given change in the ratio of prices between GM and non-GM varieties. They are seen as poor substitutes in consumption in these particular regions. Citizens in all other regions are basically indifferent, and hence the two varieties remain highly substitutable in consumption ($\sigma_G(f,k) = 5.0$).

Table 18.5. Experiment design to illustrate different representations of preference changes.

Experiment	Description
1. Base case	*Base model 1:* Consumers in all countries find GM and non-GM foods highly and equally substitutable. Elasticity of substitution between GM and non-GM varieties $\sigma_G(k)$ = 5.0 for all regions k *Shock to base model 1:* Factor productivity increase (10%) Reduced chemical use in GM oilseed and GM cereal grains (−30%)
2. Price sensitivity	*Base model 2:* Consumers in Western Europe and high-income Asia are less responsive to changes in the relative price of GM and non-GM goods. $\sigma_G(k)$ = 0.5 for k = *W.* Europe and high-income Asia. Other regions: Same as base case. *Shock to base model 2:* Same as base case.
3. Structural change	Consumers in Western Europe and high-income Asia completely shift their consumption away from GM goods. Other regions: Same as base case. *Shock to base model 2:* Same as base case plus shock the share of GM varieties in final consumption to 0.02 for Western Europe and high-income Asia.

Comparing experiments 1 and 2 corresponds to altering the curvature of the indifference curves of consumers in Western Europe and high-income Asia as illustrated in Fig. 18.1. The two curves in the figure correspond to the same level of utility, U_0. When the relative prices of GM and non-GM foods change, consumers in Western Europe and high-income Asia are, in the second experiment, assumed to be less inclined to shift consumption toward GM varieties as they were in the base case, where substitutability was high. In terms of the CES function (equation 1), this means that we are changing the $\rho_G(f,k)$ parameter by altering the substitution elasticity, in effect, changing the curvature of the CES function. The representative consumer is on the same budget line (same expenditure on the composite food product, i.e. GM plus non-GM, and hence same level of utility). The CES share parameters $\alpha_G(f,k)$ and $(1-\alpha_G(f,k))$ are unchanged by this experiment.

It is not clear, however, whether reduced price sensitivity is an appropriate interpretation of consumers' critical approach to GM foods. In some rich countries, where consumers can indeed afford to be critical of these new techniques in food production, irrespective of how cheap these products may become (relative to non-GM foods), some consumers may simply not want to consume them. In this case, we are changing the ratio of GM to non-GM foods demanded at a given (constant) price ratio, holding utility constant. This is illustrated in Fig.

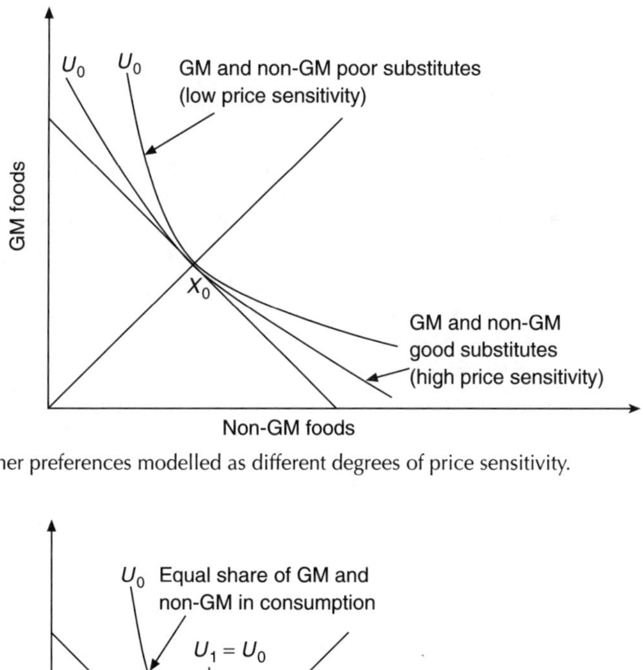

Fig. 18.1. Consumer preferences modelled as different degrees of price sensitivity.

Fig. 18.2. Consumer preferences modelled as a structural change.

18.2, where the representative consumer in Western Europe and high-income Asia is as well off as before but now with a lower share of GM foods in his/her consumption bundle. Recalibrating the model around these new shares gives us a new constant and new shares in the CES function, $\alpha_G(f,k)$ and $(1-\alpha_G(f,k))$, determining the composite use of GM and non-GM foods in consumption. In determining these new values we ensure that the total value of expenditure on each composite food item remains the same. In other words, consumers still spend the same amount on their consumption of food, but the composition hereof is changed in favour of non-GM varieties. In the experiment we reduce the GM share of foods in consumption in Western Europe and high-income Asia to 2%. The description of the three experiments is summarized in Table 18.5.

Results of Empirical Analysis

Price and trade results

Base case experiment

The increase in factor productivity and the reduced need for chemicals in the GM cereal grain and oilseed sectors causes the cost-driven prices of these crops to decline. The magnitude of this price decline in the different sectors and regions will differ, depending on the shares of primary production factors and chemicals in total production costs. In sectors and regions where these costs make up a large share of total costs, the impact of the productivity shock in terms of lower supply prices will be greater than in sectors and regions where the share is smaller. Intermediate users of GM inputs (the GM livestock and GM processed food producers) will benefit from lower input prices. The non-GM product markets will be affected by the productivity gain in the GM sectors in three ways. First, there will be increased competition for primary factors of production and intermediate inputs because GM production will increase. Second, consumers domestically might change their consumption patterns in response to the new relative prices depending on their initial consumption pattern and substitution possibilities. Third, importers will change their import pattern depending on the relative world prices, their initial absorption structures and the substitution possibilities between suppliers. In all three cases, the initial cost, consumption and import structures on the one hand, and the substitution possibilities between products for input use, final consumption and imports on the other, will determine the net impact of the productivity experiment. The net effects are theoretically ambiguous and hence must be determined empirically.

Figure 18.3 depicts for selected regions the price wedges that arise between the non-GM and GM varieties in the base case experi-

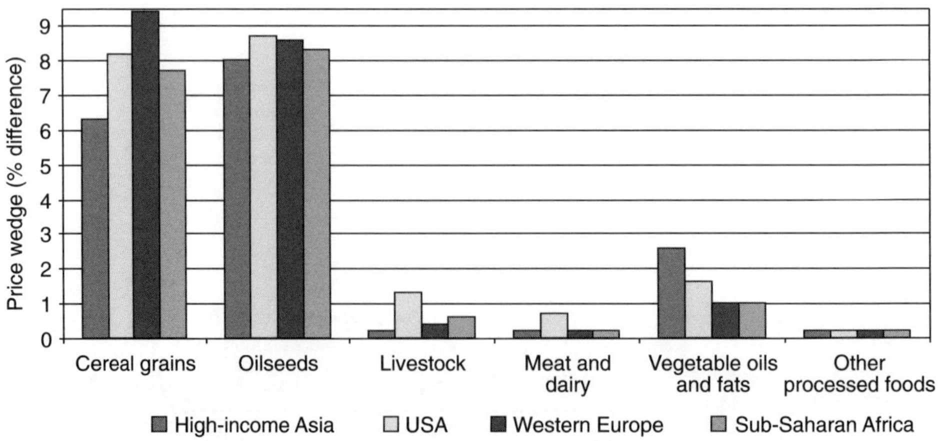

Fig. 18.3. Base case experiment: price wedges between non-GM and GM products (percentage points).

ment, where GM and non-GM foods are considered to be good substitutes in consumption in all regions. Generally, the relative price of non-GM to GM commodities rises, and the percentage point differences between the prices of non-GM and GM varieties of cereal grains and oilseeds are between 6.3 and 9.4. As described above, the price wedges vary across the regions in part because they have different shares of primary factor and chemical costs in total production costs. Hence the extent to which the individual regions benefit from the productivity increase differs.

The lower GM crop prices in turn result in lower production costs for users of GM inputs, thereby reducing those product prices relative to the non-GM varieties as well. As can be seen in Fig. 18.3, the price wedges that arise between the GM and non-GM livestock and processed food products are of course much smaller than the price wedges between GM and non-GM primary crops because the cost reduction concerns only a part of total production costs. Relatively speaking, oilseeds constitute a large share of production costs in vegetable oils and fats production (compared with oilseed and cereal grain use in other food production), and hence the spillover effect is largest here. As the base data revealed, a rather large share of cereal grains is used as intermediate input to livestock production in the USA compared with other regions. For this reason Fig. 18.3 shows a relatively large price wedge for livestock, meat and dairy products in the USA compared with the other regions.

The lower GM crop prices mean improved international competitiveness for exporters of these crops. Hence, as Table 18.6 shows, the largest exporters of cereal grains and oilseeds, the Cairns Group and the USA, increase their exports of GM crops in this base case by between 8.6% and 14.9%. Due to the reduced relative competitiveness of non-GM crops, exports of this variety decline. High-income Asia and Western Europe increase their imports of the cheaper GM varieties. This is particularly so in the case of oilseeds because these two regions are highly dependent on imported oilseeds from countries that are enthusiastic GM-adopters. Imports of the non-GM varieties decline slightly due to the reduced relative price competitiveness of non-GM products when consumers find GM and non-GM food varieties to be good substitutes.

Table 18.6. Selected trade results of base experiment (percentage changes).

	Cairns group	High-income Asia	Low-income Asia	USA	Western Europe	Sub-Saharan Africa
Exports						
Non-GM cereal grains	−4.7	−3.4	−8.0	−2.4	−3.0	−4.1
GM cereal grains	14.9	17.4	22.6	9.0	16.5	23.0
Non-GM oilseeds	−4.3	−3.0	−9.1	−2.1	−2.9	−3.0
GM oilseeds	13.0	13.5	16.7	8.6	17.6	20.7
Non-GM vegetable oils and fats	−2.0	−1.9	−2.2	−1.3	−1.0	−0.6
GM vegetable oils and fats	4.4	6.4	3.7	3.4	3.9	3.6
Non-GM other processed food	−0.2	−0.2	−0.4	−0.3	−0.2	−0.1
GM other processed food	0.7	0.8	0.9	0.7	0.8	1.1
Imports						
Non-GM cereal grains	−4.2	−0.2	−12.3	−1.8	−0.3	−4.8
GM cereal grains	5.6	1.7	19.7	2.7	0.8	32.8
Non-GM oilseeds	−9.1	−3.0	−14.8	−4.3	−1.7	−5.5
GM oilseeds	10.2	10.7	16.4	5.1	9.2	27.4
Non-GM vegetable oils and fats	−2.0	−1.9	−2.2	−1.3	−1.0	−0.6
GM vegetable oils and fats	4.4	6.4	3.7	3.4	3.9	3.6
Non-GM other processed food	−0.2	−0.2	−0.4	−0.3	−0.2	−0.1
GM other processed food	0.7	0.8	0.9	0.7	0.8	1.1

Price sensitivity experiment

The price wedges resulting from the price sensitivity experiment are not markedly different from the ones reported in the base case experiment. It may be mentioned, however, that the prices for GM cereal grains and especially oilseeds are slightly lower on the Western European and high-income Asian markets when consumers are critical (i.e. less price sensitive): larger price reductions are required in order to sell GM varieties in GMO-critical markets. Conversely, demand for non-GM crops is relatively stronger, and hence the prices of non-GM oilseeds, for example, are higher. Hence we find that the price wedges for especially oilseeds, but also cereal grains, are larger in high-income Asia and Western Europe in the price sensitivity experiment. In large oilseed-producing markets such as the USA, the price of the non-GM variety falls slightly more and the price of the GM variety falls less as compared with the base case. Compared with the base case, the increase in GM oilseed and cereal grain exports from the Cairns group and the USA is smaller when consumers in their important export markets are less responsive to the GM/non-GM price difference. Consequently, on the import side, the results show that the declines in imports of the more expensive non-GM oilseeds into high-income Asia and Western Europe are smaller. The decreases in non-GM cereal grain imports have even turned into minor increases. High-income Asia and Western Europe still increase their GM oilseed imports in this price sensitivity experiment (although at lower rates) because of their high dependence on importing from GM-enthusiastic regions. This result is due to the fact that there is a symmetry in the trade dependence concerning oilseeds: US oilseeds make up a large share of oilseed imports into high-income Asia and Western Europe, and these regions make up a large share of US exports. Hence changes in consumer preferences in these countries will have an impact on the trading conditions for US producers.

Structural change experiment

In this final experiment consumers in Western Europe and high-income Asia simply turn against GM foods. Compared with the previous experiment, final demand in these regions is not only very insensitive to relative price differences between GM and non-GM food varieties. Consumers in Western Europe and high-income Asia are assumed to simply shift their consumption patterns away from GM varieties and in favour of non-GM varieties, regardless of the relative price decline of GM foods. This shift is measured relative to the experiment, in which price sensitivity in these regions is low to begin with. Hence the effects of this structural shock are an addition to the second experiment. The results show that this rejection is clearly a much more dramatic change compared with reduced price sensitivity. Critical consumers simply do not want to eat GMOs. The price of GM varieties in the GMO-critical countries declines further because of the almost complete rejection of these products, whereas the price of non-GM foods increases. This leads to substantially larger price wedges in the GM-critical regions as compared with the previous experiments, as is evident from Fig. 18.4. By the nature of this model, the larger price wedges between GM and non-GM primary crops follow through the entire food-processing chain. The price increase for non-GM foods is, however, moderated by the fact that there are indeed markets for non-GM products in all regions in the model, so these consumers are not closing themselves off to necessary goods nor are they required to produce all the non-GM goods themselves. The model allows all countries to produce both varieties and hence supply both GMO-indifferent and GMO-critical consumers. Total US GM cereal grain and oilseed exports fall by no less than −17% and −33%, respectively (Table 18.7). Instead, exports of the non-GM varieties increase by 10% and 16%, respectively. These changes are a direct reaction to the relative prices obtainable on their key export markets, namely high-income Asia and Western Europe. The price of GM cereal grains and oilseeds on these markets plummets and the price of non-GM varieties increases slightly. The price decline for GM cereal grains in Western Europe is not as large as for oilseeds because this region is less dependent on imports of these crops, relatively speaking. This explains the larger price wedge for oilseeds compared with cereal grains in Western Europe as depicted in Fig. 18.4.

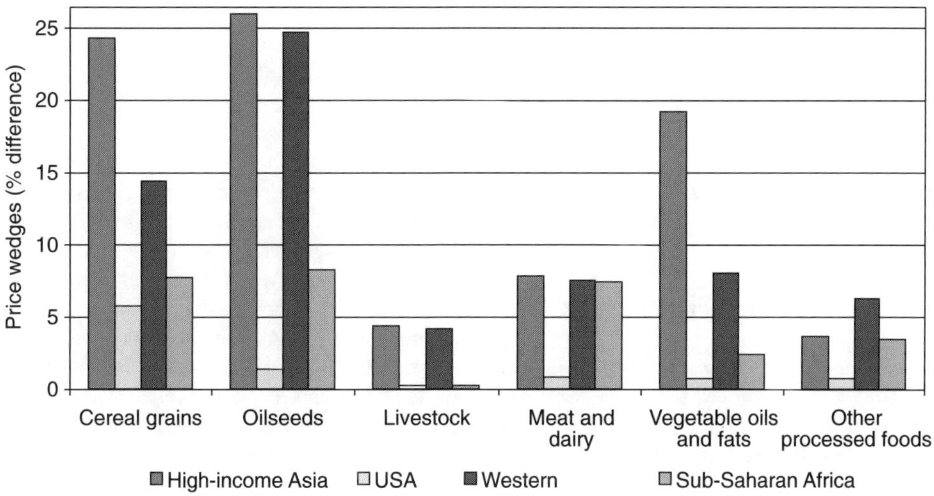

Fig. 18.4. Structural change case: price wedges between non-GM and GM products (percentage points).

Table 18.7. Selected trade results of structural shift experiment (percentage changes).

	Cairns group	High-income Asia	Low-income Asia	USA	Western Europe	Sub-Saharan Africa
Exports						
Non-GM cereal grains	1.4	4.0	−1.7	8.1	3.8	0.8
GM cereal grains	1.4	−42.5	4.1	−17.4	−30.7	−2.0
Non-GM oilseeds	10.7	12.9	1.6	14.5	2.2	5.1
GM oilseeds	−31.8	−45.8	−9.6	−33.3	−33.7	−5.9
Non-GM vegetable oils and fats	5.4	11.3	6.5	3.7	6.5	6.4
GM vegetable oils and fats	−13.9	−29.7	−35.6	−10.6	−29.2	−50.2
Non-GM other processed food	7.4	11.1	7.5	5.4	8.0	6.9
GM other processed food	−31.9	−39.6	−35.3	−30.3	−37.4	−50.6
Imports						
Non-GM cereal grains	−3.9	18.9	−12.6	−0.1	9.8	−4.0
GM cereal grains	4.4	−70.7	21.2	0.7	−59.1	34.8
Non-GM oilseeds	−6.4	23.5	−14.1	−6.0	10.3	−4.2
GM oilseeds	12.0	−56.8	28.8	22.7	−60.4	40.3
Non-GM vegetable oils and fats	3.0	24.8	0.8	2.6	9.5	5.4
GM vegetable oils and fats	−3.3	−72.4	1.5	−1.7	−59.5	−11.2
Non-GM other processed food	2.9	15.4	3.4	2.6	8.5	3.9
GM other processed food	−2.8	−66.8	−1.0	0.2	−60.2	−10.5

Turning to the import results, Table 18.7 shows that imports of GM cereal grain and oilseeds into Western Europe and high-income Asia decline substantially (between −57% and −71%). These decreases in quantities are accompanied by import price declines in the order of −21% to −26%. Conversely, imports of non-GM crops increase substantially, at slightly higher prices. The sourcing of these non-GM crop imports is spread across all regions, because in the model all regions are assumed to be able to produce both varieties and to be able to credibly verify this characteristic to importers. Clearly, this is a simplification of reality, and one can easily imagine that for some regions, living up to the principles of identity preservation and verifying this is very costly, thereby putting them at a cost disadvantage. Such effects are not captured in this model. The increases in non-GM cereal grain and oilseed imports are supplemented by increases in own production in both high-income Asia and Western Europe[9].

Production results

Being a major exporter of both crops, the increased demand for GM cereal grains and oilseeds in the base case experiment filters through to an increase in production of these crops in the USA. The effect is dampened, however, by the fact that its major destination regions (high-income Asia and Western Europe) have much larger non-GM sectors (relative to their GM sectors), which are required to use only non-GM inputs.[10] This also means, for example, that the production of non-GM crops does not fall as markedly in the USA as it does in e.g. low-income Asia, a region that is not very heavily engaged in international trade in these particular crops. Figure 18.5 compares the impact on production in the USA of the different and changing assumptions made about consumer preferences in Western Europe and high-income Asia. Since exports make up a relatively large share of the total value of production in these sectors, particularly for oilseeds, we see that there is a marked effect on the composition of production. Production of GM crop varieties increases in the first two experiments, whilst production of non-GM varieties declines somewhat. The impact is slightly less when consumers in high-income Asia and Western Europe are less sensitive to the GM/non-GM price difference.

In the structural shift experiment, however, the production of GM oilseeds in the USA declines by 15% in spite of the factor productivity gain and the reduced chemical requirements. This is because the USA is so highly dependent on exporting especially oilseeds to the GM-critical markets and

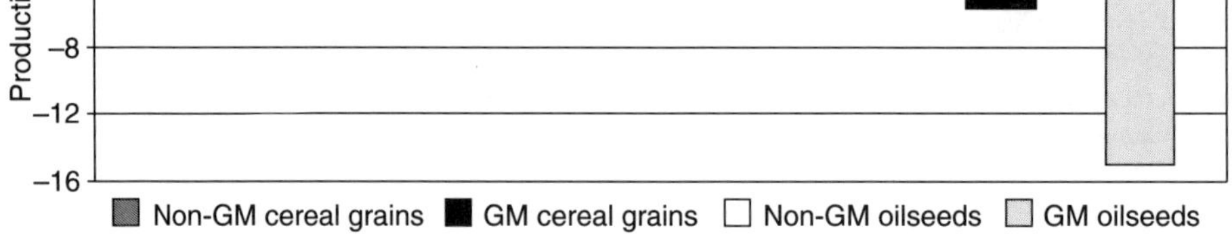

Fig. 18.5. Production effects in the USA (%).

[9] Note that Western Europe might be restricted by the Blair House agreement in terms of increasing acreage for oilseed production and so the reported production increase may not be allowed.

[10] Comparing these production effects with the results of our previous analysis, which did not have the identity preservation (IP) requirement in place (Nielsen *et al.*, 2000), we see that the effects reported here are substantially smaller. This is precisely because the IP requirement introduces much stronger restrictions on intermediate input choice for livestock producers and food processors. In our previous analysis intermediate users had a free choice between GM and non-GM varieties and could therefore benefit fully from the lower GM prices. In this model, however, intermediate users are required to use only GM or non-GM inputs.

because a structural consumer preference change has much more of an impact on this region's trading opportunities compared with the reduced price sensitivity experiment. The production of non-GM oilseeds, on the other hand, increases by 10% – another direct reflection of the importance of the GMO-critical export markets for this region.

As was discussed above, the Cairns group is not as dependent on its exports of cereal grains and oilseeds as the USA is. Furthermore, its exports of cereal grains are more evenly spread across trading partners. For these reasons, the productivity difference between GM and non-GM varieties shows up more directly in the production results for the Cairns group in the first two experiments (Fig. 18.6). The percentage changes are larger in the Cairns group compared with the USA, but these changes are from a smaller initial base. Furthermore, the structural shift experiment does not change the production results for the Cairns group as much as it does for the USA because the former group of countries is relatively less dependent on exports of these particular crops. For cereal grains production in the Cairns group countries, the changes are not large enough to switch the signs as they did in the USA because the Cairns group exports only very little to Western Europe and high-income Asia.

Although the export results for processed foods show similar compositional changes between GM and non-GM varieties, the importance of exports in total production is small, and therefore these changes only show up as small compositional changes in the production of livestock, meat and dairy products, vegetable oils and fats, and other processed food products in the USA. Similar compositional changes are reported for the Cairns group, but because exports of processed foods are relatively more important to this region, the magnitude of change in terms of production changes is slightly larger than for the USA.

An interesting question is whether these changing preferences in Western Europe and high-income Asia can open opportunities for developing countries to export non-GM varieties of cereal grains and oilseeds to these regions. Sub-Saharan Africa has some production of oilseeds, for example, and although exports of these crops do not account for a significant share of total production value at present, they might if niche markets for non-GM crops develop in Western Europe. Similarly, low-income Asian countries might look into expanding their production of e.g. non-GM oilseeds if nearby niche markets in high-income Asian countries develop. Although the differences are very small, comparing the export and production results of the three experiments indicates that this might be a path to follow if the price premiums obtainable for non-GM varieties are large enough to outweigh the relative decline in productivity and any identity preservation and

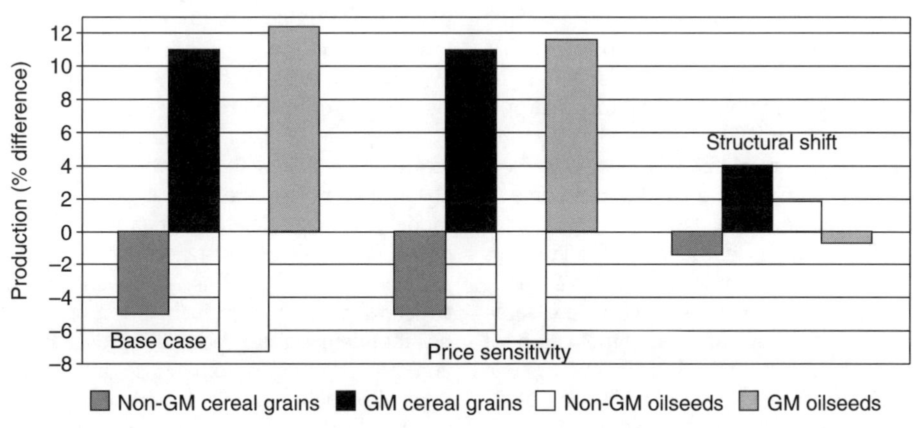

Fig. 18.6. Production effects in the Cairns group (%).

labelling costs. But even more significant in value terms for these countries are exports of processed foods, i.e. vegetable oils and fats, meat and dairy products, and other processed foods. Important in that respect are existing trade patterns, proximity of markets, historical ties, etc. which will determine whether or not producers will choose to forego productivity increases and lower costs in GM production in order to retain access to their traditional export markets by selling non-GM products. For a region like sub-Saharan Africa with strong ties to Western Europe, changing consumer attitudes toward GM foods are expected to be an important determinant of future decisions regarding genetic engineering in food production.

Absorption results

In this modelling framework, where we are operating with a representative consumer, we are implicitly aggregating over two consumer types – those who are indifferent about GM products and those who are concerned about potential hazards of consuming GM products. We have considered two changes in preferences concerning GM-inclusive foods. First, attitudes harden. The size of the two groups does not change, but those who are concerned about GM products become more price sensitive. As described above, this changes the curvature of the indifference curve, as shown in Fig. 18.1. Second, we have considered the effects of a structural preference shift – more people perceive that there are health hazards from consuming GM foods and choose to consume less, i.e. the share of consumption of GM foods drops, regardless of relative price changes. In essence, the group of GM-sensitive consumers expands. This causes the indifference curve to shift, as depicted in Fig. 18.2.

As discussed above, the level of utility stays the same when the indifference curve shifts. The representative consumer is on the same budget line with a different combination of GM and non-GM foods. We do not assume that the consumer obtains additional utility from his/her decision to increase the share of non-GM products he/she consumes. With this assumption, real absorption is an appropriate welfare measure. It indicates the change in the total amount of goods and services consumed following a change in preferences. The results of the experiments show that global absorption increases by US$7.4 billion in the base case, where consumers are assumed to find GM and non-GM foods to be good substitutes. Increasing the price sensitivity of GM-critical consumers in high-income Asia and Western Europe lowers this gain in total absorption marginally to US$7.2 billion. As the previous results have shown, the structural shift experiment is a much more dramatic change in preferences, and hence we find that the global absorption gain is only US$0.02 billion in that experiment.

The absorption results are reported for selected regions in Fig. 18.7 for the three experiments. The changes are reported in billions of US dollars and it should be noted that the percentage changes are very small. It is clear from this figure that the Cairns group, low-income Asia and the USA are the main beneficiaries of the productivity increase given that these are the regions assumed to be intense adopters of the GM crop varieties. All other regions also experience an increase in total absorption, albeit at a lower absolute level. Reducing the price sensitivity of consumers in high-income Asia and Western Europe reduces the increase in global absorption only marginally and does not change the distribution of the gains across regions. Most importantly, all regions still gain in terms of aggregate absorption from the productivity increase and hence lower product prices in spite of the increased aversion towards GM foods in high-income Asia and Western Europe.

Interpreting consumer preference changes as a structural shift, however, alters the absorption results dramatically. Because our model has completely segregated GM and non-GM production systems restricting input use to either GM or non-GM varieties, the structural preference shift has a strong effect on the demand for non-GM intermediates, and not all regions experience increases in total absorption in this experiment. Despite the productivity gain in the large GM crop sectors in the Cairns group and the USA,

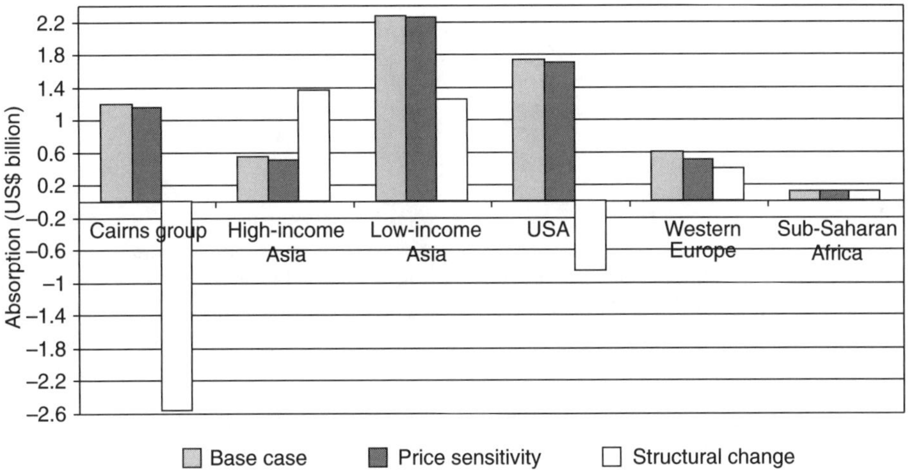

Fig. 18.7. Changes in total absorption.

these results reveal that aggregate absorption declines in these regions when consumers in important export markets turn against their main product and there is little diversion to other markets. Total absorption declines by US$2.6 billion in the Cairns group and US$0.9 billion in the USA. Although these declines amount to percentage changes of just −0.05% and −0.007%, respectively, they illustrate how different interpretations of preference changes will have very different impacts on total absorption results.

Furthermore, the impact on aggregate absorption in the GM-critical regions is ambiguous. Comparing these results with our previous work (Nielsen *et al.*, 2000), in which market segregation concerned only the primary cereal grain and oilseed sectors and preference changes were represented just as reduced price sensitivity, we find that modelling preference changes as a structural shift in this model changes not only the magnitude, but potentially also the direction of our absorption results. In our previous model, livestock and food producers that used cereal grains and oilseeds as inputs had flexibility in their choice of GM and non-GM varieties. In this model, we have restricted production decisions substantially by limiting input use to only GM or non-GM varieties throughout the food production chain. Combined with a structural shift in preferences in important export markets, amounting in effect to a rejection of GM foods, this means that regions that are major exporters of GM-potential crops as well as being enthusiastic GM-adopters risk losing out on overall absorption if consumer attitudes turn against GM foods.

Concluding Remarks

This chapter has analysed the price, production and trade consequences of changing consumer attitudes toward the use of genetic engineering techniques in food production using an empirical global general equilibrium model, in which the food-processing chain is segregated into GM and non-GM lines of production. Clearly, the present analysis relies on some simplifying assumptions, in particular those made about the productivity impact of adopting GM crops. Improved data would ideally provide information about how the GM productivity effects differ across sectors and regions. Another limitation of this analysis is that it does not explicitly take account of the costs of having to preserve the identity of a crop or food product throughout the production and marketing chains, or the costs of any testing and labelling requirements at national borders. Experience from identity preservation of speciality crops today reveals that these costs can potentially increase the price of such products by between 5 and 15% (Buckwell *et al.*, 1999). It is argued by, for example, Frandsen and Nielsen (1999) and Runge and Jackson (1999) that such a cost-price premium will – in a free market, and in the absence of unsympathetic political reactions – emerge on the guaranteed non-GM products. Whether consumers in Western Europe and elsewhere are in fact willing to pay such a premium is yet to be determined empirically. To the extent that evidence becomes available of a threshold beyond which consumers are unwilling to pay for their non-GM preferences, for example, this could be incorporated explicitly in the model used here.

Whilst keeping these caveats in mind, the empirical analysis described in this chapter does bring attention to two very important aspects of the GMO debate: (i) segregation of GM and non-GM production and marketing systems, and (ii) the power of consumer sentiment. The analysis has shown that when production and marketing systems are segregated into GM and non-GM lines all the way from primary crops through livestock feed to food processing, changing consumer attitudes towards GMOs will have substantial effects on trade, production and prices not only for the crop sectors that benefit directly from the new technology, but also for the sectors that use these crops as inputs in production. Interpreting consumer dislike of GM foods as a reduced sensitivity to relative price changes dampens the impact of the productivity difference between the two varieties.

If consumer preference changes are in fact more a matter of rejection rather than reduced price sensitivity, the effects on prices, production and trade flows are much more dramatic and the direction of effects reverses in some cases. Countries that are heavily dependent on exporting GM-potential crops to the GMO-critical regions find themselves increasing exports and hence production of non-GM varieties and reducing production of GM varieties in spite of the productivity benefit. Furthermore, total absorption results are also dependent on how preference changes are interpreted and hence modelled. Clearly, the results depend crucially on the extent of GMO rejection by consumers and the size of the productivity gain foregone in comparison with the relative price premium obtainable on non-GM varieties. For some countries the development of segregated GM and non-GM food markets is a way of retaining access to important export markets if and only if the non-GM characteristic can in fact be preserved and verified throughout the marketing system at reasonable costs.

References

Buckwell, A., Brookes, G. and Bradley, D. (1999) *Economics of Identity Preservation for Genetically Modified Crops*. Final report of a study for Food Biotechnology Communications Initiative (FBCI). With contributions from Peter Barfoot, Stefan Tangermann and Jan Blom. Mimeo.

Economic Research Service (ERS) (2000a) Biotech corn and soybeans: changing markets and the government's role. Available at http://ers.usda.gov/whatsnew/issues/biotechmarkets/. US Department of Agriculture, Washington, DC.

Economic Research Service (ERS) (2000b) Biotechnology: U.S. grain handlers look ahead. *Agricultural Outlook*. US Department of Agriculture, Washington, DC.

Frandsen, S.E. and Nielsen, C.P. (1999) *Derfor skal de GMO-frie varer mærkes* [This is why non-GMO products should be labelled], Chronicle (in Danish), *Jyllandsposten*, 20 December.

James, C. (1997) *Global Status of Transgenic Crops in 1997*. ISAAA Briefs No. 5. International Service for the Acquisition of Agri-biotech Applications, Ithaca, New York.

James, C. (1998) *Global Review of Commercialized Transgenic Crops: 1998*. ISAAA Briefs No. 8. International Service for the Acquisition of Agri-biotech Applications, Ithaca, New York.

James, C. (1999) *Global Status of Commercialized Transgenic Crops: 1999*. ISAAA Briefs No. 12: Preview. International Service for the Acquisition of Agri-biotech Applications, Ithaca, New York.

Lewis, J.D., Robinson, S. and Thierfelder, K. (1999) *After the Negotiations: Assessing the Impact of Free Trade Agreements in Southern Africa*. TMD Discussion Paper No. 46. International Food Policy Research Institute, Washington, DC.

McDougall, R.A., Elbehri, A. and Truong, T.P. (eds) (1998) *Global Trade, Assistance, and Protection: The GTAP 4 Data Base*. Center for Global Trade Analysis, Purdue University, West Lafayette.

Nelson, G.C., Josling, T., Bullock, D., Unnevehr, L., Rosegrant, M. and Hill, L. (1999) *The Economics and Politics of Genetically Modified Organisms: Implications for WTO 2000*. With Julie Babinard, Carrie Cunningham, Alessandro De Pinto and Elisavet I. Nitsi. Bulletin 809. College of Agricultural, Consumer and Environmental Sciences, University of Illinois at Urbana-Champaign.

Nielsen, C.P., Robinson, S. and Thierfelder, K. (2000) *Genetic Engineering and Trade: Panacea or Dilemma for Developing Countries*. TMD Discussion Paper No. 55. Trade and Macroeconomics Division, International Food Policy Research Institute, Washington, DC.

Nielsen, C.P., Thierfelder, K. and Robinson, S. (2001) Consumer attitudes towards genetically modified foods: the modelling of preference changes. Working paper No. 01/2001. Danish Institute of Agricultural and Fisheries Economics (SJFI), Copenhagen.

OECD (1999) *Modern Biotechnology and Agricultural Markets: A Discussion of Selected Issues and the Impact on Supply and Markets*. Directorate for Food, Agriculture and Fisheries. Committee for Agriculture. AGR/CA/APM/CFS/MD(2000)2, OECD, Paris.

Pray, C.E., Ma, D., Huang, J. and Qiao, F. (2000) Impact of Bt cotton in China. Paper presented at the International Food Policy Research Institute (IFPRI), 9 May.

Runge, C.F. and Jackson, L.A. (1999) *Labelling, Trade and Gentically Modified Organisms (GMOs): A Proposed Solution*. Working Paper WP99–4. University of Minnesota, Center for International Food and Agricultural Policy.

Index

Crop Pollination by Bees
Keith S. Delaplane and Daniel F. Mayer
CABI Publishing

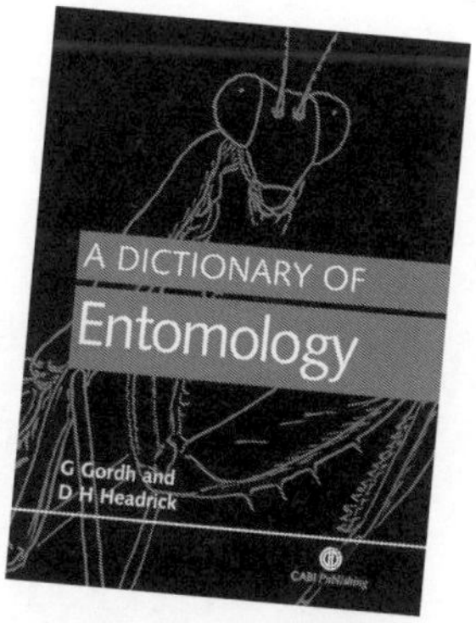
A DICTIONARY OF
Entomology
G Gordh and
D H Headrick
CABI Publishing

Principles of
CATTLE
PRODUCTION
C.J.C. Phillips

Seeds
THE ECOLOGY OF
REGENERATION
IN PLANT
COMMUNITIES
2nd EDITION
Edited by
Michael Fenner